AF802780

Driver Adaptation to Information and Assistance Systems

Driver Adaptation to Information and Assistance Systems

Edited by Alan Stevens, Corinne Brusque and Josef Krems

The Institution of Engineering and Technology

Published by The Institution of Engineering and Technology, London, United Kingdom

The Institution of Engineering and Technology is registered as a Charity in England & Wales (no. 211014) and Scotland (no. SC038698).

© The Institution of Engineering and Technology 2014

First published 2014

This publication is copyright under the Berne Convention and the Universal Copyright Convention. All rights reserved. Apart from any fair dealing for the purposes of research or private study, or criticism or review, as permitted under the Copyright, Designs and Patents Act 1988, this publication may be reproduced, stored or transmitted, in any form or by any means, only with the prior permission in writing of the publishers, or in the case of reprographic reproduction in accordance with the terms of licences issued by the Copyright Licensing Agency. Enquiries concerning reproduction outside those terms should be sent to the publisher at the undermentioned address:

The Institution of Engineering and Technology
Michael Faraday House
Six Hills Way, Stevenage
Herts, SG1 2AY, United Kingdom

www.theiet.org

While the author and publisher believe that the information and guidance given in this work are correct, all parties must rely upon their own skill and judgement when making use of them. Neither the author nor publisher assumes any liability to anyone for any loss or damage caused by any error or omission in the work, whether such an error or omission is the result of negligence or any other cause. Any and all such liability is disclaimed.

The moral rights of the author to be identified as author of this work have been asserted by him in accordance with the Copyright, Designs and Patents Act 1988.

British Library Cataloguing in Publication Data
A catalogue record for this product is available from the British Library

ISBN 978-1-84919-639-0 (hardback)
ISBN 978-1-84919-640-6 (PDF)

Typeset in India by MPS Limited
Printed in the UK by CPI Group (UK) Ltd, Croydon

Contents

About the Editors xv

Early Stage Researchers and Experienced Researchers in the ADAPTATION Project xvii

Acknowledgements xxiii

Introduction xxv

1 The ADAPTATION project **1**

Abstract 1

1.1 The motivation and objective of the project 1

1.2 ADAS and adaptation: The background 2

1.3 Scientific and technological objectives 3

1.3.1 Research methodology 4

1.3.2 The research programme 4

1.4 The partners 6

1.5 The training and research programme 6

Acknowledgement 10

References 10

2 Behavioural adaptation in response to driving assistance technologies: A literature review **13**

Abstract 13

2.1 Introduction 13

2.2 Historical background 14

2.3 Definition and assumptions 16

2.4 Theories accounting for behavioural adaptation effects 18

2.4.1 Behavioural adaptation and learning theories 18

2.4.2 Behavioural adaptation and driver risk models 19

2.4.3 Integrative models on behavioural adaptation 20

2.5 Behavioural adaptation over an extended period of time 22

2.6 Behavioural change transfer 24

2.7 A 'Joint Conceptual Theoretical Framework (JCTF) of Behavioural Adaptation in Response to Advanced Driver Assistance Systems' 26

2.8 Conclusion and further research needs 29

Acknowledgements 30

References 30

3 Methods to assess behavioural adaptation over time as a result of ADAS use **35**
Abstract 35
3.1 Introduction 35
3.2 The use of long-term studies to assess behavioural change with ADAS use 36
3.2.1 Cognitive adaptation to ADAS 36
3.2.2 Advantages of long-term studies 38
3.2.3 Inconveniences of long-term studies 39
3.2.4 Conclusion on long-term studies 40
3.3 Long-term studies using driving simulators 40
3.3.1 Introducing driving simulators 40
3.3.2 Utility of driving simulators 41
3.3.3 Advantages of driving simulators 42
3.3.4 Disadvantages of driving simulators 43
3.3.5 Conclusion on driving simulator studies 44
3.4 Long-term studies in real traffic conditions 45
3.4.1 Utility of naturalistic methods 46
3.4.2 Naturalistic driving study (NDS) versus field operational test (FOT) 47
3.4.3 Data from naturalistic methods 48
3.4.4 Conclusion on naturalistic methods 50
3.5 Conclusion 51
Acknowledgement 51
References 51

4 Effect of ADAS use on drivers' information processing and Situation Awareness **57**
Abstract 57
4.1 Behavioural adaptation and higher cognitive processes 57
4.2 Mental models 58
4.2.1 Definition 58
4.2.2 Mental model construction and update 58
4.2.3 Discussion of existing mental model measures and requirements 59
4.2.4 Development of the mental model questionnaire 60
4.3 Situation Awareness 60
4.3.1 Definition 60
4.3.2 Relationship between mental models and Situation Awareness 61
4.3.3 Situation Awareness as comprehension process 61
4.3.4 Discussion of existing Situation Awareness measures and requirements 62
4.3.5 Development of the Situation Awareness measurement technique 63

4.4 Study 1: Driving simulator study 64
4.4.1 Aims and research questions 64
4.4.2 Method and material 65
4.4.3 Results 67
4.5 Study 2: On-road study 69
4.5.1 Aims and research questions 69
4.5.2 Method and material 71
4.5.3 Results 73
4.6 Discussion and conclusions 75
4.6.1 Mental model questionnaire 75
4.6.2 Situation Awareness measurement 76
Acknowledgement 77
References 77

5 Early adopters' mental model of Adaptive Cruise Control (ACC) and its influence on behavioural adaptation to the system 81
Abstract 81
5.1 Introduction 81
5.1.1 Adaptive Cruise Control (ACC) and behavioural adaptations 81
5.1.2 Mental models 83
5.1.3 Scope of the research 84
5.2 Focus groups discussions 86
5.2.1 Procedure 86
5.2.2 Results 87
5.2.3 Discussion 89
5.3 Naturalistic Field Operational Test 90
5.3.1 Procedure 90
5.3.2 Results 92
5.3.3 Discussion 97
5.4 Discussion and conclusions 98
Acknowledgements 99
References 99

6 Intelligent driver support systems: Effects of learning and aging 103
Abstract 103
6.1 Introduction 103
6.2 Dual-task performance 104
6.2.1 Models of dual-task interference 105
6.2.2 Effects of aging on dual-task interference 107
6.3 Learning in dual-task situations 108
6.3.1 Models of learning 108
6.3.2 Effects of aging on learning in dual-task situations 110

6.4 An empirical study 111
6.4.1 Hypotheses 111
6.4.2 Experimental setup 112
6.4.3 Summary of major findings 114
6.5 Conclusion 116
Acknowledgment 116
References 117

7 Researching safety issues with intersection assistance systems for the older driver 123
Abstract 123
7.1 Introduction 123
7.2 Road traffic safety 124
7.3 Intersections and the older driver 127
7.3.1 Demographic change and its consequences 127
7.3.2 Age-related impairments, diseases/disorders and intersection performance 129
7.3.3 Crossing an intersection: A multi-level task 131
7.3.4 New developments, potential benefits and problems 132
7.3.5 Previous research relevant to intersection assistance 134
7.3.6 Drawbacks, unanswered questions and the need for further research 136
7.4 Summary and conclusion 138
Acknowledgement 138
References 139

8 Motivational factors when investigating ADAS impacts on driver behaviour 145
Abstract 145
8.1 Understanding driver behaviour: Models and approaches that attempt to explain driving behaviour 145
8.1.1 The Hatakka hierarchical model 148
8.1.2 How performance on the particular levels may influence drivers' performance on other levels 150
8.2 Motivational factors 152
8.2.1 Physiological needs 153
8.2.2 Safety needs 154
8.2.3 Social needs 154
8.2.4 Esteem needs 154
8.2.5 Self-actualisation needs 154
8.3 Motivation and driving 155
8.3.1 Physiological needs and driving 155
8.3.2 Safety needs and driving 156
8.3.3 Social needs and driving 156
8.3.4 Esteem needs and driving 156
8.3.5 Self-actualisation needs and driving 156

8.4 Motivational driver behaviour models 157
8.5 Motivational influence factors based on the theory of planned behaviour 158
8.6 The role of motivation in behavioural adaptation 159
8.6.1 Vehicle manoeuvring level and ADAS 162
8.6.2 Mastering traffic situations level and ADAS 162
8.6.3 Goals and context of driving and ADAS 163
8.6.4 Goals for life and skills for living and ADAS 163
8.7 A short view on impacts of various ADAS on driving 164
8.8 Conclusion 165
Acknowledgement 166
References 166

9 Drivers' adaptation to mobile phone use: Interaction strategies, consequences on driving behaviour and potential impact on road safety 173
Abstract 173
9.1 Introduction 173
9.2 Regulating the exposure to phone use while driving 175
9.2.1 Exposure regulation strategies 175
9.2.2 Factors influencing exposure regulation 177
9.3 Seeking after compatibility of driving and mobile phone use 179
9.3.1 Driving situation 180
9.3.2 Assisted driving 182
9.3.3 Use of hands-free devices 183
9.4 Adapting driving behaviour in response to distractive effects of mobile phone use 185
9.4.1 Impairment of driving performance due to distraction 185
9.4.2 Compensation strategies to mitigate effects of distraction 187
9.5 Conclusions on potential safety impact 188
Acknowledgement 189
References 189

10 Evaluating the potential for workload based driving assistance systems from a psychological, technological and physiological perspective 197
Abstract 197
10.1 Introduction 197
10.2 Theoretical foundation 199
10.2.1 The Yerkes–Dodson law 199
10.2.2 The adapted Yerkes–Dodson law 199
10.2.3 The flower model 199
10.3 Measuring workload 200
10.3.1 Workload induction by secondary task: The n-back task 200
10.3.2 Measuring arousal over physiology 201

10.4 Experiments and results 202
10.4.1 Experiment 1 – Relationship between brake reaction time and workload level 203
10.4.2 Experiment 2 – Wizard of Oz simulation of a WACC in the simulator 205
10.4.3 Experiment 3 – Wizard of Oz simulation of the WACC on the road 207
10.4.4 Experiment 4 – On road study with WACC 208
10.4.5 Experiment 5 – Emotion and physiology 210
10.5 Conclusion 211
Acknowledgement 212
References 212

11 Distraction and inattention prevention by combining Behaviour-Based Safety with Advanced Driver Assistance Systems 215
Abstract 215
11.1 Time for kickoff 215
11.2 The playing field: Accident prevention 216
11.3 Team player line-up 219
11.3.1 Team Advanced Driver Assistance Systems (ADAS) 219
11.3.2 Team Behaviour-Based Safety (BBS) 221
11.4 Advancing into a new league: Uniting ADAS and BBS team players 226
11.4.1 Game tactics 228
11.5 Players debriefing 230
Acknowledgements 232
References 232

12 Effectiveness of Forward Collision Warning Systems: A contribution from the cognitive analysis combining behavioural and electrophysiological measurements 239
Abstract 239
12.1 Introduction 239
12.2 The utility of FCWSs 240
12.3 A new approach for evaluating the effectiveness of FCWSs 242
12.4 Are FCWSs effective for distracted drivers? 244
12.5 Does the reliability of the system affect its effectiveness? 246
12.6 Behavioural adaptation to the system 249
12.7 Conclusions 251
Acknowledgement 252
References 252

13 Introduction of automated platooning in traffic: What is the impact on non-automated drivers? – An analysis of the influence of the short time headways held by automated platoons **257**
Abstract 257
13.1 From ADAS to automated driving 257
13.1.1 Advantages of automated systems 258
13.1.2 Different categories of ADAS 259
13.1.3 Automated driving 260
13.2 Behavioural adaptation of EVDs to the short THWs held by automated platoons 262
13.2.1 The perception based component of behavioural adaptation 262
13.2.2 The action based component of behavioural adaptation 264
13.3 The possibility for behavioural adaptation of UVDs to the short THW held by automated platoons in the light of social psychological models 264
13.4 Investigations of behavioural adaptation of UVDs to the short THW held by automated platoons 266
13.5 Conclusion 270
Acknowledgement 271
References 271

14 User-centred design approach to model scenarios on driving simulators **275**
Abstract 275
14.1 Introduction 275
14.2 What is a driving simulator scenario? 276
14.2.1 Scenario modelling systems 276
14.2.2 Challenges with programming and authoring tools 278
14.2.3 User-centred design in scenario authoring tools 278
14.3 Method 279
14.4 User survey 279
14.4.1 Results of the user survey 280
14.5 Proposed solution 283
14.5.1 Multi-layer programming 283
14.5.2 User-interface development 284
14.6 Prototype building 286
14.7 Prototype evaluation 293
14.8 Interoperability of the approach 293
14.8.1 Development of the meta-language 294
14.8.2 Interoperability framework 295
14.9 Conclusion 296
Acknowledgements 297
References 297

15 Development of a database for storage and analysis of behavioural data **301**
Abstract 301
15.1 Database fundamentals 301
15.1.1 Databases for human factors in road transportation 303
15.1.2 Advantages and disadvantages of working with databases 304
15.2 ADAPTATION database 305
15.2.1 Aims and objectives 305
15.2.2 Legal boundaries 306
15.2.3 Development process 307
15.2.4 Small-scale testing 313
15.3 Conclusion 314
Acknowledgement 316
References 316

16 Synthesis of new knowledge on adaptation processes in response to ADAS and associated advances in research methods **319**
Abstract 319
16.1 Introduction 319
16.2 The role of system characteristics 321
16.2.1 Functionality and design 322
16.2.2 Reliability 323
16.3 Considering the specificities of driver populations 324
16.3.1 Older drivers 324
16.3.2 Drivers with Parkinson's disease 325
16.3.3 ADAS-experienced drivers 325
16.3.4 Professional truck drivers 326
16.4 User support strategies for the deployment of ADAS 326
16.4.1 System knowledge 327
16.4.2 Driver awareness 327
16.4.3 Learning and training 328
16.5 Tools for research on drivers' behavioural adaptation 329
16.6 Conclusion 331
Acknowledgement 333
References 333

17 Recommendations for research on drivers' behavioural adaptation and for the design and deployment of ADAS **335**
Abstract 335
17.1 Introduction 335
17.2 Future research prospects on drivers' behavioural adaptation 336
17.2.1 Integration of the temporal dimension in the investigation of behavioural adaptation 336
17.2.2 Investigation of behavioural adaptation through different underlying processes 338

17.2.3 Countermeasures to mitigate the effects of ADAS limitations 339
17.2.4 Behavioural change induced by automation 339
17.3 Methodological challenges in research on drivers' behavioural adaptation 340
17.4 Towards improved design and deployment of ADAS 343
17.4.1 System information and driver awareness 343
17.4.2 System design: Warning presentation and adaptability based on driver monitoring 345
17.4.3 Driver training and regular reminders 346
17.5 Summary and conclusions 347
Acknowledgement 349
References 349

Index **351**

About the Editors

Alan Stevens is Chief Research Scientist and Research Director, Transportation, at the Transport Research Laboratory (TRL), in the United Kingdom, where he has been working on the application of new technology to transport for 25 years. He is an internationally recognised expert in human–machine interaction (HMI) in the driving environment and was co-author of the 'European statement of principles on HMI' through his work within the European iMobility platform, where he co-chairs the HMI working group. He was also an active member of the HMI international standards committee, regularly participating in meetings with European, US, Canadian and Japanese colleagues. He has been appointed to the EU–US working group on driver distraction following an EU–US high-level cooperation agreement and continues to be involved in the International Harmonized Research Agenda (IHRA) group and on the management committee of International Benefit Evaluation and Costs (IBEC) group. Alan's consultancy activities focus on providing advice on policy and interoperability issues to Government, developing research programmes and carrying out specific technical and human factors studies in Intelligent Transport Systems. He participates in university teaching at MSc level, supervises PhD students and is Editor-in-Chief of an international peer-reviewed journal of Intelligent Transport Systems.

Corinne Brusque is Research Manager and has been Scientific Director in charge of the evaluation of research units and researchers at Ifsttar (French Institute of Science and Technology for Transport, Development and Networks) since 2013. Her main research topics concern the factors affecting in-vehicle technology adoption, drivers' exposure to distraction related to in-vehicle technologies and the impact of in-vehicle technologies on drivers' behaviour. From 2001 to 2012, she was head of the Laboratory Ergonomics & Cognitive Sciences applied to Transport at the French Institute for Transport and Safety Research (INRETS), in Lyon, France. Prior to this appointment, she worked for 13 years as a researcher at the French Public Works Research Laboratory (LCPC), in Paris, France. She was in charge of research on the characterisation and modelling of road users' visual environment at night and on street lighting and planning. She holds an engineering degree and PhD in image processing. She supervises several doctoral students and post-doctorate researchers. She acts as reviewer for several scientific journals and conferences. She was the initiator of the biannual European conference on Human Centred Design for Intelligent Transport Systems and provides scientific coordination for several European projects including HUMANIST, INTERACTION and ADAPTATION.

Josef Krems is Professor of Cognitive and Industrial Psychology at Chemnitz University of Technology. He graduated from the University of Regensburg, Germany in 1980 and was awarded a PhD in psycholinguistics in 1984. From 1991 to 1993 he was a Visiting Assistant Professor at Ohio State University, Columbus, United States, where he worked on computational models of decision making. Then he became an Assistant Professor at the Centre for Studies on Cognitive Complexity at the University of Potsdam, Germany. Since 1995 he has been a full professor at Chemnitz University of Technology. His current research projects focus on three areas: traffic and transport (safety, driver assistance, green driving and sustainability), human factors (HMI, usability) and user acceptance. He has published or co-edited 15 books and more than 150 papers in books, scientific journals and congress proceedings.

Early Stage Researchers and Experienced Researchers in the ADAPTATION Project

ESR	BEGGIATO Matthias
Education	Master's degree in psychology/Master's degree in educational science, University of Vienna, Austria
Hosting institution	Chemnitz University of Technology, Germany with Josef Krems
Thesis topic	The thesis focuses on higher level cognitive processes underlying behavioural adaptation. Special emphasis is placed on mental models and situation awareness, including innovative assessment methods and changes over time when using Advanced Driver Assistance Systems (ADAS).

ESR	BHATTI Ghasan
Education	M.Sc in Communication and Interactivity (Computer Science), Linköping University, Sweden
Hosting institution	OKTAL, France with Guillaume Millet Ifsttar, France with Roland Brémond and Fabrice Vienne IRIT, Université Paul-Sabatier, Toulouse, France with Jean-Pierre Jessel
Thesis topic	The aim of his thesis is to develop a user-friendly and intuitive methodology and user-interface to model traffic scenarios on driving simulators. The proposed approach is for the researchers, psychologists who do not have technical or programming skills. Another objective of my work is to propose a framework and a specification language so that scenarios developed on one platform can be executed on other simulation platforms.

ESR BUENO Mercedes

Education M.Sc in Cognitive and Behavioural Neuroscience, University of Granada, Spain

Hosting institution Ifsttar, France with Alexandra Fort
Université Victor Segalen, Bordeaux, France with Colette Fabrigoule

Thesis topic The aim of this thesis was to determine the impact of a forward collision warning system on driver's information processing by using evoked potentials. The effectiveness of the system was analysed depending on its reliability and the driver' attentional state, and the behavioural adaptation to the system was examined over time.

ESR BERTHON-DONK Vera

Education M.Sc in Psychology, University of Toulouse, France

Hosting institution Leibniz Research Centre for Working Environment and Human Factors (IfADo), Dortmund, Germany and the Technical University of Dortmund, Germany with Gerhard Rinkenauer

Thesis topic The purpose of this thesis was to explore a particular question within the research domain of dual-task driving: 'Can older people learn how to use an Intelligent Driver Support System (IDSS)?' Several experiments, combining a simulated driving task with a surrogate IDSS, have shown that indeed, older adults' performance in the dual-task condition increases with practice, hence providing first indications that practice at older age can be beneficial for driving with IDSS.

ESR DOTZAUER Mandy

Education M.Sc in Human Factors, Berlin Institute of Technology, Germany

Hosting institution University Medical Center Groningen (UMCG), The Netherlands with Wiebo Brouwer

Thesis topic The doctoral thesis focused on the investigation of long-term effects of ADAS on driving performance of older drivers with and without neurodegenerative impairments. An ADAS, more tailored to fit the needs of older drivers, was developed, implemented into a driving simulator scenario and tested with older drivers with and without a neurodegenerative disease as well as young drivers.

ESR	GOUY Magali	
Education	M.Sc in Human Factors, Technische Universität Berlin, Germany	
Hosting institution	TRL, UK with Nick Reed, Open University, UK with Alan Stevens	
Thesis topic	ADAS are increasingly present within modern vehicles, supporting the introduction of semi- and fully automated driving situations. As a consequence, a mixed traffic situation is likely to emerge where vehicles equipped with different degrees of automated systems will interact with unequipped vehicles drivers (UVDs). The overarching aim of the PhD was to investigate whether the presence of automated vehicle platoons will impact UVDs' car-following behaviour.	

ESR	HAJEK Wilfried	
Education	M.Sc in Psychology, University of Klagenfurt, Austria	
Hosting institution	BMW Group Research and Technology, Germany with Ralf Decke & Bernhard Niedermaier Chemnitz University of Technology, Germany with Josef Krems	
Thesis topic	This PhD focused on driver workload while using ADAS. In high workload conditions the driver loses sight of the monitoring task and needs more time for an appropriate reaction in a dangerous situation. A compensation strategy for the usage of active cruise control in high workload conditions has therefore been designed and evaluated in simulator as well as on-road conditions. This work led to the development of the workload-adaptive active cruise control.	

ESR	HAUPT Juliane
Education	M.Sc in Psychology (Diplom-Psychologin), Chemnitz University of Technology, Germany
Hosting institution	FACTUM OG, Austria with Ralf Risser Chemnitz University of Technology, Germany with Josef Krems
Thesis topic	This doctoral thesis aimed to investigate how safety countermeasures (driver assistance systems and pedestrian priority crossings) influence drivers' motivation to pay attention to the relevant road scene or to secondary activities. Thereby, qualitative and quantitative methods were used.

ER	HUTH Véronique
Education	PhD (Dr. rer. nat.), Engineering and Traffic Psychology, Braunschweig University of Technology, Germany
Hosting institution	Ifsttar, France with Corinne Brusque
Post-doc topic	The post-doctoral research focussed on drivers' interactions with the mobile phone and their context. They were analysed in naturalistic driving studies, traffic observations and a survey, with the aim to get insights on drivers' strategies.

ER	PEREIRA Marta
Education	PhD in Ergonomics, Technical University of Lisbon, Portugal
Hosting institution	Chemnitz University of Technology, Germany with Josef Krems
Post-doc topic	The aim of the post-doc was to develop and implement a database and a data analysis protocol to enable the storage, joint access, individual analysis and integration of different types of behavioural data collected during the driving simulator or field studies.

ESR	BIANCHI PICCININI Giulio Francesco
Education	M.Sc. in Automotive Engineering, Politecnico di Torino, Italy
Hosting institution	UNIVERSITAS, Portugal with Anabela Simões Faculdade de Engenharia, Universidade do Porto with Carlos M. Rodrigues
Thesis topic	This thesis focused on the use of ACC by early adopters of the system. Qualitative and quantitative methods were used to determine possible risks for road safety, induced by the continuous usage of the ACC. The role of drivers' trust in the system and drivers' mental models of the ACC was considered for the analysis.

ESR	WEGE Claudia
Education	M.Sc Psychology, Chemnitz University of Technology, Germany and University of Oklahoma, USA
Hosting institution	Volvo Group Trucks Technology, Advanced Technology & Research, Sweden with Trent Victor Chemnitz University of Technology, Germany with Josef Krems
Thesis topic	In this PhD, the short- and long-term adaptation of attention allocation strategies while using ADAS were analysed. Of specific interest was the role of driver behavioural feedback before distraction occurs, while it occurs in the vehicle (real-time feedback) and after it has occurred.

Acknowledgements

We hope that this book will serve as a useful reference into the programme of work conducted within the European ADAPTATION project. Principally, this comprises 10 doctoral research topics from the doctoral students and the work of two post-doctoral researchers. The book also provides a summary of the State-of-the-art, as well as scientific conclusions based on research undertaken within the PhD topics and recommendations for further work. Interested readers will be able to find more details of the studies in the scientific papers published by the members of the ADAPTATION project and other dissemination material developed and referenced throughout the book.

The Marie-Curie programme, 'Initial Training Network' (ITN), within which the ADAPTATION project is situated, had from its beginning a number of scientific and other objectives. In terms of scientific achievement, and as described in more detail in Chapter 16 of this book, ADAPTATION has been able to provide a much more integrated perspective on driver adaptation to information and assistance systems than previous research. In part this has been through consideration of adaptation over a comparatively long period of time – from first use, through development of experience, then long-term use and even post-use. Also, the influence of system characteristics, in terms of the degree and type of assistance provided, has been studied as well as the influence of the driving situation and driver characteristics. Specifically, the project has:

- developed a better understanding of drivers' adaptation processes following Advanced Driver Assistance Systems (ADAS) use on different levels (energetic, cognitive and motivational);
- identified some short-, medium- and long-term effects of information and assistance system use on drivers' behaviour;
- contributed to knowledge about age as a driver characteristic on information and assistance system use and skill acquisition and
- highlighted the effects of system performance and the level of automation as important characteristics affecting drivers' adaptation.

Together, the work has developed a much more integrated and holistic theoretical model of behavioural adaptation that can be used both to generate predictions about how drivers will adapt to new vehicle technologies and to derive guidelines for the design of such systems.

The ADAPTATION project, as described in Chapter 1, has fostered an interdisciplinary and participatory atmosphere. The Marie-Curie ITN approach was a largely novel concept for most of the project participants but was a very stimulating human and scientific experience for us all. As well as completing their individual studies, the researchers (the chapter authors) have supported each other and completed a number of collaborative scientific and training activities. Chapters 2 and 3 include a joint literature review on human behavioural adaptation definitions, models and investigation methods. Chapters 16 and 17 are the result of collaborative activity where our conclusions and recommendations are presented.

This book would not have been possible without the support of many contributors.

Foremost, we gratefully acknowledge the financial contribution from the European Community's Seventh Framework Programme (FP7/2007–2013) under grant agreement no. 238833/ADAPTATION project.

Then, of course, we thank our doctoral students Matthias Beggiato, Vera Berthon-Donk, Ghasan Bhatti, Mercedes Bueno, Mandy Dotzauer, Wilfried Hajek, Juliane Haupt, Magali Gouy, Giulio Piccinini and Claudia Wege; and our post-doctoral researchers Véronique Huth and Marta Pereira. They have all contributed to this book with energy and enthusiasm, despite competing academic and organisational pressures.

We also thank ADAPTATION's senior researchers and their organisations for their contributions, both in supporting the students and to the project more widely. Many senior colleagues were involved including: Serge Boverie, Wiebo Brouwer, Ralf Decke, Colette Fabrigoule, Alexandra Fort, Gilles Gallee, Marc Grosjean, Fabienne Janin, Jean-Pierre Medevielle, Guillaume Millet, Bernhard Niedermaier, Christopher Patten, Anke Popken, Gerhard Rinkenauer, Ralf Risser, Anabela Simoes, Andrea Saroldi and Trent Victor.

We are indebted to our external reviewers for their helpful comments which have undoubtedly improved the chapters during development, particularly Gary Burnett, Cyriel Diels, Jeff Hickman, Samantha Jamson, Britta Lang, Andrew Parkes, Nick Reed and Stephen Skippon. Similarly we thank Valda Stevens for her final editing and Ting Ting Liu at the IET publisher for her patience and support.

Finally, we wish the PhD students and post-doctoral researchers tremendous success in their future careers.

Alan Stevens
Corinne Brusque
Josef Krems

Introduction

Corinne Brusque, Alan Stevens and Josef Krems

Advanced Driver Assistance Systems (ADAS) have emerged as an integral part of the traffic system. They make use of the newest information technologies in order to enhance driver's safety and comfort, and thereby avoid driver errors and accidents. Several systems and functions available on the market have the potential to meet these objectives. However, concerns have been raised that drivers adapt to these new systems in unexpected ways that can compromise safety. Nevertheless, it can also be expected that the positive effects of ADAS on drivers' behaviour can last after system use and contribute to long-term safety-oriented changes.

Adaptation is a human ability to adapt to novel situations in order to increase the chances of survival by modifying behaviour to suit new conditions of life. In road safety research, behavioural adaptation refers to drivers' changes in response to the introduction of a change in the traffic system. These behavioural changes may have been anticipated (or not) by the persons responsible for the change in the traffic system and they can induce positive or negative effects on road safety.

Studying drivers' behavioural adaptation in response to ADAS use was the main focus of the European ADAPTATION project and the aim was to boost the career perspectives of young researchers by allowing them to conduct doctoral or post-doctoral research projects within the framework of a European research network. All these individual studies are a contribution towards a better understanding of the range of adaptation processes to ADAS, including not only observable behavioural changes, but also changes in energetic, cognitive and motivational processes.

The aim of this book, which is organised into 17 chapters, is to present the main achievements of the ADAPTATION project both at the level of each doctoral or post-doctoral research project and also at the level of the network by offering a comprehensive picture of the project outcomes. A short description of each chapter is given below.

In Chapter 1, Krems, Brusque and Stevens set the background to the ADAPTATION Initial Training Network, funded by the European Commission that supports the individual doctoral and post-doctoral research and integrates them within a comprehensive framework to increase knowledge of drivers' behavioural adaptation in response to ADAS use.

In Chapter 2, Wege, Pereira, Victor and Krems examine the concepts and the theoretical models as well as the empirical research regarding drivers' behavioural changes due to ADAS use. The authors introduce a 'Joint Conceptual Theoretical

Framework on Behavioural Adaptation to ADAS' including the most relevant variables of interest to understand the behavioural response to in-vehicle technologies and the changes in energetic, cognitive and motivational processes underlying behaviour.

Chapter 3 concerns the methods to assess behavioural adaptation over time as a result of ADAS use. Dotzauer, Berthon-Donk, Beggiato, Haupt and Piccinini shed light on the driving simulator studies and the on-road studies generally used to investigate behavioural changes, pointing out not only their respective advantages and disadvantages but also their complementarity.

Chapter 4 focuses on the elaboration and validation of methods to investigate higher cognitive processes underlying behavioural adaptation in response to ADAS use. Beggiato designs a method to assess the mental model of the system created by the driver and its update through ADAS experience, and a method to assess situation awareness, that is, the mental situation model that the driver has of the driving situation while using the system.

In Chapter 5, Piccinini, Simões and Rodrigues investigate behavioural adaptation in response to adaptive cruise control (ACC) use by focusing on early adopters of the system. Through both focus groups and behavioural observations in natural settings, the authors highlight that an inappropriate driver's mental model of ACC might be a cause of both inappropriate usage of ACC and engagement in secondary activities while using ACC.

Interacting with in-vehicle technology induces dual-task situations that can affect driving performances, specifically for older drivers; Chapter 6 deals with this issue. Taking into account cognitive models of multi-task performance and learning, Berthon-Donk, Grosjean and Rinkenauer investigate the effect of practice on drivers' performance related to the both driving and interaction tasks, the effect of age on the learning rate and the different strategies implemented by older and younger drivers to cope with the dual-task situation.

In Chapter 7, Dotzauer focuses on the special needs of older drivers related to ADAS in order to increase their safety and comfort while driving. Due to deterioration in vision, cognition and motor functions, older drivers face difficulties in dealing with complex driving situations where they have to take decisions under time pressure. For this driver population, intersections represent a major safety issue that can be addressed by intersection assistance systems. Based on an intensive literature review, the author identifies knowledge gaps within this research topic.

Chapter 8 highlights the important role of motivational factors underlying drivers' behavioural adaptation in response to ADAS use. In the past, researchers have focused only on trust and acceptance. From a review of motivation definitions and driver models, Haupt and Risser identify motivation factors that are of key importance in understanding ADAS use and impacts on driver behaviour. These include risk perception and risk homeostasis, feelings of security and protection, social needs, and esteem and self-actualisation needs.

Drivers can implement behavioural strategies in response to mobile phone use while driving. In Chapter 9, Huth and Brusque distinguish three kinds of strategies to deal with the risk taken: reduction of the exposure to phone use while driving, reduction of the combined demands of the phone and driving tasks to ensure their compatibility and increase of the safety margins to mitigate the effects of distraction. The implementation of these strategies and their effects are discussed.

In Chapter 10, Hajek investigates the opportunity for adaptive driving assistance systems, based on real-time monitoring of the driver's workload derived from physiological data, in order to compensate for a driver's arousal impairment. This approach is studied from a theoretical perspective and has been implemented in the design and evaluation of a prototype of Workload-Adapting ACC (WACC). This research shows the feasibility of the approach and has identified a range of new issues to be resolved in implementing this kind of assistance system.

In Chapter 11, Wege and Victor focus on the possibility for ADAS to contribute to a long-term change in drivers' behaviour, taking as illustration a Visual Distraction Alert System (VDAS). VDAS is a distraction prevention system based on head/eye movement recordings which provides immediate feedback to the driver when looking away from the road for too long or too frequently. The authors postulate that the VDAS benefit will be increased by combining it with drivers' involvement in Behaviour-Based Safety (BBS) programs. BBS programmes aim to coach or educate drivers by providing retrospective or cumulative feedback targeting safety-critical behaviours. The interest in the BBS–ADAS approach and guidelines for its implementation are highlighted.

In Chapter 12, Bueno, Fabrigoule and Fort present an overview of behavioural studies on the effectiveness of Forward Collision Warnings systems (FCWS) conducted on the road and in driving simulators. The authors also present results from recent research using electroencephalography and the associated Event-Related Potentials in order to analyse brain activity in response to warnings. This work permits the following issues to be addressed: the effectiveness of the system when drivers are distracted, the consequences of driving with a system that is not completely reliable and behavioural adaptation to the FCWS over time.

The increasing proportion of assisted vehicles, associated with the introduction of automated vehicles, leads to consideration of a new research prospect: the behavioural adaptation of unequipped vehicle drivers facing equipped vehicle drivers. Based on behavioural models of social psychology, Gouy, Reed and Stevens, in Chapter 13, point out that drivers may imitate the observed behaviour of other drivers surrounding them. The case of platoons of vehicles maintaining a short inter-vehicle time headway is used by the authors as an illustration.

Two specific pieces of research, described in Chapters 14 and 15, have addressed some challenges related to experimental research on drivers' behaviour. First, Bhatti, Bremond, Jessel, Vienne and Millet have implemented a user-centred approach to design an intuitive and user-friendly interaction environment for modelling scenarios on driving simulators to address the lack of technical and programming skills of behavioural researchers. The authors propose also a meta-language and an interoperability framework so that scenarios can be easily ported from one simulator to another. Second, Pereira, Lietz and Beggiato have used user-centred design to develop and implement a database and a data analysis protocol to enable the storage, joint access, individual analysis and integration of different types of behavioural data, which were collected during driving simulator and field studies. This eases data sharing and secondary analysis.

An overview of the major findings obtained within the ADAPTATION project is provided in Chapter 16. Huth, Bueno, Fort and Brusque synthesise and organise the results of each individual research topic into a comprehensive picture considering the role of system characteristics, the differences between driver populations and the user support strategies for deployment of ADAS. Finally, a synthesis is given of research methods to study behavioural adaptation in the driving context.

Chapter 17 complements the previous one by presenting three types of recommendations based on the results and experience gained within the ADAPTATION project. Fort, Huth and Brusque first identify future research prospects concerning drivers' behavioural adaptation to ADAS, then give methodological recommendations to address the challenges related to this research and finally, derive operational suggestions towards improved design and deployment of ADAS.

We hope that the innovative contents of this book will be of interest to a broad audience including researchers in human factors related to in-vehicle technologies, system designers and policy makers concerned with system deployment.

Chapter 1

The ADAPTATION project

Josef Krems[1], *Corinne Brusque*[2,3] *and Alan Stevens*[4]

Abstract

ADAPTATION is the short name of 'Drivers' behavioural ADAPTATION over the time in response to ADAS use', which was a *Marie Curie Initial Training Network (ITN)* funded between 2010 and 2013 under the European Commission Call: FP7-PEOPLE-ITN-2008. The major objective of ADAPTATION was to improve the career perspectives of young researchers by taking part in a research programme aiming to investigating drivers' behavioural adaptation and its underlying processes over time in response to Advanced Driver Assistance Systems (ADAS) use. Within ten PhD projects, accompanied by two post-doctoral projects, ADAPTATION has studied drivers' behavioural adaptation from an integrated perspective working towards an integrated theoretical model of behavioural adaptation.

1.1 The motivation and objective of the project

The objective of ADAPTATION was to improve the career perspectives of young researchers by taking part in an innovative and ambitious European research programme aiming to investigating drivers' behavioural adaptation and its underlying processes over time in response to Advanced Driver Assistance Systems (ADAS) use. The research programme integrated, under a joint theoretical framework and a joint longitudinal methodological design, a set of individual projects dealing with various aspects of the adaptation process.

In addition to training-through-research, personalised training actions were implemented to extend the skills of this future generation of academic and industrial researchers. The training programme implemented during the ITN lifetime aimed to accelerate acquisition of skills in human factors applied to ADAS design, favour multi-disciplinary approaches, strengthen abilities to disseminate research results and widen career prospect with complementary skills of team and project management.

[1] Technische Universität Chemnitz, Chemnitz, Germany
[2] IFSTTAR-LESCOT, Bron, France
[3] University of Lyon, Lyon, France
[4] Transport Research Laboratory, Wokingham, UK

1.2 ADAS and adaptation: The background

ADAS have emerged as an integral part of modern road vehicles in order to support the driver while driving. ADAS make use of the latest information technologies in order to enhance driver safety and comfort, and thereby avoid driver errors and accidents. Today a number of systems are available on the market (see Chapters 7 and 13) that have great potential to meet these safety objectives. However, concerns have also been raised that drivers adapt to these new systems in unexpected ways that can compromise safety (see [1] for an overview).

The term 'behavioural adaptation' has emerged in this regard and refers to 'those behaviours which may occur following the introduction of changes to the road-vehicle-user [...] They create a continuum of effects ranging from positive increase in safety to a decrease in safety' [2, p. 23]. Although interest in and research on behavioural adaptation goes back to the 1970s (see, e.g., [3]), the topic has recently re-emerged as a key issue in research.

Considerable research has been conducted on drivers' behavioural adaptation in response to ADAS (see Chapter 13). Those studies have yielded a diversity of results, which can be briefly summarised as follows:

- Behavioural changes sometimes occur and sometimes do not
- These changes affect different aspects of behaviour
- The behavioural changes differ in magnitude and direction; even with regard to one particular system, the results reported are sometimes similar and sometimes contradictory

Apparently, behavioural changes do not occur in an 'all-or-nothing' fashion, but evolve from the complex interplay of a number of factors. Three of these factors are summarised in the next paragraph.

First, they depend on the assistance system being studied. ADAS vary with respect to the type of driving function they support and hence influence driver control processes in different ways [4]. ADAS also differ in their design of the driver–system–vehicle interaction, e.g. the degree to which they automate particular driving functions [5]. The level of automation, the system's performance (e.g. reliability) and the design of the human–machine interface have all been found to affect human cognitive, affective and motivational processes underlying behaviour [6]. Behavioural adaptation has been found to be mediated by changes in drivers' trust [7], situation awareness [8], fatigue [9], mental workload [10] and perceived risk [11]. Second, behavioural adaptation occurs in a specific context, which, in its narrowest sense, refers to the momentary driving situation (driving task demands), along with the travel conditions (such as purpose and duration); and, in its widest sense, refers to the social and cultural background (including driver population needs, habits, attitudes and legislation). Third, behavioural adaptation is influenced by individual driver characteristics, such as driving experience, age, gender, personality traits, attitudes and motives [12, 13].

A number of theories have been proposed to account for drivers' adaptive changes in response to safety regulations and new technologies (e.g. for a review for risk models, see [3]). They place emphasis on different types of processes and mechanisms that influence behavioural adaptation, and are able to explain the empirical findings to some extent. However, none of them allow specific predictions of how ADAS will influence drivers' behaviour and underlying processes dependent on the type of system studied, the design of the driver–system interaction, the context of use or individual driver characteristics.

To summarise, when the ADAPTATION project started, there was a lack of research programmes that had systematically investigated the conjoint short-, medium- and long-term effects of system properties on drivers' cognitive, affective and motivational processes underlying behaviour. Moreover, the adaptation effects found in previous studies showed considerable variability and a lack of coherence, largely due to the differences in system characteristics studied and the influence of situation variables and driver characteristics. The ADAPTATION project aimed at addressing these identified research deficiencies by studying drivers' behavioural adaptation to ADAS from an integrated perspective, in order to work towards an integrated theoretical model of behavioural adaptation.

1.3 Scientific and technological objectives

The ultimate objective of the ADAPTATION approach is to study the whole range of adaptation processes in response to ADAS including behavioural adaptation, risk compensation and changes in information processing from a comprehensive perspective, in an integrated way. We have adopted a wide view regarding the nature of adaptation processes, emphasising that adaptation includes not only observable behavioural changes, but also changes in energetic, cognitive and motivational processes. Therefore, the project work programme was established to focus not only on behavioural changes, but also on the various processes underlying those observable behaviours and especially the interaction of those processes.

Four main scientific objectives were assigned to the research projects:

- A better understanding of drivers' adaptation processes following ADAS use on different levels (energetic level, cognitive level, motivational level)
- Distinguishing short-, medium- and long-term effects of ADAS use on drivers' behaviour
- Studying the impact of age as a driver characteristic on ADAS use and skill acquisition
- Assessing the effects of system's performance and the level of automation as important system characteristics for drivers' adaptation

More technologically oriented objectives also underlie the definition of the ADAPTATION research programme; the main intended outcomes were:

- An empirical baseline for the design of ADAS
- A shared database from different studies that can be used for further testing
- Empirical evidence for safety and usability evaluation procedures of ADAS

- A set of recommendations and guidelines for the design and development of ADAS
- A set of recommendations to heighten public awareness for safe use of ADAS
- A set of methodological recommendations concerning the study design for investigations of adaptation processes

1.3.1 Research methodology

In response to these scientific and technological objectives, a joint methodological framework was developed to investigate the temporal effects of ADAS use on drivers' behaviour and underlying processes benefiting from the different disciplines present in ADAPTATION (see Chapters 2 and 3). It was based on different research contexts according to the systems' maturity and their distribution among European drivers, for example surveys and observation of ADAS use in everyday life where systems were already available, or laboratory experimentation with driving simulators in other cases.

With regard to experimental design, the following dimensions were implemented as independent variables:

- Level of experience/training with ADAS: short-, medium- and long-term effects
- Age and gender as major driver characteristics
- System performance and level of automation

Dependent variables were defined for each type of adaptation process: energetic/regulatory mechanisms, higher cognitive processes and motivational processes. Basic behavioural parameters essential for driving (longitudinal and lateral control) were also measured. Moreover, drivers' reliance on ADAS was assessed with performance-based measures (Figure 1.1).

The methodological approach adopted by the ADAPTATION project, as a whole, allowed the integration, analysis, interpretation and validation of the scientific results concerning different levels of adaptation across the individual PhD topics. This approach has contributed significantly to research on drivers' behaviour and ADAS design by providing a more comprehensive understanding of drivers' behavioural adaptation and its underlying processes.

1.3.2 The research programme

The research programme of ADAPTATION was structured around five work packages:

1. ADAS function selection and research hypothesis generation

   The major objectives of this work package were to perform a review of existing theoretical models and empirical studies on behavioural adaptation; to classify results according to ADAS types or functions studied and the drivers–system function allocation implemented; to decide on a restricted set of ADAS applications that could be used in the empirical studies of each individual PhD project; to develop a joint conceptual model of drivers' adaptation processes including changes in cognitive, affective, motivational processes underlying

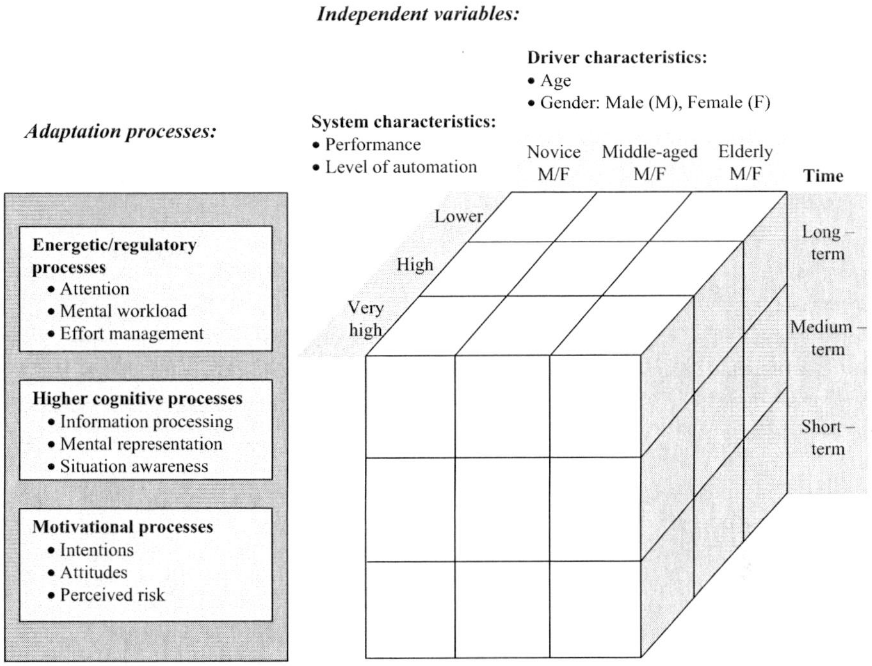

Figure 1.1 The general framework for the ADAPTATION project (influenced by the work of Popken [14])

behaviour; and to specify the outlines of the individual PhD projects including research hypotheses.

2. Definition of a joint methodology to investigate long-term changes in drivers' behaviour and underlying processes in response to ADAS

   The major objectives of this work package were to develop a joint methodology that allows for the complementary and comprehensive investigation of drivers' behavioural adaptation while focusing on different types of processes underlying behavioural changes; to develop and implement a joint data analysis protocol in order to enable the storage and joint access of data collected within each individual PhD project; to implement procedures for the common use of distributed data sets; to support the dissemination of data and results between individual projects; and to monitor the compliance of the empirical test regimes with ethical standards.

3. Establishing joint driving simulator and field studies and associated data collection

   The major objective of this work package was the joint design and realisation of driving simulator and field studies in order to collect data concerning drivers' long-term adaptation processes. The aim was to appreciate and take advantage of the relative strengths and weaknesses of these two methodological approaches and to allow for the validation of results obtained in the two test environments.

4. Analysis of data collected regarding drivers' adaptive changes
   The major objective of this work package was analysis of the data collected in driving simulator and field studies concerning changes in energetic/regulatory processes (attention, vigilance, mental workload, effort regulation), higher cognitive processes (information processing, mental representation, situation awareness) and motivational processes (attitudes, intentions and risk perception) that accompany or result in changes in drivers' behaviour.
5. Integration of results on drivers' long-term adaptation processes
   The major objectives of this work package were integration and synthesis of results obtained in the individual PhD topics regarding the different types of adaptation processes; the specification of design recommendations for ADAS in order to limit potential negative effects on safety; and the definition of a minimum set of requirements for conducting future studies on drivers' adaptation processes in response to ADAS.

1.4 The partners

The Initial Training Network, ADAPTATION, comprised 10 full partners, each being well-recognised European organisations working on drivers' behaviour research and ADAS design. The consortium gathered researchers/specialists from different disciplines (including neurophysiology, experimental psychology, psycho-sociology, ergonomics, engineering), coming from various working environments (three universities, two research institutes on road safety, a research institute on occupational health, a Small and Medium Business (SMB) carrying out traffic and social analyses, a driving simulator developer and two automotive manufacturers). Therefore, the research was not disconnected from industrial needs and this facilitates the transfer of knowledge into practice. The co-ordination of the ADAPTATION network was carried out by Ifsttar. A technical manager (from the Technische Universität Chemnitz) participated in general management of the project.

In addition to this partnership with substantial industry involvement, several European stakeholders from the automotive sector (suppliers and car manufacturers, public authorities and research associations) participated on the supervisory board as associated partners. They took part in the scientific and strategic planning of the network and in definition of the skill requirements for future generations of professionals. The network partners and associated partners are shown in Table 1.1.

1.5 The training and research programme

To support the research programme, a training programme was implemented for the early stage researchers (ESR) performing their PhD within the framework of the ADAPTATION project and for the two experienced researchers (ER) who were also involved. This training programme took into account the academic background of researchers, the requirements of the ADAPTATION research domain in human

Table 1.1 Partners and associate partners of ADAPTATION

Partners	Institut Français des Sciences et Technologies des Transports, de l'Aménagement et des Réseaux (Ifsttar) Technische Universität Chemnitz (CUT) BMW Research and Technology FACTUM Chaloupka & Risser OHG IfADo – Leibniz Research Centre for Working Environment and Human Factors OKTAL SA Transport Research Laboratory (TRL) UMCG Universitair Medisch Centrum Groningen UNIVERSITAS, Cooperativa de Ensino Superior e Investigaçao Cientifica, C.R.L. Volvo Technology Corporation
Associated partners	Continental Automotive France SAS Centro Ricerche FIAT Statens väg-och transportforskningsinstitut HUMANIST association

factors applied to ADAS and the needs of ESRs and ERs in terms of skills acquisitions necessary to boost their future researcher careers. Three main objectives were assigned to the training programme:

1. To favour multi-disciplinary approaches to investigate drivers' adaptation processes
2. To strengthen the abilities of researchers to disseminate research results
3. To widening the career prospects of researchers

During the 48 months of the project, complementary actions for training were implemented at both local and network-wide levels, to meet these three main objectives (Figure 1.2).

For each objective of the training activities, actions were carried out at both local and network-wide level, by seeking complementarities and synergies between the two levels. The distribution of activities between these local and network-wide actions was based both on the nature of the actions (e.g. specific for one researcher or relevant for several researchers) and on the added value of exploiting the scientific and technical competencies of ADAPTATION partners to deliver network-wide training (compared to what could be offered at a local level).

The institutions hosting the ESRs and the universities supporting the PhDs were fully in charge of training at the local level. Their actions included scientific supervision through regular meetings between the ESRs and their supervisors, access to the facilities necessary for performing their PhD work, integration into a research team and participation in the scientific life of the hosting institution and university. The training programme provided by partners for their own staff or students was open to the ESR. To ensure the quality of supervision offered by the scientists in charge of the project, the number of PhD students was limited to one per partner.

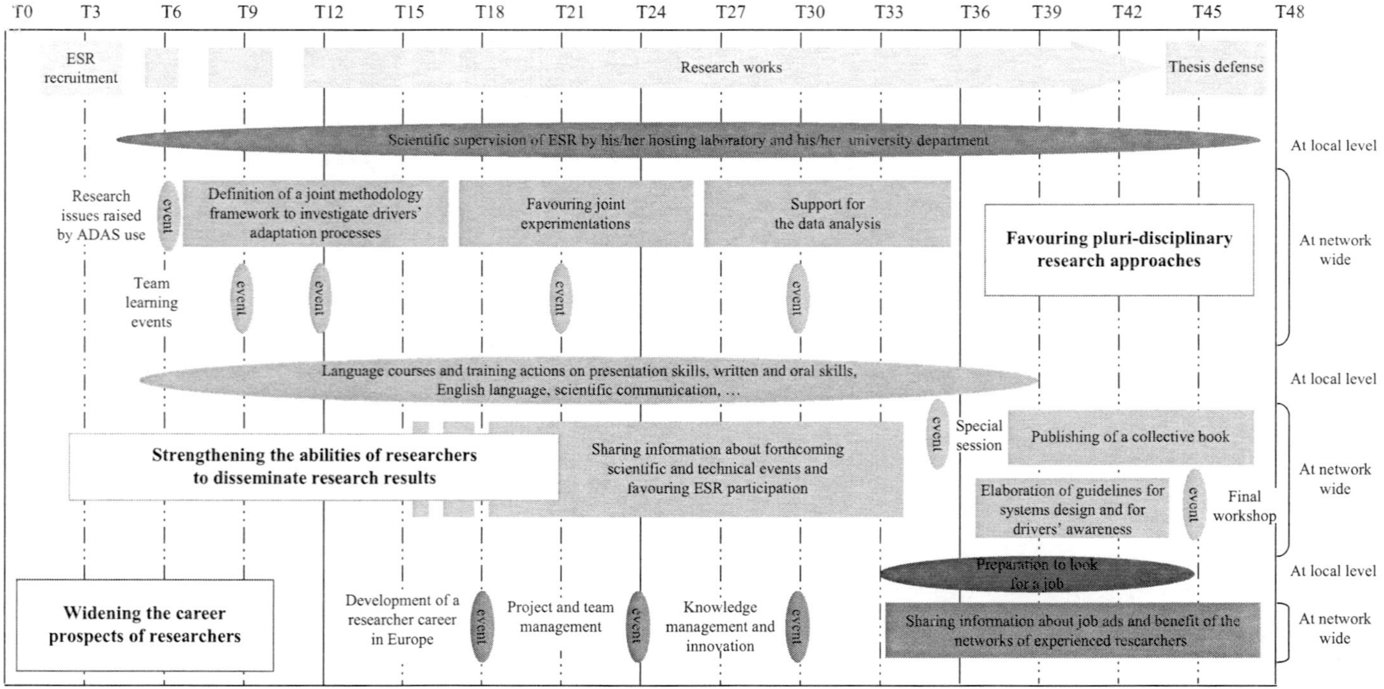

Figure 1.2 The timetable of ADAPTATION

Training activities performed at the network-wide level mobilised the scientific and technical competencies and know-how of the ADAPTATION partners. These were based on transfer of knowledge between senior scientists and young researchers and also between the young researchers themselves. Indeed, ADAPTATION had the benefit of being a multi-disciplinary research project with various disciplines present among ESRs. In this context, team learning was a challenging but rewarding method for training young researches in interdisciplinary skills. Team learning consisted of co-operative exchanges between ESRs with different backgrounds. ESRs, being experts in their specific domain, acted both as teacher and learner, and developed both these roles continuously during their discussions. The team learning sessions were overseen by senior scientists guiding and supervising the exchange between the learning partners.

Increasing the employability of young researchers and the competitiveness of European Industry was a key motivation for this training programme. These objectives were addressed through strong involvement of industry members of the supervisory board in the definition of skills requirements and the training actions needed. Industry representatives from both human resources departments and innovation and research departments were involved in setting up the training programme. As well as acting as advisers they also contributed as trainers.

Each training activity at the network-wide level was oriented towards the production of a common result, such as an integrated methodology, a joint publication or a joint dissemination event. In reaching these goals, the actions included courses and lectures given by researchers and specialists, the sharing of scientific and technical literature, discussion and debates of scientific results, and other collaborative work.

Training was implemented through joint meetings and seminars, and through partners' visits and stays abroad. Between each physical meeting, partners were able to have regular exchanges to share information, to debate research issues and to work together via a dedicated collaborative working environment (CWE). A private area of the project website included a forum for ESR/ER exchanges, an electronic library (for gathering articles and documents related to the different training actions) and also a calendar for announcing upcoming events organised inside and outside of the network, such as scientific conferences, workshops, technical seminars and training events.

Training-through-research was, nevertheless, the most crucial part of the training programme: it was through their participation in the ADAPTATION research programme, and the collaborative work performed with other ESR, that the PhD students developed their research skills. For ERs recruited into the framework of ADAPTATION, their involvement in coordination of training activities for ESR was a challenging and effective way to acquire complementary skills of project and team management through a training-by-doing process, while also developing their own personal research (Table 1.2).

In elaborating the programme of training, we aimed to identify, a priori, training actions that would be relevant and useful for the whole of the ADAPTATION researcher community. Nevertheless, each ESR/ER was unique in terms of their

Table 1.2 Partners of ADAPTATION and research topics

Partners	Individual research topics
Ifsttar	Impact of confidence in forward collision warning system during driving
CUT	Short- and long-term changes in drivers' higher level cognitive processes when driving with assistance systems
BMW	Adaptive driver assistance based on monitoring of driver's workload derived from physiological data
FACTUM	Drivers' changes in intentions to pay attention to the driving task related to the use of ADAS
IfADo	Adaptation in different age groups to ADAS: effects of practice over time
OKTAL	Scenario modelling to evaluate driver assistance systems with driving simulators
TRL	Behavioural adaptation of drivers of unequipped vehicles to short time headways observed in a vehicle platoon
UMCG	Long-term effects of ADAS on driving performance of older drivers with and without mild cognitive or physical impairments
UNIVERSITAS	The driver's mental representation of ADAS functions and its influence on his/her behaviour
VTEC	Integrating ADAS feedback and behaviour based safety approaches to improve drivers' attention allocation

academic background, working environment and research topic. So, from the beginning, the acquisition of some specific skills was understood to be necessary due to the requirements of each research topics. To take account of these individual requirements, a Personal Career Development Plan (PCDP) was established for each ESR/ER in order to aid the development and provision of a suitable research training programme. These PCDPs were established immediately following recruitment, updated on a regular basis and used throughout the duration of ADAPTATION.

Acknowledgement

ADAPTATION has received funding from the European Community's Seventh Framework Program (FP7/2007-2013) under grant agreement no. 238833/ADAPTATION project (www.adaptation-itn.eu).

References

1. Rudin-Brown M., Jamson S. (eds.). *Behavioural Adaptation and Road Safety: Theory, Evidence, and Action*. London: CRC Press; 2013
2. OECD. *Behavioural Adaptations to Changes in the Road Transport System*. Paris: Organization for Economic Co-operation and Development; 1990
3. Brown C. M., Noy Y. I. 'Behavioural adaptation to in-vehicle safety measures: Past ideas and future directions' in Rothengatter T., Huguenin R. (eds.). *Traffic and Transport Psychology: Theory and Applications (Proceedings of the ICTTP 2000)*. Elsevier; 2004. pp. 25–46

4. Engström J., Hollnagel E. 'A general conceptual framework for modelling behavioural effects of driver support functions' in Cacciabue P. C. (ed.). *Modelling Driver Behaviour in Automotive Environments. Critical Issues in Driver Interactions with Intelligent Transport Systems*. London: Springer; 2007. pp. 61–84
5. Nilsson L. 'Automated driving does not work without the involvement of the driver' in Underwood G. (ed.). *Traffic and Transport Psychology: Theory and Application. Proceedings of the ICTTP 2004*. Elsevier; 2005. pp. 273–301
6. Cacciabue P. C. (ed.). *Modelling Driver Behaviour in Automotive Environments. Critical Issues in Driver Interactions with Intelligent Transport Systems*. London: Springer; 2007
7. Stanton N. A., Young M. S. 'Driver behaviour with adaptive cruise control'. *Ergonomics*. 2005;48(10):1294–1313
8. Baumann M., Krems J. F. 'Situation awareness and driving: A cognitive model' in Cacciabue P. C. (ed.). *Modelling Driver Behaviour in Automotive Environments. Critical Issues in Driver Interactions with Intelligent Transport Systems*. London: Springer; 2007. pp. 253–265
9. Matthews G., Desmond P. A. 'Task-induced fatigue states and simulated driving performance'. *The Quarterly Journal of Experimental Psychology*. 2002;55A(2):659–686
10. Ward N. J., Shankwitz C., Gorgestani A., Donath M., De Waard D., Boer E. R. 'An evaluation of a lane support system for bus rapid transit on narrow shoulders and the relation to bus driver mental workload'. *Ergonomics*. 2006;49(9):832–859
11. Rajaonah B., Tricot N., Anceaux F., Millot P. 'The role of intervening variables in driver-ACC cooperation'. *International Journal of Human-Computer Studies*. 2008;66(3):185–197
12. Fuller R. 'Motivational determinants of control in the driving task' in Cacciabue P. C. (ed.). *Modelling Driver Behaviour in Automotive Environments. Critical Issues in Driver Interactions with Intelligent Transport Systems*. London: Springer; 2007. pp. 165–188
13. Summala H. 'Towards understanding motivational and emotional factors in driver behaviour: Comfort through satisficing' in Cacciabue P. C. (ed.). *Modelling Driver Behaviour in Automotive Environments. Critical Issues in Driver Interactions with Intelligent Transport Systems*. London: Springer; 2007. pp. 189–207
14. Popken A. *Drivers' Reliance on Lane Keeping Assistance Systems as a Function of the Level of Assistance*. Doctoral thesis, Technische Universität Chemnitz; 2009

Chapter 2

Behavioural adaptation in response to driving assistance technologies: A literature review

Claudia A. Wege[1], Marta-Sofia Pereira[2], Trent W. Victor[1] and Josef Krems[2]

Abstract

Adaption processes become important each time a driving situation embodies one or several unfamiliar components. These processes involve a behavioural change emerging into previously established behavioural patterns. Research shows that behavioural changes due to Advanced Driver Assistance Systems (ADAS) are on a continuum ranging from an increase to a decrease in safety. This chapter reviews concepts, theoretical models as well as empirical research regarding these behavioural changes. The literature reviews showed the need for a Model capturing the most relevant factors inducing behavioural adaptation which resulted in the development of a '*Joint Conceptual Theoretical Framework (JCTF) of Behavioural Adaptation in Response to Advanced Driver Assistance Systems*'. Alongside, the traditional OECD definition of behavioural adaptation to driving assistance technologies is critically discussed by investigating its main assumptions and its adequacy for current on-market and future ADAS applications.

2.1 Introduction

The philosopher Khalil Gibran (1883–1931) once said, 'It takes a minute to have a crush on someone, an hour to like someone, and a day to love someone... but it takes a lifetime to forget someone.' Is this idea of 'adaptation to suit new life conditions' transferable to modern transport analysis research? How long does it take to get familiar with driving assistance technologies? How lasting is the behavioural adaption effect? And, how persistent is an effect once the technology is taken away from the driver? After getting used to a certain routine, is it almost impossible to adapt to changes in the traffic system? This chapter provides a

[1] Volvo Group Trucks Technology (GTT), Sweden
[2] Chemnitz University of Technology, Germany

literature review of the concept of behavioural adaptation in response to driving assistance technologies. In particular, the chapter presents the scope of the state-of-the-art knowledge that guided the ADAPTATION project. It was found that a conceptual framework including the most relevant characteristics of behavioural adaptation (short-, medium- and long-term) is still missing. Therefore, a framework that collects evidence of the most important factors and underlying psychological processes that affect behavioural adaptation was developed and is introduced in this chapter.

The Oxford Dictionary [1] defines *Adaptation* as 'the action or process of adapting or being adapted'. To adapt is defined as 'to become adjusted to new conditions'. The ability to adapt to novel situations, meaning to modify behaviour to suit new conditions, is intrinsic to human nature [2] and from an evolutionary perspective improves human chances of survival. As Smiley [2, p. 47] states 'Adaptation ... is one of our most valuable characteristics and the reason that a human presence is desirable to monitor even the most highly automated systems – to deal with the unexpected.' She further argues that 'Adaptation is a manifestation of intelligent behavior' and '... that adaptation will occur is predictable – we should be more surprised by its absence' [2, p. 48].

The introduction of ADAS to the (car, truck and bus) market has been implemented at an increasing rate and speed in recent years. According to drivers' needs and requirements, the general purpose of ADAS is to enhance safety, comfort, entertainment and awareness of the driver by optimising the driving task with the overall objective of avoiding drivers' errors and accidents. Albeit the overall purpose of ADAS is a positive effect on safety and comfort, drivers adapt to new systems in unexpected ways that can compromise safety. According to Rudin-Brown [3] it is these negative effects of behavioural adaptation that are of most interest to road safety professionals. These negative effects have not been studied to a full extent yet, especially in regards to the diversity of safety systems. As research has demonstrated, there is a tendency for behavioural adaptation to develop differently for different ADAS [4–7] (for an overview, see Rudin-Brown [3]). Despite recent efforts in developing theoretical models and experimental studies, research is still at an early stage. This chapter outlines the progress of research from its early stages to the present and aims to advance the scientific debate for future researchers.

2.2 Historical background

Although the number of fatalities caused by road accidents has decreased in Europe over the last two decades (from 75,426 fatalities in 1991 in the 27 member states to about 39,300 in 2011 [8]), the current number of traffic deaths and injuries is still regarded as unacceptable. Various measures have been implemented to increase road safety, e.g. Vision Zero initiative [9]. However, both manufacturers and researchers have already identified potential problems with current countermeasures, as the safety effects of certain measures have been lower than initially expected. Aiming to identify the phenomenon responsible for this discrepancy, the interaction between the human and the different elements of the road transport system has been brought into focus. To convey that the implementation of safety changes might be inducing modifications in driver behaviour, terms like risk compensation [10], human behaviour feedback [11], behavioural adaptation [12], danger compensation [13], adverse behavioural change [14], and driver-behavioural barrier [15] were used. These phenomena are now frequently jointly called behavioural adaptation. Evans [11] stated that the earliest explicit reference to behavioural adaptation was by Gibson and Cooks in 1938 [16]. Gibson and Cooks mentioned in a footnote that 'except for emergencies, more efficient brakes on an automobile will not in themselves make driving the automobile any safer. Better brakes will reduce the absolute size of the minimum stopping zone, it is true, but the driver soon learns this new zone and, since it is his field-zone ratio which remains constant, he allows only the same relative margin between field and zone as before' (p. 458). Over a decade later, Smeed [17] considered the existence of a 'regressive tendency' that led drivers to drive faster as a result of improved sight distances and better roads. Wilde's work on the risk homeostasis theory contributed to further explain and theoretically support this phenomenon [10]. Wilde's work was based on the idea that road users tend to maintain the same level of risk, independent of changes made to the road system. To achieve the same level of risk, drivers compare their perceived subjective risk with the target level of risk (i.e. the level of risk desired to accept) and adjust their behaviour accordingly. In other words, individuals are hypothesised to modify their behaviour when perceived risk deviates from the target risk level in order to keep risk at the preferred level. The target level of subjective accident risk is assumed by Wilde as very important, as it is the variable capable of influencing accident rate, thus matching the expected safety effects with the real ones. A model following Wilde's homeostasis theory [10] is the task–capability interface model by Fuller [18]. It focuses on the possibility of task demands to exceed the driver's available capability. Where capability exceeds demand, the task is easy; where capability equals demand, the driver is operating at the capability limits, which makes the (driving) task difficult. As in Wilde's risk homeostasis theory, individuals are hypothesised to modify their behaviour when task demand exceeds capability in order not to fail at a task that would result in the loss of control, potentially precipitating a near-crash.

Evans [11] collected information on different approaches that tried to understand and reduce the toll from traffic accidents. Evans included the notion that, when a system is modified, the user usually does not ignore it, but instead responds with some type of behavioural change, which he defined as 'alteration in behaviour in response to changing external physical conditions' (p. 558). This change can be on a continuum of possibilities, suggesting that the outcome of a safety program can vary between two extremes: from (a) changes even greater than expected, to (b) safety changes in opposition to what was expected, i.e. decrease in safety. The term 'feedback' was used as a key element that not only characterises the degree to which there is feedback provided by a system, but also defines the difference between the estimated safety change and the actual safety benefit. This means that a change that is perceived by road users will result in a change of behaviour, and might lead to different safety outcomes when compared with what was originally planned. However, Evans reported not to have addressed the important issue of feedback variations over time.

2.3 Definition and assumptions

In 1990 a report prepared by the Organisation for Economic Co-operation and Development (OECD) further developed the understanding of behavioural adaptation in a compilation of research [12]. The OECD report aimed at examining the 'evidence of road user behaviours that occur in response to road safety programmes' (p. 13). The term 'behavioural adaptation' was presented and defined as follows: 'Behavioural adaptations are those behaviours which may occur following the introduction of the changes to the road-vehicle-user system and which were not intended by the initiators of the change. Behavioural adaptations occur as road users respond to changes in the road transport system such that their personal needs are achieved as a result, they create a continuum of effects ranging from a positive increase in safety to a decrease in safety' (p. 23). When considering the OECD definition, one aspect is clearly different from the formalism presented by Evans. Not all behavioural changes are covered, but only the ones that are *not* consistent with the purpose of the change. It is defined that both positive and negative effects are included in the definition, not only negative effects. The important condition to be considered a behavioural adaptation is to be *unexpected*, either unexpectedly positive or negative. The fact that the existence of 'unexpected behaviour' is dependent on the intentions of the initiator has been criticised. Grayson [14], for example, stated that the intentions behind the road safety changes are not always clear and often all that can be said with any degree of certainty is that measures aim to reduce accidents, an aspect that is not easily linked with behavioural changes. Furthermore, any estimates on what will happen after a change is complex and might neither be predicted nor understood by the initiator of the change [19]. 'Only if reasonable estimates of what should happen are available, would it be possible to compare what actually did happen with these expectations, and hence determine whether there had been any adverse effect' ([14], p. 6). The fact that the change is not necessarily subject to be predicted or understood by the initiator shows that it is yet unclear whether or not behavioural adaptation is a conscious or a subconscious process. In order to evaluate this phenomenon, studies need to investigate the perceived and actual performance change of ADAS users. A further discussion on the level of drivers' awareness of the behavioural change is presented in Section 2.4.2 in connection with compensatory behavioural adaptation.

The adequacy of the OECD definition can be a matter of discussion regarding not only the predictability of the events, but also the type of consequences. Grayson [14] and Rudin-Brown [3] argue that although the OECD definition covers positive, negative or neutral effects on crash risk and overall road safety, the negative consequences of behavioural adaptation are usually the focus of research. Grayson emphasised that a less vague term should be used ('adverse behavioural consequences' was his proposition), focusing on the literal use of the word 'adverse' as opposed to safety. Grayson's vision to some extent has been supported in recent publications as, in certain cases, a reformulation of the term presented by the OECD was used. For example, Dragutinovic, Brookhuis and Marchau [4] focused exclusively on the negative aspects and mentioned that, in the framework of their study, the term 'behavioural adaptation' refers to 'unintended and unwanted changes in driver behaviour'. Likewise, Rudin-Brown [3] stated that for the purpose of the discussion elaborated in her article, behavioural adaptation would be defined as 'unintended behaviour that arises following a change in the road traffic system that has negative consequences on safety' (p. 252).

In spite of the discussions about the adequacy of terminology, the OECD definition has been used by several authors in the context of transportation psychology for studying the occurrence of such phenomenon after the introduction of in-vehicle technology [4, 20–26]. Therefore, it is worth explaining its boundaries in further detail. Additional assumptions about explanatory characteristics regarding the occurrence of behavioural adaptations contribute to a clearer understanding of the proposed concept. The first is the notion of *feedback*. Similarly to the work presented by Evans [11], for behavioural adaptation to occur, feedback from the road system change has to exist and has to be perceived by the road user (though not necessarily consciously). The speed at which behavioural adaptation occurs might be dependent on the feedback, as changes that are perceived quickly might result in faster behavioural changes. Different levels of feedback can be distinguished. For a discussion on the different time characteristics of feedback, see Section 2.5.

Apart from feedback, two other explanatory assumptions were integrated in the behavioural adaptation definition proposed by the OECD: the driver has to be *able to change the behaviour* and also has to have the *motivation to act* upon it. Both assumptions are straightforward. A driver that perceives a change in the system but is not able to change his/her behaviour will not show signs of behavioural adaptation. The same applies to motivation. For behavioural adaptation to occur, users must be motivated to behave differently; for example, they have to recognise the benefit of the behaviour modification. Furthermore, the OECD definition does not identify a temporal or spatial range of behavioural adaptation. The authors justify this due to the limited amount of empirical and theoretical work conducted on this matter. When the OECD definition was formulated, only the existence of the phenomenon and its different effects were able to be gathered. It cannot be expected that behavioural adaptation identified following a certain change will also appear when changes of different nature occur in the road transport system. For example, behavioural changes following the introduction of anti-lock braking systems will be different from behavioural changes following speed control systems. Behavioural adaptation is not only highly system specific but it also varies for different elements of the road transport system. This means that changes to the infrastructure (e.g. lane width, shoulder width edge line markings), changes related to education and enforcement (e.g. publicity campaigns, education and training, legislation and enforcement) and non-automated vehicular modifications in the vehicles (e.g. high-mounted braking lights, studded tyres, seat belts) might lead to such phenomena. Studies have shown that in response to increases in road and lane width, people drive faster [27–29], engage in more erratic manoeuvres (e.g. centre line crossings or steering corrections) [30] and drive closer to the road edge [31]. Other studies have found changes in driver speed related to street lighting [32, 33]. With regard to the previous studies, the introduction of new in-vehicle technologies is expected to change the driving task either by the automation of specific sub-tasks of driving, or by providing extra information to the driver (related or unrelated to the driving task) [34, 35]. The expected consequence of the technical development is the related change of the role of the driver. In the past the driver's role was to control the vehicle; in the future the driver's task will be to monitor the function of the ADAS and to resume manual control in certain situations such as an emergency. On one hand, ADAS replace some of the driver's tasks (e.g. speed choice, distance keeping, detection of relevant traffic information etc.) while on the other hand the management of ADAS is imposed on the driver. The driver needs to gradually develop this 'new role', because otherwise, unintended ADAS effects due to drivers' human capabilities and limitations could emerge. System developers can help to prevent unintended effects by considering the 'human factor' within the complex interaction of driver, vehicle and environment when new systems are introduced.

2.4 Theories accounting for behavioural adaptation effects

A number of theories have been proposed to account for drivers' adaptive changes in response to new technologies [36–40]. The theories place varying amounts of emphasis on different types of processes and influencing factors. They all try to explain, to some extent, the underlying mechanisms of behavioural change found in empirical research. As stated above, researchers [e.g. 3, 5, 7] have reviewed the 'behavioural adaption' debate. All attempts to explain the phenomena can be classified into either *learning theories* or *driver risk theories*. Both approaches are described in more detail below. In order to investigate a more conclusive approach to contribute to the understanding of behavioural adaption, three *integrative models* are presented as well.

2.4.1 Behavioural adaptation and learning theories

Behavioural adaptation is strongly connected to learning effects. In the literature, it is assumed that the exponential learning curve is applicable for changes in human behaviour over time. In the context of skill learning and traffic research, Elvik [41] has identified statistical regularities that refer to 'laws of accident causation'. One of these is the *universal law of learning*, which implies that accident rate per unit of exposure will decline as the amount of exposure increases. Although Elvik refers to a driver's ability to detect and control traffic hazards as the amount of travel increases, the universal law of learning is also appropriate for explaining adaptation effects when exposure to ADAS increases. Thus, the more exposure to ADAS the more learning is involved, which results in fewer accidents.

Another theory in connection to behavioural adaptation and learning is the *Skill-Rules-Knowledge model* [42], which explains how human behaviour changes with practice. This model was developed further from the original model by Fitts and Posner [43], which has also resulted in simulation models such as ACT-R [44, 45]. Fitts and Posner termed the initial stage of skill learning as the cognitive phase followed by the associative phase and finally the autonomous phase. It is assumed that experts can execute certain tasks following automated routines that demand little or no conscious control and/or attention. However, the three processing stages highly depend on the task at hand. As such, an expert driver can all of a sudden become almost novice due to the introduction of an ADAS which might change the task at hand.

The *change in mental representations* is another example of an issue worth mentioning. During interaction with ADAS, drivers build mental representations of the system. These mental representations are internal models formed in a specific context and for a specific aim [46]. They directly influence the modes of cooperation between the driver and the system because they reflect the users' understanding about the functioning principles and the usage conditions of the system. Mental representations can change over time and as they are not true copies of objective reality they may diverge from it considerably. Therefore, when mental representations are not properly formed, they can trigger misuses of the system, leading to possible dangerous situations. Finally, behavioural adaptation and learning is also connected to the *enhancement of perception and attention*. Thus, Smiley [2] suggests that the primary motivations for behavioural adaptation are the intelligent re-allocation of attention and effort. This will not necessarily lead to constant accident rates but may lead to trade-offs between mobility and safety.

2.4.2 Behavioural adaptation and driver risk models

The OECD scientific expert group [12] suggested that Wilde's *risk homeostasis theory* [10] provides the most complete explanation for behavioural adaptation, and although controversial and heavily criticised, this theory has received the greatest amount of attention from researchers. Its basic assumption is that people have a target level of risk that they accept, tolerate, prefer, desire or choose. Whenever road users perceive a discrepancy between the target level of risk and experienced risk in one direction or the other, they will attempt to restore the balance through some kind of behavioural adjustment. This is referred to as risk compensation or behavioural compensation performed by drivers in response to any change in the perceived risk. Two examples of behavioural compensation are first, drivers compensating for the increased cognitive demand of driving while using a mobile phone by driving more slowly and thereby increasing safety margins [47]. And second, Lewis-Evans and Charlton [48] found that drivers reduce speeds on a narrowed road accompanied by increased ratings of risk. The results support a zero perceived risk model of behavioural adaptation to road width as an implicit perceptual process rather than explicit, conscious process. In their study Lewis-Evans and Charlton [48] compared perceived and actual performances in response to traffic changes but not in response to ADAS. When evaluating driver's level of consciousness and self-awareness for compensatory behaviour when using ADAS, a study design should investigate driver's perceived and actual response to ADAS.

Evolving from research on compensatory behaviour is the concept of *counterproductive behavioural adaptation*. Heijer, Brookhuis, Van Winsum and Duynstee [49] define counterproductive behavioural adaptation as a phenomenon where drivers start to behave in riskier ways because they are supported by an ADAS. Contrary to compensatory behavioural adaptation, counterproductive behavioural adaptation only targets behavioural changes that affect safety negatively. Supporting theories on counterproductive behavioural adaptation is a study by Comte [6] comparing measures of safety when driving with and without an 'Intelligent Speed Adaptation System', restricting drivers to the posted speed limit. Results show that drivers were more inclined to engage in riskier behaviour when driving with the 'Intelligent Speed Adaptation System'.

2.4.3 Integrative models on behavioural adaptation

The OECD work [12] criticises most general attempts to model behavioural adaptation as too vague, overly general and only indirectly related to behavioural adaptation. Therefore, recently there were attempts to specifically model behavioural adaption in order to draw more definitive conclusions. One example is the '*Qualitative model of behavioural adaption*' by Rudin-Brown [3] and Rudin-Brown and Noy [38]. Instead of establishing a hierarchical model (static) that would explain behavioural changes due to the use of ADAS, the model is a dynamic model that helps to understand behaviour and the associated changes over time. Through its iterative character, the model offers the opportunity to explore and explain behavioural adaptation over time periods, which can also include changes from an earlier negative displayed behaviour back to a positive driving behaviour. The model includes psychological concepts and their contribution to driver's behaviour and proposes that the driver's dimensions locus of control and sensation-seeking, both being personality factors, contribute to the development of behavioural adaptation. Drivers with an internal locus of control may rely more on their own skills and abilities and, no matter how reliable a safety system, always maintain more direct involvement with the driving task than drivers with an external locus of control. Conversely, drivers with an external locus of control may be more likely to give up control to an external system, relying on it completely, resulting in reacting more slowly when the system fails to perform the task it was designed to do. Furthermore, according to the model, drivers who are high sensation seekers, compared to low sensation seekers, may demonstrate more behavioural adaptation due to their preference for a higher level of risk (assuming that the intended effect of an ADAS is to reduce the level of risk). Additionally, personality factors directly and indirectly (through trust in automation) influence the driver's 'mental model'. Generally, false trust in an ADAS is the primary concern of road safety researchers. False trust is where an operator trusts and uses poor quality automation and shows reduced alertness due to less monitoring of the device. The 'Qualitative model of behavioural adaption' [38] predicts that the degree of behavioural adaptation will be related to the amount of trust a driver has in the system (wherein trust includes 'reliability' and 'competence' of the system), which is determined by the system's characteristics such as feedback timing (immediate vs. delayed), amount of usage (amount of exposure) and persistence.

An example for a static rather than a dynamic model is the '*Process Model of behavioural adaptation*' by Weller and Schlag [40]. In the model behavioural adaptation only occurs when the driver is aware of the new possibilities offered by the system. This may depend on several factors such as an indirect feedback (e.g. media), a direct feedback (e.g. haptic or visual) or driver's characteristics (e.g. experience, age). Furthermore, it is likely that behavioural change may occur even though this change in behaviour is not consciously perceived by the driver. An example of this may be the earlier case of a driver with an external locus of control who subconsciously is more likely to give up control to an external system, resulting in reacting more slowly when the system fails to perform the task it was designed to do. Because this model does not take the effect of time into account, it is questionable whether the variables in the model influence the direction and magnitude of behavioural adaption.

A model that aims to draw attention to the interdependences of relevant variables influencing the driver's behaviour is the '*Conceptual model of driver appropriation*' presented by Cotter and Mogilka [36]. As an extension of existing conceptual models [e.g. 38, 40] an attempt was made to consider the full range of driver's behaviour processes, including behavioural adaption, risk compensation, and changes in information processing in response to the introduction of ADAS. The model takes into account the timely interaction of these processes as well as mediating factors and the relationship between them. The model not only accounts for observable behaviour but also the underlying cognitive, energetic and motivational factors. The central focus of the model is a driver's mental representation of an ADAS. During the initial phases of interaction with a new ADAS, the driver begins to build up a mental representation of the system's behaviour and functioning. This mental representation is continually elaborated and refined, thus getting more and more comprehensive and sophisticated over the time of system use. After some learning and a familiarisation period, the driver will have developed a relatively stable mental representation of the system that subsequently governs behaviour. Nevertheless, the consequences of driver's behaviour will provide feedback and cause the mental representation to be revised and refined accordingly.

In sum, generally a number of theories have been proposed to account for driver's adaptive changes in response to new technologies [36–40]. The theories place varying amounts of emphasis on different types of processes and influencing factors. They all try to explain, to some extent, the underlying mechanisms of behavioural change found in empirical research. As stated earlier, all attempts [e.g. 3, 5, 7] to explain the phenomena can be classified into either *learning theories, driver risk theories* or a more conclusive approach, which is *integrative models of behavioural adaptation.*

In the future, in order to investigate underlying behavioural adaptation processes it is necessary to carefully examine behavioural adaptation effects. Survey questionnaires can be an appropriate tool to study the subjective modification of behaviour, which might give valuable insight into less obvious processes. One example of a less obvious process is a user not fully attending to the road ahead when using an in-vehicle navigation system [50]. Questionnaire data can also show ADAS perception and acceptance. For example, questionnaire results of 130 Adaptive Cruise Control system users showed that the longer drivers use the Adaptive Cruise Control system, the more aware of its limitations they become [51].

2.5 Behavioural adaptation over an extended period of time

Long-term ADAS effects can only be studied during long-term usage (e.g. days, weeks, months or years of exposure). One important aspect of ADAS, often asked by system developers, is, how long do drivers take to adapt to a system, e.g. to use a system as intended and to understand a systems functions?

Evans [11] suggested that *the time* taken for behavioural changes to occur depends on the *ability of road users to detect changes*. Using his interpretation, easily perceived changes will result in relatively quick behavioural adaptation (hours, days or weeks). For more subtle changes of a system (e.g. road width modification), it takes longer for both the detection and manifestation of the behavioural adaptation (months or years). The OECD expert group [12] views behavioural adaptation differently. As mentioned before, the OECD postulated three prerequisites for behavioural adaption to occur: (1) the presence of system feedback to the driver, (2) the driver's ability to change the behaviour and (3) the driver's motivation to act accordingly. Though it is possible to distinguish between different levels of feedback and timings for the behavioural change, the OECD report highlighted that the initial response after a change in the road system is not included in the proposed behavioural adaptation definition. The initial response might be required and expected by the promoter of the system change, to elicit the planned goals. Therefore, behavioural adaptation occurs only *after the initial response* and can be characterised by the process during which the driver integrates the behavioural change in their normal behaviour. A similar notion is followed by Saad and colleagues [52], addressing that behavioural adaptation to in-vehicle systems may not always appear immediately when the driving context is changed, but it usually appears after a familiarisation period. Viti, Hoogendoorn, Alkim and Bootsma [53] regarded a 'two' week learning phase to Adaptive Cruise Control systems as 'fast', referring to a study by Weinberger, Winne and Bubb [54] that estimated that 67% of the drivers learn how and when to use an Adaptive Cruise Control system and when to overrule it within two weeks.

Previous studies [55, 56] give evidence that there are different *behavioural adaptation stages*. Overall, the timing of these stages seems as unclear as the

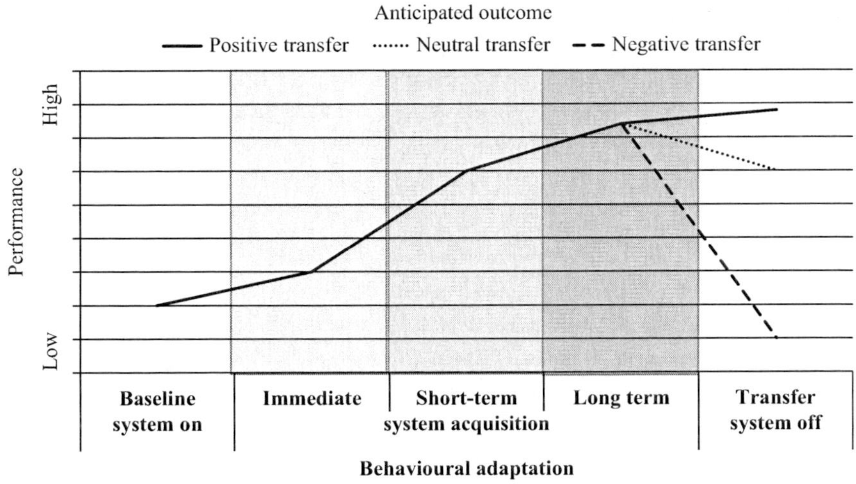

Figure 2.1 Stages of adaptation relative to system use (adapted and printed with permission [55])

definition of these stages. Manser, Crease and Boyle [55] consider adaptation in three stages: immediate (immediately after a driver experiences a change in a safety system), short term (hours, days or weeks after a change in a safety system) and long term (months or years following the change). Manser *et al.* updated the original figure [55] as shown in Figure 2.1. These stages may be considered when examining behavioural adaption relative to safety system use as the quality of the performance is a result of the introduction of a system and its continuous use. According to Manser *et al.* there is always an improvement in performance after the system is activated, a fact that has not always been supported by other researchers.

Some researchers have shown negative behavioural changes that could compromise safety [e.g. 4, 7, 38, 57, 58]. Different effects of studies [59] could be explained by differently defined adaptation time frames (behavioural adaptation stages). A study that systematically investigated two behavioural adaptation stages showed varying driver reactions depending on the time passed after drivers received an ADAS warning, and provided primary empirical evidence for various behavioural adaptation stages following changes in the driving task. Wege, Will and Victor [56] found, using naturalistic driving data, that there is a further adjustment ('post-threat recovery period') after the initial response ('threat-period') to brake-capability forward collision warnings. The use of Manser *et al.* terms and Wege *et al.* terms represents an important area that needs to be clarified in further literature. An attempt to define the stages or phases of behavioural adaption to new in-vehicle technologies over time has been made by Wheatley [60] in the context of the 'culture shock concept' that may help to understand some of the processes involved. 'Culture shock' can be defined as 'the psychological disorientation experienced by people who suddenly find themselves living and working in radically different cultural environments' [61]. Culture shock is characterised by four stages [62], which may apply in similar ways to the process of adaption to new products or systems. These four stages are: the honeymoon stage, the hostility stage, the adjustment stage and the acceptance stage. According to Wheatley the classic stages of culture shock, and its emotional reactions, may be adapted to describe the consequences of driver's behavioural stages following the introduction of ADAS. The first and fourth stages may be associated with positive pleasure experiences while the second and third are associated with negative experiences and stress.

Lately, large field operational test (FOT) studies have investigated behavioural adaption to active safety systems over an extended period of time, e.g. in the Netherlands [53] and across several European countries [63]. One of the first long-term ADAS-related studies presented was an Australian case study reporting a year-long real-world deployment of a driver inattention warning system (DSS) [64]. Results show a trend in the reduction of distraction events during the DSS-alert period. The change in the driver's behaviour is also reflected in the decrease in high acceleration and deceleration events recorded by the DSS. This trend can be attributed to a change in driver's behaviour to avoid actions that trigger distraction alerts, such as long glances away from the road.

Apart from the potential of FOTs investigating ADAS effects over an extended period of time, FOTs can also shed light on real-world examples of *behavioural adaptation to non-driving related tasks*. For instance, in one in-car study the use of speed regulation assistance systems and mobile phone use over a time period of 4 weeks was investigated. Drivers reported strategies related to the use of speed regulation assistance systems in order to compensate for attentional resource allocation to a phone call [65]. This corresponds to the results of a focus group study that suggested that the use of Adaptive Cruise Control systems may promote the occurrence of non-driving related activities [66]. Rudin-Brown and Parker [21] found that drivers reinvested some of the spare cognitive resources into the cognitively demanding non-driving task, e.g. using a cell phone while driving. Rudin-Brown and Parker argue that the increased comfort level provided by driver support systems is similar to a currency to the driver. The mental (and physical) resources 'saved' by the safety and controllability functions of the system can be used to 'finance' behavioural adaptation by 'purchasing' other secondary activities in conjunction with the primary driving task that satisfy the driver's motives [21, 67]. Once more, naturalistic driving studies are found to be a valuable approach in studying the effects of ADAS on the primary driving task as well as on the secondary non-driving tasks.

2.6 Behavioural change transfer

Research should not only include how long it takes to change behaviour, but also whether the change continues after a safety system is no longer in use. Recently, the first theories on a more extended timespan of behavioural adaptation have been published [55]. Those theories form around the *concept of behavioural change transfer* that identifies 'carry-over' effects after ADAS removal. Three examples of behavioural change transfer are: first, a case when a driver decides to switch off a certain safety device after using it for an extended period of time, or second, when a fleet driver switches to drive from a fleet vehicle equipped with ADAS to drive a fleet vehicle not equipped with ADAS. Third, behavioural adaptation may also change as a result of a change in the configuration of the technology, e.g. improved radar capability. These cases are known from real-world examples from FOT studies [e.g. 63] and show that the experimental design of field or driving simulator studies should include the investigation of behavioural change transfer effects. In the stage schemata [Fig. 2.1], the quality of the performance is not only represented by the result of the introduction of a system and its continuous use, but also after the removal of the safety system. According to Manser *et al.* [55], after the driver has used the system for some time, the anticipated outcome in performance slowly deteriorates and can have three possible outcomes after system removal: a positive transfer (quality of performance is higher than before system use), a neutral transfer (quality of performance is on the same level as before system use) or a negative transfer (quality of performance is worse than before system use). Manser *et al.* state that 'most safety system developers hope that their products would promote neutral transfer in that any behavioural adaptations present would dissipate after safety system use has concluded' [55, p. 344]. In that regard, positive and negative transfer only relates to any carry-over effects without addressing the quality of the effect. This theory is derived from the human motor learning and control literature, wherein positive transfer suggests only those behaviours learned during skill acquisition are exhibited sometime later, whereas negative transfer suggests that opposite behaviours occur. Hence, *positive transfer* is defined as any performance, after system removal, being consistent with the systems intended goals of enhancing safety. Positive carry-over means that an ADAS continues to support safe mobility after system use is discontinued. Because the discussion on possible 'carry-over effects' is very new, the terms used to describe these effects need to be standardised and used in a way that is not contradictory. Also, a *multi-stage* description of adaptation processes will be needed in future literature. Manser and colleagues postulate a '*two-stage adaptation process*' ((1) ADAS use, (2) ADAS disuse). However, in practise there could also be a 'three-stage adaptation process' to technology ((1) ADAS use, (2) ADAS disuse, (3) ADAS re-use) and so forth.

Future research would benefit from the inclusion of 'behavioural adaptation transfer' as an experimental methodological factor, for example using an ABA experimental design (control–treatment–control experimental design). ABA-design studies allow analysing persistent intervention effects such as in a study by Carney, McGehee, Lee, Reyes and Raby [68], who found that an event-triggered video intervention system significantly reduced the number of behavioural events during the intervention and did not significantly increase during the second baseline, which was assessed after the intervention ended. Within the ADAPTATION project, behavioural change transfer was also taken into consideration when designing simulator experiments. For example Dotzauer, Caljouw and Brouwer [69] investigated the effects of an intersection assistant system for older drivers in an ABA experimental design. A similar notion of behavioural adaptation was considered in a study by Gouy [70] on behavioural adaptation of unequipped vehicle drivers to the short time headway maintained by vehicles in a platoon. Unequipped vehicle drivers were found to reduce their time headway while they were driving next to a platoon, but the effect disappeared as soon as they passed the platoon.

In sum, in order to answer the question 'How lasting is the effect of an ADAS intervention and/or assistance?', behavioural adaptation studies need to be designed accordingly in the future. It is necessary to consider the different extended time frames of behavioural change, otherwise one might interpret that no behavioural adaptation occurred while it actually takes longer to capture an existing behavioural change [e.g. 24].

2.7 A 'Joint Conceptual Theoretical Framework (JCTF) of Behavioural Adaptation in Response to Advanced Driver Assistance Systems'

The result of the literature review on behavioural adaptation in response to new technologies showed that there is no model that joins the plurality of factors into one framework. As this chapter outlines, a number of theories have been proposed and each place emphasis on different types of processes and mechanisms that influence behavioural adaptation. Although several theories try to model and explain ADAS effects, they can only partly elucidate the inconsistent results. However, none of them allows for specific predictions of how ADAS will influence the driver's behaviour and underlying processes dependent on the type of system studied, the design of the driver–system interaction, the context of use and individual driver characteristics. Currently, there is a lack of research programs that have systematically investigated the conjoint short-, medium- and long-term effects of system properties, on driver's cognitive, energetic and motivational processes underlying behaviour. Thus, one objective of the ADAPTATION project was to develop a conceptual framework [71]. This framework displays the multiple factors acting simultaneously in a complex interplay. Because research on behavioural adaptation to ADAS is still at an early stage, the development of a detailed model is challenging. The aim of the '*Joint Conceptual Theoretical Framework (JCTF) of Behavioural Adaptation in Response to Advanced Driver Assistance Systems*' (Figure 2.2) was to identify the range of relevant internal and external factors associated with behavioural adaptation. The objective of the JCTF was not to develop a model on behavioural adaptation including details on interaction mechanisms, but to show a wide view

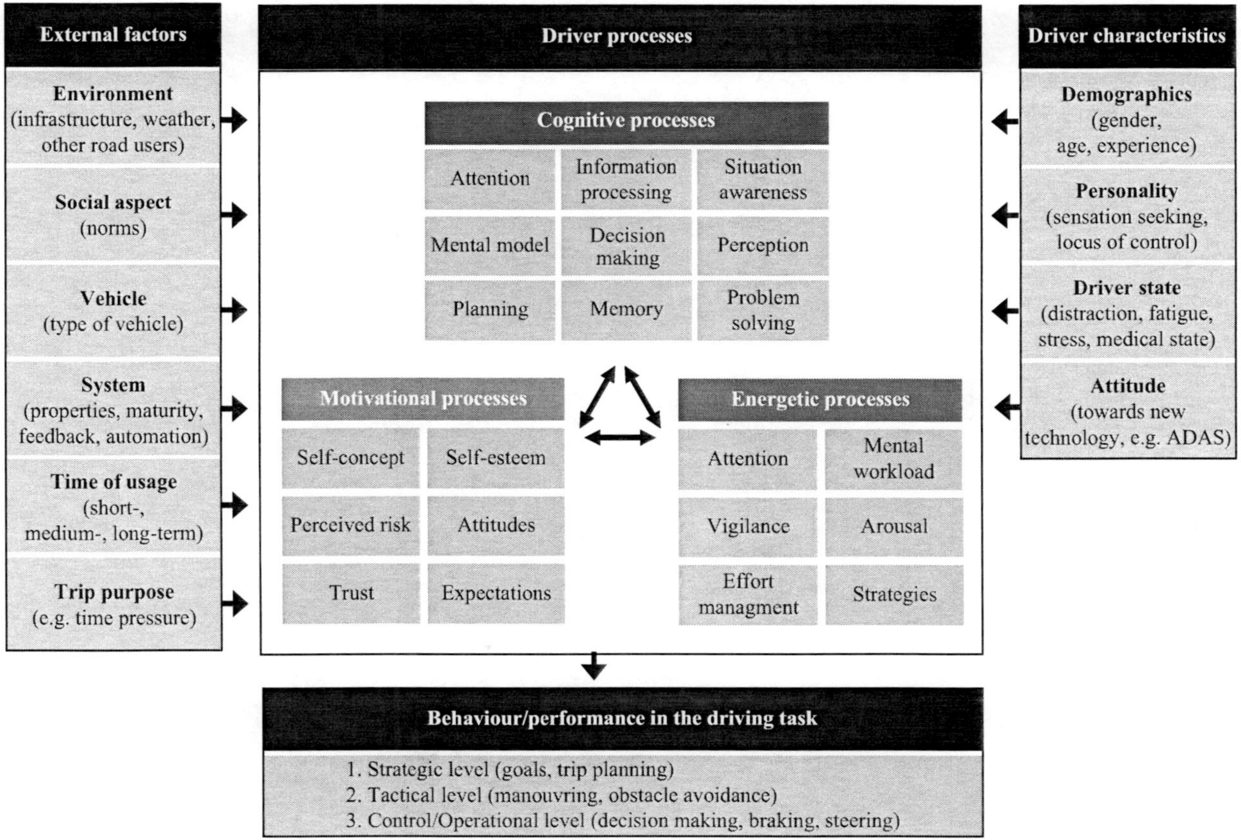

Figure 2.2 Joint Conceptual Theoretical Framework (JCTF) of Behavioural Adaptation in Response to Advanced Driver Assistance Systems

regarding the nature of adaptation processes, emphasising that adaptation does not only include observable behavioural changes, but also psychological driver processes. The definition of behavioural adaptation did not only include unintended changes but also intended changes following ADAS interaction. It is an *integrative* theoretical framework that enables researchers to generate research questions and predictions about the influence of ADAS on different behavioural levels. The JCTF does not target one specific ADAS. It is more a union of various ADAS applications. The behavioural levels, such as strategic, tactical and operational, are adapted from Michon [72]. Furthermore, the JCTF identifies the most relevant external factors and driver characteristics influencing the occurrence and magnitude of behavioural adaptation. External factors such as context variables, e.g. driving task demands, conditions of travel, and driver characteristics, e.g. personality traits such as sensation seeking, locus of control, or a general propensity to trust, as well as attitudes towards ADAS, are expected to influence different psychological processes. The outstanding and innovative characteristic of the JCTF is that it does not only focus on behavioural performance changes but also on their underlying driver internal processes. Those processes are identified as cognitive, energetic and motivational. Each process is characterised by different psychological concepts, which are interrelated with one another without following a hierarchical order.

The *cognitive process* includes:

- Attention [56, 65, 69] and this book Chapters 6, 8, 11 and 12
- Situation awareness [73] and this book Chapter 8
- Mental model [73] and this book Chapter 5
- Information processing [69, 73] and this book Chapters 6 and 12
- Perception [56] and this book Chapter 12
- Problem solving, this book Chapter 12
- Memory, this book Chapter 12
- Planning [65]
- Decision making [65, 69] and this book Chapter 6

The *motivational process* includes:

- Trust [73] and this book Chapters 1 and 5
- Expectations, this book Chapters 5 and 8
- Attitudes, this book Chapter 8
- Perceived risk, this book Chapter 8

The *energetic process* includes:

- Mental workload, this book Chapter 10
- Strategies [65]

In the following chapter of this book some of the collected factors are explained in more detail and in Chapter 17 further research recommendation in regard to the JCTF is given.

2.8 Conclusion and further research needs

In the past decades, behavioural adaptation has received an increasing amount of research attention. Since the OECD [12] definition that postulated the prerequisites of behavioural adaptation: (1) the presence of feedback of the behaviour, (2) the motivation to change behaviour and (3) the human capability to change behaviour, 24 years have passed and new in-vehicle technologies have entered the market. It is time to revisit these prerequisites, to include behavioural transfer into the definition and to advance the development of integrated theoretical models. The ADAPTATION project has expedited this development. The literature review on behavioural adaption to new technologies showed there is a need to advance the debate towards clearer terminology on behavioural adaptation processes.

With the vast progress in automobile technology and the implementation of ADAS, the driving task changes. In the future, the driver's role might become being a 'supervisor of ADAS' rather than the person controlling the vehicle. System designers assume that behavioural adaptation will occur [59] because it is intrinsic to human nature to try to modify behaviour to suit new conditions. Research must identify the nature of these changes in the complex human–machine interplay in order to enhance system design. The results of research on behavioural adaptation should be included in the design of new in-vehicle support systems and in the assessment processes of the associated user interfaces.

Until recently, most research on driving support systems only took a 'snapshot' of behaviour with the system. In order to capture instances of behavioural adaption, studies need to last over longer periods of time. In FOT studies and naturalistic driving studies, a natural behavioural response to ADAS, without the interference of an examiner, can be investigated. The challenge of assessing behavioural change in laboratory settings to naturalistic driving settings is further discussed in Chapter 3.

Empirical research does not yet reveal how lasting the behavioural adaptations during safety system exposure are, and how transient the behavioural adaption effect after system removal is. It is likely that there are immediate, short-term and long-term effects on performance. When designing a study investigating behavioural adaptation, it is necessary to consider the different extended time frames of behavioural change, otherwise one might interpret that no behavioural adaptation occurred while it actually takes longer to capture a change.

Based on the review of existing theories of behavioural adaptation, a '*Joint Conceptual Theoretical Framework (JCTF) of Behavioural Adaptation in Response to Advanced Driver Assistance Systems*' was developed in order to combine multiple aspects of the driver–vehicle–environment concept. The strength of the framework is the identification of important variables associated with behavioural adaptation. In further research, it is necessary to critically analyse the concepts and variables involved, to identify the advantages and disadvantages of a qualitative framework and to find out about the independence or overlapping of different concepts.

Is the initial question in this chapter of Khalil Gibran's idea of 'adaption to new life conditions' transferable to transport research? As this chapter outlined, the answer is complex. However, one fact seems to be certain: 'Change is constant'.

Acknowledgements

This research received funding from the European Community's Seventh Framework Programme (FP7/2007-2013) under grant agreement no. 238833/ ADAPTATION project (www.adaptation-itn.eu). The authors would like to thank all ADAPTATION partners for developing the JCTF and for conducting a literature review on ADAS effects.

References

1. http://oxforddictionaries.com [Accessed 20 February 2013]
2. Smiley A. 'Behavioral adaptation, safety, and intelligent transportation systems'. *Transportation Research Record: Journal of the Transportation Research Board*. 2007;1724(2007):47–51
3. Rudin-Brown C. "Intelligent" in-vehicle intelligent transport systems: limiting behavioural adaptation through adaptive design'. *IET Intelligent Transport Systems*. 2010;4(4):252–261
4. Dragutinovic N. M., Brookhuis K. A., Marchau V. A. W. J. 'Behavioural adaptation in response to advanced driver assistance systems' in de Waard D., Brookhuis K. A., Weikert C. M. (eds.). *Human Factors in Design*. Maastricht, the Netherlands: Shaker; 2004. pp. 47–51
5. Rudin-Brown C., Jameson S. (eds.). *Behavioural Adaptation and Road Safety: Theory, Evidence, and Action*. Boca Raton, FL: CRC Press; 2013
6. Comte S. 'New systems: new behaviour?'. *Transportation Research Part F*. 2000;3:95–111
7. Carsten O. 'Behavioural adaptation: friend or foe?' in de Waard D. (ed.). *Human Factors: Security and Safety*. Maastricht, the Netherlands: Shaker; 2009. pp. 208–232
8. European Commission – Directorate General for Mobility and Transport. *EU road fatalities*. [online]. 2013. Available from http://ec.europa.eu/transport/road_safety/pdf/observatory/trends_figures.pdf [Accessed 16 April 2013]
9. Whitelegg J., Haq G. *Vision Zero: Adopting a Target of Zero for Road Traffic Fatalities and Serious Injuries*. [online]. Stockholm Environment Institute; 2006. Available from http://sei-international.org/mediamanager/documents/Publications/Future [Accessed 30 January 2013]
10. Wilde G. 'The theory of risk homeostasis: implications for safety and health'. *Risk Analysis*. 1982;2:209–225
11. Evans L. 'Human behaviour feedback and traffic safety'. *Human Factors*. 1985;27:555–576
12. Organisation for Economic Co-operation and Development (OECD) (Paris, France) Scientific Expert Group. *Behavioural Adaptations to Changes in the Road Transport System*. Paris: OECD; 1990
13. Evans W. N., Graham J. D. 'Risk reduction or risk compensation: the case of mandatory safety-belt use laws'. *Journal of Risk and Uncertainty*. 1991;4(1): 61–73

14. Grayson G. B. *Behavioural Adaptation: A Review of the Literature.* TRL report 254; 1996
15. Marchau V. A. W. J., van der Heijden R. E. C. M. 'Introducing advanced electronic driver support systems: an exploration of market and technological uncertainties'. *Transport Reviews: A Transnational Transdisciplinary Journal.* 2000;20(4):421–433
16. Gibson J., Crooks L. 'A theoretical field-analysis of automobile-driving'. *The American Journal of Psychology.* 1938;51(3):453–471
17. Smeed R. J. 'Some statistical aspects of road safety research'. *Journal of the Royal Statistical Society.* 1949;Series A:1–34
18. Fuller R. 'The task-capability interface model of the driving process'. *Recherche – Transports – Sécurité.* 2000;66:47–57
19. Jiang C., Underwood G., Howarth C. I. 'Towards a theoretical model for behavioural adaptations to changes in the road transport system'. *Transport Reviews.* 1992;12(3):253–264
20. Vincent A., Noy I., Laing A. 'Behavioural adaptation to fatigue warning systems'. [online]. *Proceedings of the 16th International Technical Conference on the Enhanced Safety of Vehicles; Ontario, Canada, May–June 1998*; Paper Number 98-S2-P-2 1. Available from http://www-nrd.nhtsa.dot.gov/pdf/nrd-01/Esv/esv16/98S2P21.PDF [Accessed 4 January 2013]
21. Rudin-Brown C. M., Parker H. A. 'Behavioural adaptation to adaptive cruise control (ACC): implications for preventive strategies'. *Transportation Research Part F.* 2004;7(2):59–76
22. Hoedemaeker M., Brookhuis K. A. 'Behavioural adaptation to driving with an adaptive cruise control (ACC)'. *Transportation Research Part F.* 1998;1:95–106
23. Saad F. 'Some critical issues when studying behavioural adaptations to new driver support systems'. *Cognition, Technology & Work.* 2006;8(3):175–181
24. Hjälmdahl M., Várhelyi A. 'Speed regulation by in-car active accelerator pedal-effects on driver behaviour'. *Transportation Research Part F.* 2004;2(7):77–94
25. Fuller R. 'Towards a general theory of driver behaviour'. *Accident Analysis and Prevention.* 2005;37:461–472
26. Ward N. J. 'Automation of task processed: an example of intelligent transportation systems'. *Human Factors and Ergonomics in Manufacturing.* 2000;10(4):395–408
27. Godley S. T. *A Driving Simulator Investigation of Perceptual Countermeasures to Speeding.* Doctoral thesis, Department of Psychology, Monash University; 1999
28. De Waard D., Steyvers F. J. J. M., Brookhuis K. A. 'How much visual road information is needed to drive safely and comfortably?' *Safety Science.* 2004;42:639–655
29. Leong H. J. W. 'The distribution and trend of free speeds on two-lane rural highways in New South Wales'. *Proceedings, Australian Road Research Board.* 1968;4:791–814

30. Messer C. J., Mounce J. M., Brackett R. Q. *Highway Geometric Design Consistency Related to Driver Expectancy*. Washington, DC: Federal Highway Administration. FHWA-RD-81-035; 1981
31. van Driel C. J. G., Davidse R. J., van Maarseveen M. F. A. M. 'The effects of edgeline on speed and lateral position: a meta-analysis'. *Accident Analysis and Prevention*. 2004;36:671–682
32. Rockwell T. H., Lindsay G. F. 'Driving performance, Part II' in *Effects of Illumination on Operating Characteristics of Freeways.* Washington, DC: NCHRP Report, Highway Research Board; 1968;60:51–71
33. Huber M. J., Tracey J. L. *Effects of Illumination on Operating Characteristics of freeways.* Washington, DC: NCHRP Report No. 60; 1968
34. Engström J., Hollnagel E. 'A general conceptual framework for modeling behavioural effects of driver support functions' in Cacciabue P. C. (ed.). *Modeling Driver Behavior in Automotive Environments: Critical Issues in Driver Interactions with Intelligent Transport Systems*. London: Springer; 2007
35. Cacciabue P. C., Saad F. 'Behavioural adaptations to driver support systems: a modelling and road safety perspective'. *Cognition, Technology & Work.* 2008;10(1):31–39
36. Cotter S., Mogilka A. *Methodologies for the assessment of ITS in terms of driver appropriation processes over time.* HUMANIST Project Deliverable E.6, 2007
37. Rudin-Brown C. M., Parker H. A. 'Behavioural adaptation to adaptive cruise control (ACC): implications for preventive strategies'. *Transportation Research Part F: Traffic Psychology and Behaviour.* 2004;8(2):59–76
38. Rudin-Brown C. M., Noy I. 'Investigation of behavioural adaptation to lane departure warning'. *Transportation Research Records*. 2002;1803:30–37
39. Saad F. 'Dealing with behavioral adaptations to advanced driver support systems' in Cacciabue P. C. (ed.). *Modeling Driver Behavior in Automotive Environments*. London: Springer; 2007. pp. 147–161
40. Weller G., Schlag B. 'Verhaltensadaptation nach Einfuehrung von Fahrerassistenzsystemen: Vorstellung eines Modells und Ergebnisse einer Expertenbefragung' in Schlag B. (ed.). *Verkehrspsychologie: Mobilität – Verkehrssicherheit – Fahrerassistenz.* Lengerich: Pabst Science; 2003
41. Elvik R. 'Laws of accident causation'. *Accident Analysis and Prevention*. 2006;38:742–747
42. Rasmussen J. 'Skills, rules, knowledge; signals, signs, and symbols, and other distinctions in human performance models'. *IEEE Transactions on Systems, Man and Cybernetics.* 1983;13:257–266
43. Fitts P. M., Posner M. I. *Human Performance*. Belmont, CA: Brooks-Cole; 1967
44. Anderson J. R. 'Acquisition of cognitive skill'. *Psychological Review.* 1982;89:369–406
45. Byrne M. D., Anderson J. A. 'Enhancing ACT-R's perceptual-motor abilities'. *Proceedings of the Nineteenth Annual Conference of the Cognitive Science Society*. Stanford University, Mahwah, NJ: Erlbaum; 1997. p. 880

46. Bellet T., Bailly-Asuni B., Banet A., Mayenobe P. 'A theoretical and methodological framework for studying and modeling drivers' mental representations'. *Safety Science*. 2009;47(9):1205–1221
47. Haigney D. E., Taylor R. G., Westerman S. J. 'Concurrent mobile (cellular) phone use and driving performance: task demand characteristics and compensatory processes'. *Transportation Research, Part F*. 2000;3:113–121
48. Lewis-Evans B., Charlton S. G. 'Explicit and implicit processes in behavioural adaptation to road width'. *Accident Analysis and Prevention*. 2006;38(3):610–617
49. Heijer K., Brookhuis K., van Winsum W., Duynstee L. *Automation of the Driving Task*. SWOV Institute for Road Safety Research, Leidschendam, Report R-98-9; 1998
50. Forbes N. L. *Behavioural Adaptation to In-Vehicle Navigation Systems*. Doctoral thesis, University of Nottingham; 2009
51. Larsson A. F. 'Driver usage and understanding of adaptive cruise control'. *Applied Ergonomics*. 2012;43(3):501–506
52. Saad F., Hjälmdahl M., Canas J., Alonso M., Garayo P., Macchi L., Nathan F., Ojeda L., Papakostopoulos V., Panou M., Bekiaris A. *Literature Review of Behavioural Effects*. AIDE Project, Deliverable D1.2.1; 2004
53. Viti F., Hoogendoorn S. P., Alkim T. P., Bootsma G. *Driving Behaviour Adaptation under ACC: Results from a Large Field Operational Test in the Netherlands*. [online]. Available from http://www.academia.edu/1091192/Driving_behavior_adaptation_under_ACC_results_from_a_large_Field_Operational_Test_in_the_Netherlands [Accessed 8 February 2013]
54. Weinberger M., Winne, H., Bubb H. 'Adaptive cruise control field operational test – the learning phase'. *Society of Automotive Engineers of Japan Review*. 2001;22:487–494
55. Manser M. P., Crease, J., Boyle L. 'Methodological and measurement issues in behavioural adaptation' in Rudin-Brown M., Jamson S. (eds.). *Behavioural Adaptation and Road Safety: Theory, Evidence, and Action*. Boca Raton, FL: CRC Press; 2013. pp. 339–358
56. Wege C., Will S., Victor T. 'Eye movement and brake reactions to real world brake-capacity forward collision warnings – a naturalistic driving study'. *Accident Analysis and Prevention*. 2013;58(9):259–270
57. Brookhuis K. A., Waard De. D., Janssen W. H. 'Behavioural impacts of advanced driver assistance systems – an overview'. *European Journal of Transport and Infrastructure Research*. 2001;1(3):245–253
58. Gossner O., Picar P. 'On the consequences of behavioral adaptations in the cost-benefit analysis of road safety measures'. *Journal of Risk & Insurance, The American Risk and Insurance Association*. 2005;72(4):577–599
59. Smiley A. 'Behavioural adaptation, safety and ITS'. Presented at the ITS Safety Test and Evaluation Workshop, IHRA-ITS; Washington, DC, 2000
60. Wheatley D. J. 'Culture shock and adaptation to new in-vehicle telematics systems'. *Proceedings of the International Conference on Affective Human Factors Design*; Singapore, June 2001. London: Asean Academic Press; 2001. pp. 480–486

61. Ferraro G. P. *The Cultural Dimension of International Business*. 3rd Edn. Upper Saddle River, New Jersey: Prentice Hall; 1998
62. Ober K. 'Culture shock – adjustment to new cultural environments'. *Practical Anthropology*. 1960;7/8:177–182
63. Malta L., Ljung Aust M., Gustafsson D., Dobberstein J., Benmimoun M., Pütz A., Saint Pierre G., Metz B., Freek F. *Final Results: Impacts on Traffic Safety*. euroFOT deliverable 6.4. Brussels; 2012
64. Heinzman J., Zelinsky A. 'Automotive safety solutions through technology and human-factors innovation'. Presented at the 12th International Symposium on Experimental Robotics; Delhi, India, December 2010
65. Huth V., Lancelle V., Gabel C., Bonnard A., Brusque C. 'Exploring mobile phone usage and its context with naturalistic driving observations' in de Waard D., Brookhuis K., Dehais F., Weikert C., Röttger S., Manzey D., Biede S., … Terrier P. (eds.). *Proceedings HFES Europe Chapter Conference:* Toulouse, France, October 2012. Available from http://hfes-europe.org [Accessed 28 June 2013]
66. Bianchi Piccinini G., Simoes A., Rodrigues C. M. 'Effects on driving task and road safety impact induced by the usage of Adaptive Cruise Control (ACC): a focus groups study'. *The International Journal of Human Factors and Ergonomics*. 2012;1(3):234–253
67. Manser M. P., Kuge N., Boer E. 'Driver behavioural adaptation in response to a novel haptic driver support system'. *Proceedings of the Human Factors and Ergonomic Society 49th Annual Meeting;* Orlando, FL, 2005. pp. 1950–1954
68. Carney C., McGehee D. V., Lee J. D., Reyes M. L., Raby M. 'Using an event-triggered video intervention system to expand the supervised learning of newly licensed adolescent drivers'. *American Journal of Public Health*. 2010;100(6):1101–1106
69. Dotzauer M., Caljouw S., Brouwer W. H. 'Intersection assistance: a safe solution for older drivers'. *Accident Analysis and Prevention*. 2013;59:522–528
70. Gouy M. *Behavioural Adaptation of Drivers of Unequipped Vehicles to Short Time Headways Observed in a Vehicle Platoon*. Doctoral thesis. Milton Keynes: Open University; 2013
71. Popken A. *Drivers' Reliance on Lane Keeping Assistance Systems as a Function of the Level of Assistance*. Doctoral thesis. Chemnitz: Chemnitz University of Technology; 2009
72. Michon J. A. 'A critical view of driver behaviour models: what we know, what should we do?' in Evans L., Schwing R. C. (eds.). *Human Behavior and Traffic Safety*. New York, NY: Plenum Press; 1985. pp. 485–520
73. Beggiato M., Krems J. F. 'The evolution of mental model, trust and acceptance of adaptive cruise control in relation to initial information'. *Transportation Research Part F*. 2013;18:47–57

Chapter 3

Methods to assess behavioural adaptation over time as a result of ADAS use

Mandy Dotzauer[1], Vera Berthon-Donk[2], Matthias Beggiato[3], Juliane Haupt[4] and Giulio Piccinini[5]

Abstract

In recent decades, major technological advances have allowed a large number of Advanced Driver Assistance Systems (ADAS) and In-Vehicle Information Systems (IVIS) to be introduced that intend to improve road and driver safety. With the introduction of ADAS and IVIS, there have also been unintentional potential driver distractions and other safety effects. These effects can be of a short-, medium- or long-term nature. Driving simulator studies and on-road studies offer opportunities to investigate behavioural change as a result of the use of ADAS. In order to design the most suitable approach, it is important to know more about each investigation method, contrasting the methodological approaches with regard to utility, potential research questions, data collection and validity. Moreover, the approaches can also be viewed as complementary for assessing behavioural adaptation effectively and efficiently. This chapter will shed some light on the different methods, pointing out the advantages and disadvantages of realistic driving settings and simulated settings for the investigation of behavioural change.

3.1 Introduction

Over the last 20 years, major technological advances have occurred in the field of road traffic, in general, and specifically in the domain of automobile driving. An increasing number of cars are now equipped with Advanced Driver Assistance Systems (ADAS) and In-Vehicle Information Systems (IVIS) that are designed to help drivers perform better in their task by providing information, warnings and even (partially) take over control. As such, they provide longitudinal (e.g. Adaptive Cruise

[1] University Medical Centre Groningen, Groningen, the Netherlands
[2] IFADO, Dortmund, Germany
[3] Chemnitz University of Technology, Chemnitz, Germany
[4] FACTUM OHG, Vienna, Austria
[5] ISEC, Lisbon, Portugal

Control) or lateral support (e.g. Lane Departure Warning), aid in navigating through an unknown city, or warn about upcoming critical events. As a result, driving safety may be enhanced, in general, and the driving behaviour of populations at risk (e.g. older drivers, novice drivers) may become more efficient and safe [1].

One of the drawbacks of all those systems, however, is that they can potentially become a serious source of in-vehicle distraction. Substantial research has shown that the secondary task represented by observing, attending to, or monitoring ADAS has an impact on the performance of the primary driving task. However, as discussed in Chapter 2, often those conclusions are mainly based on usability studies in which ADAS inexperienced participants were observed during the first-time use of a system in a single session. Such studies allow only limited conclusions to be drawn concerning the adaptation of users to new or unknown systems. In contrast, longer-term experimental designs have the benefit of allowing the examination of efficacy of an ADAS system when freshly introduced as well as after continued use. The continued use allows examination of whether drivers continually adapt to ADAS and whether such adaptation is associated with performance improvements or it declines over time. Furthermore, long-term evaluations are useful to examine retention effects and post-use transfer effects concerning other systems [2]. Several methods allow the investigation of behavioural effects of ADAS use over time, such as driving simulator studies and naturalistic driving studies (NDS).

In this chapter, we will first look at the use of long-term studies. Based on a model describing the acquisition of skills with regard to driving and ADAS, we consider the advantages and inconveniences of this type of research. Then we will look at both simulator studies and NDS, carefully contrasting each methodological approach with regard to utility in behavioural adaptation research, potential research questions, data collection, validity, advantages and disadvantages.

3.2 The use of long-term studies to assess behavioural change with ADAS use

To understand the utility of long-term studies in examining (potential) behavioural change induced with ADAS use, it is important to first look at the cognitive processes that take place when such systems are used while driving. In this section, we will first look at a theoretical model aiming at explaining the processes that influence long-term behavioural adaptations, before looking at the advantages and the inconveniences of the use of such studies.

3.2.1 Cognitive adaptation to ADAS

3.2.1.1 Defining a theoretical framework

Within the framework of the European project Adaptive Integrated Driver-vehicle InterfacE (AIDE) [3], a comprehensive literature study was conducted to define processes influencing long-term behavioural adaptations. The project identified some temporal factors that affect behavioural adaptation and that can be observed while using long-term experimental designs. First, they identified a *learning phase*. During this phase, drivers discover the system and explore how to operate it, identify precise system limits and define domains of utility. This learning process plays an important role in the construction of a mental representation, the driver's confidence in the system and the definition of its optimal use. The speed of the learning process is highly influenced by the behaviour and interface of the system. If a system is easy to use and transparent concerning different system states (such as working, warning, alarm), the driver will more easily adapt to it rather than to a system that is ambiguous. Furthermore, individual factors, such as familiarisation with new technologies, in general, and the driver's experience with other in-vehicle systems, might influence this learning process.

The AIDE project then defined an *integration phase*, which is identified by a reorganisation of the driving task in order to include system functionalities. Michon [4], based on the work by Allen, Lunenfeld, and Alexander [5], argues that the driving task can be defined by a hierarchical structure of subtasks. As such, the most basic tasks are those that take place at a so-called *operational* (or *control*) level that refers to direct control of the vehicle such as pushing the gas or brake pedal. Higher up in the hierarchy, tasks take place on the *manoeuvring* level (or *tactical* level), which refer to the interaction with the direct road environment, for example, interacting with other road users and within the driving environment by respecting safety distances, taking turns and respecting traffic signs.

Finally, the *strategic* level represents higher-order tasks such as trip planning or navigating to reach certain destinations in the most efficient way. What makes this model interesting is the *factor* time that is inherent to each level. Decisions/actions on the operational level are immediate reactions to danger and operate in the millisecond to second zone. Decisions on the tactical level are not immediate reactions to danger but reactions to cues or contexts that predict danger. They operate in the second to minute zone and can strongly reduce the probability of ending up in time-pressured operational level tasks. Drivers can also make adjustments on the strategic level, which in turn influences task accomplishments on the operational level and minimises the probability of decision making under time pressure. While planning a trip, drivers can decide on the time of travel and the route, for example, avoiding rush hours, highly complex intersections or routes with high traffic density [6].

3.2.1.2 Expected behavioural adaptations with the introduction of ADAS

There are few reported long-term studies in the literature that examine behavioural changes under the influence of ADAS. It can be expected that while using such studies to observe changes in driver behaviour, the learning phase will be foremost represented by changes on the operational level, for example, an increase in facility of use while activating or deactivating the assistant system. When the driver is used to the basic functionalities of the system, the integration phase starts; the ADAS becomes integrated within the mental model of the driving task represented by changes on the manoeuvring level. Changes at on the manoeuvring level can be positive, as shown by Caird, Chisholm, and Lockhart [7], who found that advanced in-vehicle signs giving information about an upcoming intersection reduced the velocity towards that intersection. However, negative changes on the manoeuvring level have also been observed, as shown by Vollrath, Schleicher, and Gelau [8] who found that drivers reacted more slowly to critical events while using Cruise Control (CC) or Adaptive Cruise Control (ACC). Finally, when the system is fully integrated within the mental model, it might alter tasks on the strategic level as well. Some drivers will use the ADAS as a 'reference tool' [9] in order to receive feedback about their driving performance and improve when necessary (for an example see [10]). However, some other drivers might use the ADAS as a 'slave' [9] offering them the opportunity to allocate their attention to other in-vehicle tasks or adapt their behaviour in ways that is not beneficial for traffic safety, for example smaller minimum time-headway, higher mean speeds or stronger brake forces while using an ACC [11]. It should be noted that the aforementioned studies were all single-session studies, and that long-term studies could potentially shed more light onto the underlying processes involved in learning to use an ADAS.

3.2.2 Advantages of long-term studies

3.2.2.1 Modelling behavioural adaptation

Repetitive training trials allow the calculation of learning curves (i.e. mathematical models used to estimate efficiencies gained when an activity is repeated) which can be used for predictions of speed and ease of learning of a new system. One example of such a mathematical model is the power law of practice. Fitting a power law of practice to repeated performance observations can provide useful insights into the degree of learning needed for the proper use of an ADAS or its human–machine interface (for an example see [12]). For more information on the use of the power law of practice in driving situations, see Chapter 6.

3.2.2.2 Retention effects

Another advantage of using long-term driving studies to study behavioural changes while using ADAS is the possibility to study retention effects. What happens once a driver gets used to a system and then has to drive without it? What impact does the use of one system have on the use of another, similar system, or a combination of systems? Manser and Creaser [2] found that participants who had been using a Co-operative Intersection Collision Avoidance System – Stop Sign Assist (CICAS-SSA) on three consecutive days showed positive transfer effects on the fourth day when the system was turned off. Drivers seemed to continue to benefit from the system use (showing enhanced intersection crossing behaviour, such as shorter wait times before crossing, as compared to intersection crossing behaviour before system use) even when the system was no longer available. Unfortunately, the opposite effect has also been observed. Stanton, Young, and McCaulder [13] found in a driving simulator study that, due to increased automation from using an ACC, some drivers had difficulties reclaiming quick control of the car in case of system failure. Stanton, Young, Walker, Turner, and Randle [14] reported similar findings from a study in which they compared the effect of driving with or without

ACC and/or Active Steering (AS) in a driving simulator study where participants using automation were less likely to reclaim control in an emergency-braking scenario and therefore were more likely to be involved in collisions than when driving the car under manual control. The authors attribute these negative effects on behaviour from driving with an ADAS to over-reliance on the technology.

In conclusion, it can be said that the examination of potential (positive or negative) retention effects on behavioural changes, due to ADAS use, can be of interest when examining the effect of behavioural changes due to ADAS use.

3.2.3 Inconveniences of long-term studies

3.2.3.1 Definition of optimal length

Although long-term studies allow us to gain insight into strategy changes and retention effects when using ADAS, a few inconveniences are paired with long-term studies. It might be a problem to define the length of the experimental period, the number of experimental sessions within this period and the interval time between those experimental sessions. A literature research showed that there seems to be no real consensus for these issues and experimental periods and sessions and intervals vary between experiments. Shina, Tractinsky, and Compton [15] used five sessions extended over 14 days, with one to four days between sessions, to study the effect of practice on the number of errors in a mathematical task while driving. Chisholm, Caird, and Lockhart [16] studied the practice effect of repetitive MP3 use while driving with six experimental sessions each within a one-week interval and over a two-month period. Finally Brookhuis, de Vries, and de Waard [17] tested 15 participants on every (working) day for three weeks to assess the effect of operating a mobile phone in different traffic conditions. All these studies found behavioural adaptation effects over time, but a few methodological questions remain:

- Would the observed behavioural effects have been different with an increase or decrease in the experimental period?
- Are there any confounding factors that could explain behavioural effects observed over time, which were not within experimental control, such as boredom with the task, lack of motivation, fatigue, increased familiarity, over-learning or over-stimulation [9]?
- When do we know that a performance plateau is reached and that no further testing is needed?

All these questions can be hard to answer before starting a long-term study as they will largely depend on the system studied, the system interface and individual factors such as motivation, curiosity and familiarity with new technologies.

3.2.3.2 Motivational issues

Participants need to stay motivated to perform (often) repetitive manoeuvres over days, weeks or even months in simulated driving environments where no real driving risks are present. Therefore, performance incentives will often be needed. Ranney [18] describes the use of performance incentives such as feedback and reward/penalty systems in order to increase the psychological fidelity and enhance the perception of risk in driving simulators. He defines good driving performance incentives to include clear instructions on the performance space and the driving task at hand. As such, in a long-term study, participants should be informed that performance depends on completing all sessions. Caution should however be taken when using rewards (e.g. money or course credits) especially when those rewards consist of refunds of expenses and are independent of the participant's performance. This said, incentive will not guarantee motivation until the end of the experiment due to increased familiarity, over-learning or over-stimulation. The experimenter should try to control those confounding factors in choosing the right experimental duration.

3.2.3.3 Costs

Finally, long-term studies might be costly, not only at a monetary level (if participants need to be rewarded for extended participation) but also at an organisational level.

3.2.4 Conclusion on long-term studies

In conclusion, we can state that long-term behavioural observations have the advantage of allowing the examination of behavioural adaptation to newly introduced systems at several points in time and at different levels of the driving task. The observation of repetitive use of a system allows the definition or fit of learning curves that can be useful in predicting future use by novice ADAS users. Nonetheless, when designing long-term studies, it is necessary to keep in mind the difficulties and challenges that come with time-consuming and cost-intensive research.

3.3 Long-term studies using driving simulators

The next section focuses on driving simulators and simulated testing environments, and provides information about what can be done with simulations and the robustness of results and conclusions.

3.3.1 Introducing driving simulators

Driving simulators were developed in order to avoid costly and time-consuming field studies. Because of flexible configurations of simulators, they offer cost savings and allow investigating a wide range of research questions. Driving simulators also increase experimental control and allow presentation of hazardous situations in a safe environment [19]. Starting around 1950, driving simulators were used to study the driver, vehicle and the environment. In the late 1970s, the development of driving simulations had evolved into five key areas [20]:

1. Visual and auditory display generation
2. Kinaesthetic cues
3. Driving scenario programming
4. Computational vehicle dynamics
5. Vehicle cab including controls and instruments

Thirty years ago, throughout Europe and the United States only 20 simulators were in use. They were mainly utilised for training and licensing purposes. Throughout the years, technological difficulties, especially with time delays, have been overcome by overstepping from analogue electronics to digital computation. The presentation of the simulation improved over the years as well. Initially, film-based motion picture simulators were used for research, whereas, today, simulations use digital graphics, providing a realistic driving experience [21].

Driving scenarios are projected onto a big screen and participants are either seated on a simple seat in front of a desk with a steering wheel attached to it, or in a real car including a fully equipped dashboard and instrument cluster. Participants are instructed to drive as they would normally and the experimenter and the participant are often connected through a microphone for further instructions during the drive. Participants' behaviour and performance can be recorded to allow for more in-depth post-analysis [22].

Simulators range from low- and medium- to high-fidelity simulators. Low-fidelity simulators are simple desktop simulators with a videogame-type steering and pedal input device. Medium fidelity simulators are more advanced with participants usually seated in either part of a vehicle or an actual vehicle cab. High-fidelity simulators are the most sophisticated types of driving simulators [23]. The two common types of high-fidelity simulators used are dynamic and stationary. Dynamic driving simulators mimic movements such as acceleration, deceleration and declination. These types of simulators provide the most realistic driving experience. Stationary simulators, on the other hand, are fixed-based and do not simulate movements. Participants do not receive any haptic feedback on their driving style. It makes the driving experience less realistic than in a dynamic simulator [22].

3.3.2 Utility of driving simulators

Driving simulators can be used for many different applications. Common broad research areas are highway and road design, testing planned constraints on drivers, and investigating driver distraction and cognitive load [24].

3.3.2.1 Testing road design

Driving simulators provide the opportunity to conduct studies that would not be realisable on concrete roadways due to time and cost constraints. As an example, Kantowitz [25], by comparing orthogonal combinations of curve radius and deflection angle in a driving simulator, retrieved valid data and results that could not be obtained by observational studies. Other research for which driving simulators have been of use include studies related to road tunnel design. Specific tunnel and road design concepts can be tested in a simulator before the road/tunnel has been constructed [24].

3.3.2.2 Investigating planned constraints on drivers

Driving simulators can also be used to investigate the effects of planned constraints placed on the driver, for example, in order to gain more insight into potential safety benefits. One example is a study for the European Commission, in which variations of Intelligent Speed Assistance (ISA) were tested in a simulated environment [26]. Different modes of the system were programmed and tested before implementing road speed management methods [22].

3.3.2.3 Researching driver distraction and cognitive load

Driving simulator studies also provide a safe environment to test and investigate driver distraction and cognitive workload. Secondary tasks, which can either be simulations of real systems [27] or surrogate tasks [28], can be programmed to induce different types of workload and distraction. It is important to keep in mind that perceived priorities between the driving and the secondary task might shift, as no real accident risk is associated with driving in a simulator [29]. Nonetheless, driving simulators are a valuable tool to investigate the effects of distraction and workload on the driving performance as long as validation studies are also performed on the road. Using simulators can also help to efficiently eliminate aspects of workload and distraction that do not influence the driving performance and therefore do not require further testing in a real-world setting [24].

3.3.3 Advantages of driving simulators

Driving simulator studies have several advantages over NDS and FOTs (field operational tests). They are less time and cost intensive and they allow for a vast amount of data collection. Another benefit of driving simulations is that they provide a standardised, controllable, reproducible and safe environment for different driving manoeuvres.

3.3.3.1 Controllability, reproducibility and standardisation

An advantage of driving simulator research is that almost all variables can be controlled and manipulated thereby offering high experimental control. Identical driving scenarios can be presented across participants including identical virtual traffic (e.g. traffic density, behaviour of other traffic), weather conditions and road layout. Those factors can be further manipulated and configured with respect to specific research questions and research aims. Using simulators, it is possible to test participants in physically different locations applying the same scenario. Driving simulators allow for standardised driving tests and produce comparable as well as reproducible results [19, 22].

3.3.3.2 Dangerous traffic conditions

Driving simulator studies provide a safe environment in which crashes and hazardous situations can be simulated without harming participants and other road users. One example is measuring driving performance under the influence of drugs or alcohol, which would not be possible in real road settings. Dangerous driving situations such as police pursuits or driving in extreme weather conditions can also be tested in a simulator [19]. Simulated driving scenarios can also be used to prepare novice drivers handling unpredictable or safety-critical tasks to which they cannot be exposed in the real-world traffic (e.g. collision avoidance). Considering that the number of crashes among younger drivers is significantly higher during the first year of driver's license possession and with increasing driving experience the number of crashes decreases [30], simulators can be used to create training curricula that allow novice drivers to improve driving skills in a safe and simplified environment, before being transferred to real traffic conditions [31].

3.3.3.3 Time and money saving

Executing experiments in a driving simulator can be a cost- and time-efficient alternative to naturalistic or test track driving. Once programmed and configured, scenarios can be easily reproduced and spread [32]. Variables of interest can be systematically manipulated and resulting driving performance and behaviour measured [23] to quickly obtain vast data with regard to a precise research question.

3.3.3.4 Accurate and efficient data collection

In driving simulators, vast amounts of driving performance parameters can be recorded accurately as well as efficiently. Examples are longitudinal control (e.g. speed, speed variability, time headway and the distance to a stopping line), lateral control (e.g. lateral position, standard deviation of lateral position, lane departure, time-to-lane crossing) and hazard-response detection (e.g. perception response time, brake response time, time-to-collision). In real-world driving, information about lateral positioning can be difficult to obtain because it requires visible lane markings that are sensitive to weather conditions, reflections and shades [33].

The laboratory setting of driving simulations also allows for controlled conditions that make it possible to record other measures such as eye movements or physiological measures. Furthermore, it provides the opportunity to assess subjective workload.

3.3.4 Disadvantages of driving simulators

Using driving simulators for testing a variety of research questions is compelling, but they are not beneficial in every situation. Low fidelity, lacking validity and simulator sickness are examples of negative side effects that may influence participants' driving behaviour and performance in a simulated driving environment.

3.3.4.1 Fidelity

Fidelity plays an important role in how realistic a driving task is perceived; the lower the fidelity of a driving simulator, the greater the chances of being perceived as unrealistic. Participants often do not believe in the authenticity of driving simulators and therefore their responses and reactions might differ from those they would have in a real driving environment. Invalid research results may be produced [34]. The safety aspect that has been mentioned as an advantage can also be considered a disadvantage of driving simulator studies. Real danger and consequence of actions is not conveyed in a driving simulator, and therefore, a false sense of safety, responsibility or competence is implied [31], influencing driving behaviour.

3.3.4.2 Validity

Questions regarding validity of results produced in driving simulator studies are frequent and difficult to answer. Validity addresses whether results/measures of a simulator study are consistent with results found in the field. Researchers need to deal with a multi-level correspondence problem referred to as simulation validation. This is defined as the replication of simulator and on-road studies to determine the extent to which measures correspond across contexts [19]. A main concern is the internal and external validity of driving simulator studies. Due to the high controllability of simulator studies, the internal validity is high compared with NDS. However, external validity, the transfer of results to the 'real world', is rather low. For a more detailed discussion on driving simulator studies and validity, see [19, 36–39].

3.3.4.3 Simulator sickness

Simulator sickness is a persistent challenge faced by researchers when conducting driving simulator studies. The occurrence of simulator sickness symptoms often leads to high dropout rates. Being aware of the persistent issue of simulator sickness, researchers normally screen their participants for tendencies towards motion sickness and test proneness to simulator sickness before the actual experiment. They also design driving scenarios in a way that helps to reduce symptoms of simulator sickness (e.g. limiting driving time and avoiding curvy or winding roads). Research shows that older participants or medically impaired participants are more prone to simulator sickness than younger drivers [40, 41]. Even though simulator sickness has not been fully understood, evidence suggests that a combination of system design, technological deficiencies of the simulation and susceptibility to motion sickness result in the occurrence of simulator sickness symptoms [32]. For a detailed description of the different factors contributing to simulator sickness and technological advances helping to reduce symptoms, see Stoner and colleagues [42].

In summary, studies conducted with simulators have the advantage of being time- and cost-efficient, reproducible, controlled and standardised, and they allow for gathering precise data in a safe driving environment. Low-fidelity simulators might provoke unrealistic driving behaviour and no valid conclusions can be drawn from driving simulator studies concerning driving behaviour in real-world driving situations.

3.3.5 Conclusion on driving simulator studies

Regarding the assessment of behavioural change over time, one of the main advantages of a simulator study is the higher controllability (i.e. higher internal validity) of the whole setting that makes hypotheses-driven assessment of behavioural change possible. Influencing factors can be manipulated or controlled systematically; data acquisition is cheaper, easier and admits a broader range of measurements with good data quality. The reduced risk for participants and other road users in a simulated setting also allows for testing premature systems, critical situations as well as the influence of risk-increasing factors on driver behaviour. However, the highly standardised setting imposes some restrictions. One major drawback for the assessment of behavioural change with driving simulators is the artificiality of the setting. Even with high-fidelity simulators, the external validity of the results obtained in a simulator study can be questionable and results cannot automatically be transferred to reality.

3.4 Long-term studies in real traffic conditions

This section focuses on investigating driver behaviour by means of naturalistic methods: naturalistic driving studies (NDS) and field operational tests (FOT). The information is based on the report by Lietz and colleagues [22]. In contrast to simulator settings, in most realistic driving settings, participants drive a real car in real traffic situations (this does not apply to test track settings). Realistic settings investigating driver behaviour can be divided into four types:

1. Naturalistic driving studies (NDS): Participants usually drive an instrumented car for a certain period of time on their usual routes without any limiting instructions. Data are recorded continuously.
2. Field operational tests (FOT): In contrast to an NDS, an FOT aims at investigating the effect of one or more independent variables (e.g. assistant systems, different groups, different conditions) on driving behaviour. Natural driving behaviour is recorded continuously such as in the NDS. The experimental design of an FOT allows for limited hypothesis testing and manipulation of conditions.
3. Experimental road tests: These tests concern experimentations carried out with instrumented cars in real traffic conditions on a predefined test route. In order to cover different experimental conditions, participants often have to drive the same test route several times. Generally, a researcher accompanies the participants on the test route in order to give them the experimental instructions and to monitor their driving behaviour.
4. Test tracks: In contrast to the three former tests, participants of test track studies drive the car on specifically designed tracks and not public roads. This allows for higher control over the route and testing of, for example, infrastructure measures. Although the conditions are more artificial than in an NDS or FOT, the real road setting provides greater external validity than simulator studies.

A characteristic of naturalistic methods for acquiring driver behaviour data is driving in naturalistic conditions. Thereby, participants drive their own vehicle or one which they can use for the whole period of acquiring data. Participants do not receive manipulative instructions with respect to *how* they shall drive. In general, during the test no investigator is present. The data collected will not be subject to any experimental control. Since this produces a high variance in the data, in general a large number of participants and/or kilometres driven are considered for the analysis in order to enhance the power of the study. Normally, when used in the study, the vehicles are unobtrusively equipped with the necessary devices for data acquisition.

3.4.1 Utility of naturalistic methods

Naturalistic methods illustrate naturalistic driver behaviour. Accordingly, naturalistic methods are suitable to gain externally valid statements, to explore driver behaviour and to generate unexpected or not anticipated hypothesis that can be later tested in controlled study settings. They are suitable for validating already existing results that have been collected and analysed using other paradigms such as driving simulator studies.

3.4.1.1 Deriving hypotheses and explorative behaviour description

Naturalistic methods offer the opportunity to record actual driver behaviour. Behaviour that occurs spontaneously as a response to an unforeseen event is recorded and is not evoked explicitly. This approach to behaviour is especially promising when investigating behaviour that has not been observed before and expectations have not been formulated. Hypotheses of driver behaviour and reactions can be derived from the data gained during observations and the descriptive analysis [43]. These newly generated hypotheses may subsequently be tested in an inference statistical way. This post-hoc procedure is characterised by data that were ascertained without a concrete objective (vs. experimental studies that follow a concrete research question and accordingly align the study design). Thus, one disadvantage of naturalistic driving studies is that conclusions drawn from gained results are rather limited. Hence, outcomes of such post-hoc analysis should be investigated in target-oriented follow-up studies.

3.4.1.2 Development of methods

Recording naturalistic driver behaviour serves as a good base to develop methods that may simulate the driving task, parts of the driving task or other tasks that correspond with naturalistic driving [44]. This means that here, at least, two methods are used; recording naturalistic behaviour data as well as the replication of the found results by means of new developed methods in laboratory settings. This serves to develop economic methods that can simulate the driving task or other related tasks in the laboratory in order to economise the effort of conducting naturalistic methods. Examples for methods that simulate the driving task, or parts of it, are car-following and the lane-change task [45, 46].

3.4.1.3 Avoidance of socially desired behaviour

Since recording driver behaviour data within naturalistic methods is less obvious and sometimes even without the drivers' knowledge (e.g. being observed at an intersection), naturalistic methods are suitable to avoid socially desired behaviour. Because of ethical issues, this (partially) covered observation should always be scrutinised.

3.4.1.4 Evaluation of results from the laboratory setting in the field

Naturalistic studies are utilised to investigate already existing results that were found in laboratory settings or in a realistic setting [47]. An advantage of conducting a naturalistic study is to confirm laboratory outcomes; a more internal and especially external valid statement can be derived.

3.4.1.5 Long-term effects

Naturalistic methods are suitable to investigate long-term effects in naturalistic conditions such as effects of ADAS on driver behaviour [48]. Especially in the context of research concerning the introduction of new technology (e.g. electric vehicles), the drivers have the opportunity to get more familiar with handling. Finally, more externally valid subjective and objective data referring to the usage of new technologies can be gathered. Data that is logged during the experiment can be used to create a curve that illustrates how behaviour changes over the time of usage.

3.4.2 Naturalistic driving study (NDS) versus field operational test (FOT)

Two naturalistic methods are reviewed in more detail; NDS and FOT. In both study types, data are collected in the same way. The two methods differ in their design and objectives. FOTs investigate the effects of ADAS use on driver behaviour (e.g. [49]). NDSs follow no experimental control in terms of group assignments or control conditions. Examples for NDS are observing driver behaviour via cameras or persons in intersection areas [50] or investigating a predefined number of participants for a certain time period (e.g. six months) in which they drive an equipped vehicle that records relevant data [51].

Experimental control decreases with a higher level of reality of the driving task. Hence the more realistic the data collected, the less comparable are the conditions in which particular participants drive. For example, two persons participating in an NDS may undertake a journey to work of 45 min but in different conditions. One person might drive daily on the motorway whereas the other might drive only in city traffic. The routes to work are not experimentally controlled and cannot be considered as equivalent. In contrast, each participant in a driving simulator study experiences the same conditions with the same vehicle, route and weather. The different paradigms (simulator studies and naturalistic methods) work under different conditions, which affects the reliability and validity of the studies. Accordingly, two NDS that investigate the same research question are less likely to match (less reliable and less valid) in their results compared with two driving simulator studies that investigate the same research question.

Because of the different characteristics of research paradigms for every research question, different advantages and disadvantages arise that need to be taken into account when designing for a study. Therefore, the study design of NDS and FOT and the quality and quantity of data are discussed below.

3.4.2.1 Naturalistic driving study

As discussed earlier, NDS can seldom be experimentally controlled. This means that participants are not grouped and variables are not actively manipulated.

Study designs do not provide manipulative instructions and participants are not asked to alter their behaviour. As no obvious limitations in its naturalistic driving behaviour are forced onto the participants, they show their actual behaviour to a very high degree. Hence, data that are recorded within an NDS are very realistic. Conclusions can be formulated concerning general driving, which means that NDS is externally highly valid.

A disadvantage of the missing experimental control is that many factors may influence driver behaviour. These factors can neither be ascertained completely nor be controlled. For example, a car driver may experience a 10 min city drive with optimal visibility, many green lights and less traffic. On another day on the same route the same driver may experience many red lights, high traffic density and bad weather conditions that influence his/her visibility negatively. These uncontrollable factors make it difficult to derive sound statements about causality. Whether the route or one of the numerous other factors causes the indicated behaviour cannot be concluded. NDS is internally not valid. Replication studies only produce the same results in very few cases.

3.4.2.2 Field operational test

In FOTs, a number of participants may drive a vehicle that is equipped with at least one ADAS. Times during which they use the system serve as experimental time periods; the rest of the time serves as baseline. Alternatively, an FOT can be designed as a between-participant study in which one group drives a car with an activated system and another, control group, drives with a deactivated system. Hence, FOT offers better experimental control than NDS. However, due to the lack of specific instructions and the naturalistic driving, the internal validity is not as good as in laboratory studies. The behaviour displayed by participants during an FOT is very similar to typical driving behaviour. The external validity of FOT is higher than the external validity of driving simulator studies. FOT can be utilised to investigate the effect of ADAS in the field and to generalise results as long as there is no potential danger for participants and other road users and the ADAS functions properly.

3.4.3 Data from naturalistic methods

By using naturalistic methods, five types of driver data can be collected: (1) driving data, (2) glance data, (3) physiological data, (4) subjective data and (5) observer data. The quality (the reliability of measuring actual driver behaviour) and the quantity of the gained data depend on what technology is used and the kind of data the technology is able to deliver. The variance within paradigms can be larger than between different methods depending on the technology used and the conscientiousness with which a study is planned and conducted.

In general, naturalistic studies deliver a huge amount of data. Such data offers the opportunity to investigate a wide range of different behaviour, more specifically, behaviour in different traffic and driving situations. Even though data collected during FOT are characterised by great variance and it is difficult to derive clear causal statements, the repertoire of behaviour data in different traffic situations is substantial, so that a lot of research questions can be analysed, unexpected hypothesis can be generated and subsequently investigated. Driving data are usually, depending on the data loggers, of good quality.

The substantial amount of data also has its drawbacks. This data is often very complex. Relevant information has to be filtered from irrelevant. The variance within and between persons can differ significantly. Deriving clear causal statements is often difficult due to the above-mentioned constraints. Huge amounts of data lead to another potential problem; that is statistical tests of significance may show significant results, although the effect is very small. This may be caused by the big sample size and lead to biased representations of the results. Thus, the effect size should be stated in order to interpret a statistical significant outcome in the correct way.

It is not possible to control the infrastructure and complexity of the traffic situation such as road characteristics, the number of other road users, but sensor systems (e.g. GPS, cameras) can provide information about the present traffic situations and contribute towards assessing particular driving situations.

The most commonly used method to observe drivers is video recording. The viewer of such recordings can analyse and document what a driver does during his/her journey. Video recordings can also be used to determine glance directions of the driver. However, it is important to bear in mind that glance direction data retrieved from ordinary video recordings are not as precise as glance data recorded with an eye tracker. Even though eye tracking data are more precise, in the context of naturalistic driving, technical limitations are soon reached and the fact that the driver is being observed is more obvious than using video cameras. Moreover, installing video cameras is not as cost-intensive as installing eye trackers.

3.4.3.1 Driving data

Driving data that can be recorded directly are raw data such as steering wheel angle, lateral position, speed, pedal and gear handling, time headway. From these data, parameters (e.g. mean standard deviation of lateral position, mean distance to lead the vehicle) can be calculated. When high quality technology is used, naturalistic as well as laboratory methods are suitable to collect a huge amount of driving data.

3.4.3.2 Glance data

Records of glance data include the coordinates or angles of horizontal and/or vertical and/or spatial depth of the drivers glance direction. Ascertaining glance data via eye tracker is possible by means of naturalistic methods and laboratory studies. In general, the quality of eye tracking data gathered in naturalistic studies is not as good as data collected in a laboratory setting due to the driving task. Also glances can be collected by means of ordinary video recordings. In regard to quality of the data, obtained glance behaviour is not as accurate and precise as data collected by means of an eye tracker but due to cost constraints and applicability are often used in naturalistic settings to collect glance behaviour.

3.4.3.3 Physiological data

Physiological data include the measure of electric activity of the brain, electrodermal activity, heartbeat rate and spatial illustration of activated brain structures. Recording physiological data becomes more difficult, the more realistic the research paradigm, due to difficulties in setting up the equipment for the experiment. It depends on the kind of data to be explored. For example, a functional magnetic resonance imaging (fMRI) that visualises activated structures inside the human brain (in order to visualise neuronal changes induced by cognitive processes during driving for example) is basically only possible for merely watching a traffic simulation [52].

3.4.3.4 Subjective data

Questionnaires, interviews and focus groups are methods to collect subjective data. In general, they are used in order to gain some knowledge about the drivers' personality traits, general driving behaviour and their attitudes (e.g. towards new technologies). Since these measurements are independent of the paradigms, they can be carried out in parallel to laboratory as well as to naturalistic studies.

3.4.3.5 Observer data

Observing drivers can be done either directly or by means of video cameras. The most desired setting is for drivers not to feel observed and to act as naturally as possible. When drivers are aware that they are observed, they tend to react in a socially desired manner. Even though observations run the potential risk of causing socially desired behaviour, they are still used.

3.4.4 Conclusion on naturalistic methods

With regard to the assessment of behavioural changes, the pros and cons of naturalistic methods can be summarised as follows. A major advantage of NDS and FOT is that participants drive in their real everyday life context without experimental restrictions on how they shall drive. This allows assessment of naturalistic behaviour with a high external validity. Particularly for assessing long-term behavioural changes, naturalistic methods offer advantages such as several participants can drive unobtrusively instrumented vehicles in parallel for a long period of time – months or even years. Continuous recording of different behavioural data allows detailed analysis of gradual changes over a long period of time. However, one challenge of naturalistic methods is to handle and filter the huge amount of data to extract meaningful indicators. This issue is connected to one of the major drawbacks of naturalistic methods — the absence of experimental control. Under naturalistic driving conditions, many factors such as road type, weather conditions and traffic density may influence driver behaviour and can neither be controlled nor completely assessed. Therefore, conclusions drawn from post-hoc analysis are rather limited to a mainly explorative character in contrast to hypotheses-driven experimental studies.

Hence NDS and FOT are not the panacea to investigate all research questions in the area of driver behaviour and attitudes. NDS and FOT are only two of many ways to collect driver behaviour and subjective data. The challenge is to be well informed of the variety of existing methods and to choose an appropriate method for a specific research question. However, beside NDS and FOT, other empirical paradigms such as laboratory studies and more experimental field studies play an important role.

3.5 Conclusion

Based on the literature review in this chapter, we can state that the use of long-term studies for the study of behavioural changes with the introduction of ADAS might be a powerful tool, allowing the modelling of behavioural adaptation over time and looking at potential retention or transfer effects. Driving simulator studies as well as naturalistic driving studies (NDS) offer the opportunity to look into behavioural change inherent in ADAS use. In order to depict the most suitable approach for long-term research issues, we reviewed and compared the use of driving simulators as well as naturalistic methods. Both methods can be powerful tools for long-term studies as long as the advantages and inconveniences that are inherent to each method are taken into account. One major advantage of driving simulator studies over NDS and FOT is the high controllability that allows specific manipulation of conditions. In NDS and FOT studies, data acquisition is cheaper and easier than data collection. Driving simulator studies also permit a broader range of measurements with good data quality. Alternatively, NDS and FOT are characterised by their naturalistic and realistic driving experience, ensuring high external validity and allowing assessment of naturalistic driving behaviour. One methods' inherited inconvenience is the other methods' advantage, making both methods complementary, an aspect to keep in mind when assessing behavioural adaptation effectively and efficiently on different levels. Understanding the benefits and drawbacks of each method will ensure experimental designs that go beyond the capabilities provided by each method individually and provide useful insights into behavioural adaptation.

Acknowledgement

The research leading to these results has received funding from the European Community's Seventh Framework Program (FP7/2007-2013) under grant agreement no. 238833/ADAPTATION project (www.adaptation-itn.eu).

References

1. Brookhuis K. A., De Waard D., Janssen W. H. 'Behavioural impacts of Advanced Driver Assistance Systems – an overview'. *European Journal of Transport and Infrastructure.* 2001;1:245–253
2. Manser M., Creaser J. 'Assessing driver behavioral adaptation to a rural intersection driver support system'. Presented at the Sixth International

Driving Symposium on Human Factors in Driver Assessment, Training and Vehicle Design; Olympic Valley, Lake Tahoe, CA, 2011

3. Engström J., Arfwidsson J., Amditis A., Andreone L., Bengler K., Cacciabue P. C., et al. 'Towards the automotive HMI of the future: mid-term results of the AIDE project' in Valldorf J., Gessner W. (eds.). *Advanced Microsystems for Automotive Applications*. Berlin: Springer; 2006. pp. 379–405
4. Michon J. A. 'A critical view of driver behavior models: what do we know, what should we do?' in Evans L., Schwing R. C. (eds.). *Human Behavior and Traffic Safety*. New York, NY: Plenum Press; 1985. pp. 485–520
5. Allen T., Lunenfeld H., Alexander G. 'Driver information needs'. *Highway Research Record*. 1971;366:102–115
6. Brouwer W. H., Withaar F. K., Tant M. L. M., Van Zomeren A. H. 'Attention and driving in traumatic brain injury: a question of coping with time-pressure'. *Journal of Head Trauma Rehabilitation*. 2002;17:1–15
7. Caird J. K., Chisholm S. L., Lockhart J. 'Do in-vehicle advanced signs enhance older and younger drivers' intersection performance? Driving simulation and eye movement results'. *International Journal of Human-Computer Studies*. 2008;66(3):132–144
8. Vollrath M., Schleicher S., Gelau C. 'The influence of Cruise Control and Adaptive Cruise Control on driving behaviour – A driving simulator study'. *Accident Analysis & Prevention*. 2011;43(3):1134–1139
9. Saad F., Hjälmdahl M., Cañas J. J., Alons M., Garoyo P., Macchi L., et al. *AIDE Project Deliverable D1_2_1: Literature Review of Behavioural Effects*. European Commission; 2004
10. Shinar D., Schechtman E. 'Headway feedback improves intervehicular distance: a field study'. *Human Factors: The Journal of the Human Factors and Ergonomics Society*. 2002;44(3):474–481
11. Hoedemaeker M., Brookhuis K. A. 'Behavioural adaptation to driving with an adaptive cruise control (ACC)'. *Transportation Research Part F: Traffic Psychology and Behaviour*. 1998;1(2):95–106
12. Jahn G., Krems J. F., Gelau C. 'Skill acquisition while operation In-Vehicle Information Systems: interface design determines the level of safety-relevant distractions'. *Human Factors*. 2009;51(2):136–151
13. Stanton N. A., Young M., McCaulder B. 'Drive-by-wire: the case of driver workload and reclaiming control with adaptive cruise control'. *Safety Science*. 1997;27(2-3):149–159
14. Stanton N. A., Young M. S., Walker G. H., Turner H., Randle S. 'Automating the Driver's Control Tasks'. *International Journal of Cognitive Ergonomics*. 2011;5(3):221–236
15. Shina D., Tractinsky N., Compton R. 'Effects of practice, age, and task demands, on interference from a phone task while driving'. *Accident Analysis & Prevention*. 2005;37(2):315–326
16. Chisholm S. L., Caird J. K., Lockhart J. 'The effects of practice with MP3 players on driving performance'. *Accident Analysis & Prevention*. 2008;40(2): 704–713

17. Brookhuis K. A., de Vries G., de Waard D. 'The effects of mobile telephoning on driving performance'. *Accident Analysis & Prevention.* 1991;23(4): 309–316
18. Ranney T. A. 'Psychological fidelity: perception of risk' in Fisher D. L., Rizzo M., Caird J. K., Lee J. (eds.). *Handbook of Driving Simulation for Engineering, Medicine, and Psychology.* Boca Raton, FL: CRC Press; 2011. pp. 9.1–9.13
19. Caird J. K., Horrey W. J. 'Twelve practical and useful questions about driving simulations' in Fisher D. L., Rizzo M., Caird J. K., Lee J. D. (eds.). *Handbook of Driving Simulator for Engineering, Medicine, and Psychology.* Boca Raton, FL: CRC Press; 2011. pp. 5.1–5.18
20. Fisher D. L., Rizzo M., Caird J. K., Lee J. D. (eds.). *Handbook of Driving Simulator for Engineering, Medicine, and Psychology.* Boca Raton, FL: CRC Press; 2011
21. Allen W. R., Rosenthal T. J., Coom M. L. 'A short history of driving simulation' in Fisher D. L., Rizzo M., Caird J. K., Lee J. D. (eds.). *Handbook of Driving Simulator for Engineering, Medicine, and Psychology.* Boca Raton, FL: CRC Press; 2011. pp. 2.1–2.16
22. Lietz H., Petzoldt T., Henning M., Haupt J., Wanielik G., Krems J., Mosebach H., et al. *Methodische und Technische Aspekte einer Naturalistic Driving Study.* FAT-Schriftenreihe 229. Berlin: Verband der Automobilindustrie (VDA); 2011
23. Rudin-Brown C. M., Williamson A., Lenne M. G. *Can Driving Simulation be Used to Predict Changes in Real-World Crash Risk?* Accident Research Centre, Monash University, 2009
24. Kantowitz B. H. 'Using driving simulators outside of North America' in Fisher D. L., Rizzo M., Caird J. K., Lee J. (eds.). *Handbook of Driving Simulation for Engineering, Medicine, and Psychology.* Boca Raton, FL: CRC Press; 2011. pp. 3.1–3.14
25. Kantowitz B. H. 'Using models for transportation human factors'. *Proceedings of the First International Driving Symposium on Human Factors in Driver Assessment, Training and Vehicle Design*; Aspen, CO, 2001. Iowa University; 2001. pp. 42–57
26. Hogema J. H., Rook A. M. *Intelligent Speed Adaptation: The Effects of an Active Gas Pedal on Driver Behaviour and Acceptance.* On request of the Dutch Ministry of Transport, Public Works and Water Management, Transport Research Centre AVV and the European Commission. TNO report TM-04-D011. TNO Human Factors Research Institute, Soesterberg; 2004
27. Vollrath M., Schleicher S., Gelau C. 'The influence of Cruise Control and Adaptive Cruise Control on driving behaviour – A driving simulator study'. *Accident Analysis & Prevention.* 2011;43(3):1134–1139
28. Wilschut E. S., Rinkenauer G., Brookhuis K. A., Falkenstein M. 'Effects of visual search task complexity on lane change task performance'. *Proceedings of the European Conference on Human Centred Design for Intelligent Transport Systems*; Lyon, France, 2008. Lyon: HUMANIST; 2008. pp. 23–33

29. Berthon-Donk V., Grosjean M., Rinkenauer G. 'Effect of feedback on performance in the Lane-Change Test'. Presented at the Sixth International Driving Symposium on Human Factors in Driver Assessment, Training and Vehicle Design; Olympic Valley, Lake Tahoe, CA, 2011
30. Mayhew D. R., Simpson H. M., Pak A. 'Changes in collision rates among novice drivers during the first months of driving'. *Accident Analysis and Prevention.* 2003;35:683–691
31. Allen R. W., Park G. D., Cook M. L., Fiorentino D. 'The effect of driving-simulator fidelity on training effectiveness'. *Proceedings of the Driving Simulation Conference North America,* Iowa City, IA; 2007. Available from http://www.nads-sc.uiowa.edu/dscna/2007/main.htm [Accessed 26 July 2013]
32. Moroney J. F., Lilienthal M. G. 'Human factors in simulation and training: an overview' in Hancock P., Vincenzi D., Wise J., Mouloua M. (eds.). *Human Factors in Simulation and Training.* Boca Raton, FL: Taylor & Francis; 2009. pp. 3–38
33. Roskam A. J., Brookhuis K. A., De Waard D., Carsten O. M. J., Rea L., Jamson S., et al. 'HASTE Deliverable 1: development of experimental protocol' [online]; 2002. Available from http://ec.europa.eu/transport/roadsafety_library/publications/haste_deliverable_1_v1_1.pdf [Accessed 28 January 2013]
34. de Winter J. C. F., van Leeuwen P. M., Happee R. 'Advantages and disadvantages of driving simulators: a discussion'. *Measuring Behaviour.* Utrecht, the Netherlands; August, 2012. Available from http://www.3me.tudelft.nl/fileadmin/Faculteit/3mE/Over_de_faculteit/Afdelingen/BioMechanical_Engineering/Organisatie/Medewerkers/Winter/doc/driving_simulators.pdf [Accessed 26 July 2013]
35. Käppler W. D. 'Views on the role of simulation in driver training'. *Proceedings of the 12th European Annual Conference on Human Decision Making and Manual Control*; Kassel, Germany, June 1993. pp. 5.12–5.17
36. Kaptein N. A., Theeuwes J., Van der Horst R. 'Driving simulator validity: some considerations'. *Transportation Research Record.* 1996;1550:30–36
37. Yan X., Abdel-Aty M., Radwan E., Wang X., Chilakapati P. 'Validating a driving simulator using surrogate safety measures'. *Accident Analysis & Prevention.* 2008;40:274–288
38. Blauuw G. J. 'Driving experience and task demands in simulator and instrumented car: a validation study'. *Human Factors.* 1982;24:473–486
39. Törnros J. 'Driving behavior in a real and a simulated road tunnel: a validation study'. *Accident Analysis & Prevention.* 1998;30:497–503
40. Godley S. T., Triggs T. J., Fildes B. N. 'Driving simulation validation for speed research'. *Accident Analysis and Prevention.* 2002;34:589–600
41. Stanney K. M., Kennedy R. S. 'Simulator sickness' in Hancock P., Vincenzi D., Wise J., Mouloua M. (eds.). *Human Factors in Simulation and Training.* Boca Raton, FL: Taylor & Francis; 2009. pp. 117–127
42. Stoner H. A., Fisher D. L., Mollenhauer Jr. M. (2011). In Fisher D. L., Rizzo M., Caird J. K., Lee J. D. (eds.). (2011). *Handbook of Driving Simulator for Engineering, Medicine, and Psychology.* Boca Raton, FL: CRC Press

43. Malenfant J. E. L., Van Houten R. 'Observations of how drivers fasten their seatbelts in relation to various startup tasks'. *Accident Analysis & Prevention*. 2008;40:309–314
44. Hirata T., Yai T., Takagawa T. 'Development of the driving simulation system MOVIC-T4 and its validation using field driving data'. *Tsinghua Science & Technology*. 2007;12(2):141–150
45. Mattes S. 'The lane-change task as a tool for driver distraction evaluation' in Rausch H., Bubb H. (eds.). *Quality of Work Products in Enterprises of the Future*. Stuttgart: Ergonomia Verlag; 2003. pp. 57–60
46. Mattes S., Hallen A. 'Surrogate distraction measurement techniques: the lane change test' in Regan M. A., Lee J., Young K. L. (eds.). *Driver Distraction, Theory, Effects and Mitigation*. Boca Raton, FL: CRC, Taylor & Francis; 2009. pp. 107–122
47. Sayer J. R., Devonshire J., Flannagan C. A. (2005). *Effects of Secondary Tasks on Driving Performance* (Report No. UMTRI-2005-29). Ann Arbor, MI: The University of Michigan Transportation Research Institute
48. Lai F., Hjälmdahl M., Chorlton K., Wiklund M. 'The long-term effect of intelligent speed adaptation on driver behaviour'. *Applied Ergonomics*. 2009;41(2):179–186
49. Young K. L., Regan M. A., Triggs T. J., Stephan K., Mitsopoulos-Rubens E., Tomasevic N. 'Field operational test of a seatbelt reminder system: effects on driver behaviour and acceptance'. *Transportation Research Part F: Traffic Psychology and Behaviour*. 2008;11(6):434–444
50. Rosenbloom T., Ben-Eliyahu A., Nemrodov D., Biegel A., Perlman A. 'Committing driving violations: an observational study comparing city, town and village'. *Journal of Safety Research*. 2009;40(3):215–219
51. Hanowski R. J., Olson R. L., Hickman J. S., Dingus T. A. *The 100-Car Naturalistic Driving Study: A Descriptive Analysis of Light Vehicle-Heavy Vehicle Interactions from the Light Vehicle Driver's Perspective, Data Analysis Results* (Report# FMCSA-RT-05-005). Washington, DC: Federal Motor Carrier Safety Administration; 2006
52. Hsieh L., Young R. A., Bowyer S. M., Moran J. E., Genik Ii R. J., Green C. C., et al. 'Conversation effects on neural mechanisms underlying reaction time to visual events while viewing a driving scene: fMRI analysis and asynchrony model'. *Brain Research*. 2009;1251:162–175

Chapter 4

Effect of ADAS use on drivers' information processing and Situation Awareness

Matthias Beggiato

Abstract

This chapter focuses on the elaboration and validation of new methods for assessing drivers' higher level cognitive processes reflecting mental models and Situation Awareness (SA). When drivers start using an advanced driver assistance system (ADAS), they acquire a mental model of the system's purpose, function and performance. These mental models should be considered in investigations of behavioural adaptation to ADAS, as they are expected to influence information processing, SA and the selection of appropriate actions. After defining mental models and SA, advantages and disadvantages of existing assessment techniques are presented as well as requirements for new approaches. A newly developed mental model questionnaire is described that was used in a driving simulator study and a field test. SA is measured in real time based on an implicit performance approach using a continuously presented secondary task. Benefits and limitations of both methods are discussed in a final summary.

4.1 Behavioural adaptation and higher cognitive processes

Advanced driver assistance systems (ADAS) aim to support drivers such that comfort and safety are enhanced. However, the positive effects of these systems may be diminished or even inverted by unintentional changes in user behaviour (behavioural adaptation). Higher level cognitive processes are considered as underlying factors for these changes in behaviour and should therefore be included in investigations of behavioural adaptation to ADAS. Within the ADAPTATION project, higher level cognitive processes mainly refer to information processing, based on mental models (i.e. being synonymous with mental representation) and Situation Awareness (SA). These processes are thought to influence behavioural adaptation (e.g. [1–3]). The relevant factors that are descriptive of behavioural

Department of Cognitive and Engineering Psychology, Chemnitz University of Technology, 09107 Chemnitz, Germany changes occurring with ADAS use have been conceptualised in the driver appropriation model developed in the HUMANIST project [4]. The model integrates and extends previous theories (e.g. [5, 6]). The broader definition of 'driver appropriation' includes cognitive, regulatory and motivational processes underlying observable behaviour, as well as temporal interactions. The driver appropriation model comprises three main interacting factors: the system, the situation and the driver. The driver creates a mental representation of the system, which is continuously elaborated and refined by experience. Despite several studies addressing the connection between mental models and changes in driving behaviour in response to ADAS use (e.g. [1, 7–9]), knowledge of the relation between the appropriateness of mental models, system trust and behavioural changes over time is lacking [2]. This chapter describes elaboration and validation of innovative methods for assessing higher cognitive processes, with an emphasis on SA and mental models.

4.2 Mental models

4.2.1 Definition

Mental models have been studied to understand how humans make decisions, perceive, know and construct behaviour in a variety of environments. In the Human–Machine Interaction (HMI) domain, mental models can be conceptualised as '... a rich and elaborate structure, reflecting the user's understanding of what the system contains, how it works and why it works that way' [10, p. 12]. Durso and Gronlund [11, pp. 297–298] define a mental model more precisely as a long-term memory knowledge structure, that is '. . . a representation of the typical causal interconnections involving actions and environmental events that influence the functioning of the system'. They point out that mental models, in the context of complex dynamic systems, differ from mental models of static objects or abstract syllogism. In transportation research, mental models are considered fruitful for explaining relevant cognitive processes [12], and a realistic mental model of ADAS functionality is considered fundamental for adequate use of these systems [1, 13, 14].

4.2.2 Mental model construction and update

To estimate and predict the effect of mental models, a theoretical concept of how mental models are constructed and updated is required. Durso, Rawson, and Girotto [15] distinguish between mental models and situation models. While the former is considered a general long-term memory knowledge structure, the latter is formed in a particular situation and is built by environmental input (bottom-up) in connection with top-down knowledge structures. One mental model can therefore give rise to several situation models associated with particular situational characteristics. When using an ADAS, a correct mental model of system functionality is considered crucial for the construction of an adequate situation model in real road situations [4, 16]. On the one hand, incorrect mental models can lead to the misinterpretation of environmental input, and on the other hand, experience updates and correct the mental model. Theories from text comprehension research [15, 17, 18] elaborate on the cognitive processes underlying the interpretation and comprehension of a given situation. The Construction–Integration Theory [19] assumes two stages in the comprehension process: In the first construction-related phase, environmental input activates knowledge stored in long-term memory (nodes) that is linked to the perceived information. Activation in this first stage is considered rather diffuse and general. In the second integration phase, bottom-up information is used to specifically select corresponding knowledge structures: compatible elements within the activated knowledge network are activated, while incompatible nodes are suppressed. Highly interconnected nodes accumulate activation while less activated nodes lose activation. As a result, less-well-connected nodes may disappear from the knowledge network over time [15]. Through these inhibitory and excitatory processes, a coherent situation model is formed. After the integration phase, the remaining network is restored in long-term memory and retrieved in future processing cycles.

4.2.3 Discussion of existing mental model measures and requirements

Tools and measures to assess users' mental models have already been established. Cherri, Nodari, and Toffetti [20] presented a list of such measures, including focus groups, in-depth-interviews, self-reported diaries, questionnaires, task analysis, decision trees, card sorting and the 'potato head' technique. These methods have some advantages and drawbacks:

- Task analysis, decision trees, card sorting and potato head techniques primarily focus on the optimal design of user-centred concrete systems. These techniques identify the most favourable design configurations for optimising user–system interactions, avoiding misleading terminology and analysing decision processes while performing tasks. Therefore, they work best in the design and evaluation process of ADAS. However, it is difficult to apply these methods in a more general research environment due to the very specific configurations of particular systems. Furthermore, most of these techniques are costly, complex and time consuming in terms of development, setup and analysis.
- Focus groups, in-depth interviews and diaries are primarily qualitative methods. They allow for a detailed, flexible and individualised assessment of mental models. On the one hand, these techniques are useful in exploring mental models, particularly in describing and explaining individual experiences and relationships. On the other hand, data analysis is mostly complex and time consuming. Direct comparisons between or within subjects as well as statistical calculations are difficult due to lack of standardisation.

A key objective in the ADAPTATION project was the development of new tools for assessing mental models in the context of behavioural adaptation. Because behavioural adaptation is considered a process, the new assessment tool should be able to track changes in the mental model over time. Moreover, it should be possible to quantify how the mental model has changed, for example, due to interaction with a system. In addition to within-subjects measurement, a comparison should also be possible between subjects. This allows statistical tests between groups with different characteristics. Furthermore, the new tool should be based on a theoretical concept and be amenable to adaptation to the different ADAS investigated within the ADAPTATION project. Also, the development, administration and analysis procedures should be as simple as possible.

4.2.4 Development of the mental model questionnaire

The simplest method meeting these requirements appears to be a standardised questionnaire approach, based on the mental model definitions stated in Section 4.2.1: A user's understanding of a system is intended as the representation of causal interconnections between situational characteristics and system functioning. Accordingly, questionnaire items were developed, focusing on concrete system functionality in specific situations. Subjects answered questions using a six-point Likert scale ranging from 1 (*totally disagree*) to 6 (*totally agree*); for example: 'The system maintains a predetermined speed in an empty lane', 'The system reacts to pedestrians in the traffic lane' or 'The system works during night time'. The final mental model questionnaire consisted of 35 items. Results of the driving simulator study, where the questionnaire was administered to track mental model changes in three experimental groups, are reported in Section 4.4.3.1.

The application of the questionnaire in a real driving context is presented in the second study (Section 4.5). This study assessed how users interpret statements given in the car manufacturer's manual concerning ACC, and how their interpretation changed over time with experience. For this study, the mental model questionnaire was slightly adapted to statements in the user's manual, but based on the same principles.

4.3 Situation Awareness

4.3.1 Definition

The concept of SA originated in aviation research in the 1980s and became widespread among several disciplines and work environments, such as power plant operations, anaesthesiology, military activity and automobile driving [21]. Increasing automation of the driving task shows similarities with previous developments in other domains, such as aviation or air traffic control, thus rendering the SA concept fruitful in connection with research into the effects of automation in driving [22]. The most common and widely used definition is from Endsley [23, p. 792]: 'Situation Awareness is the perception of the elements in the environment within a volume of time and space, the comprehension of their meaning, and the projection of their status in the near future'. Based on this definition she developed a model with three levels of SA.

Level 1. *Perception*: The perception of status, attributes and dynamics of relevant elements in the environment is fundamental to acquiring a correct picture of a situation.

Level 2. *Comprehension*: Perceived elements on Level 1 need to be integrated into a holistic picture of a situation. That is, an awareness of elements must be complemented by an understanding of their significance in, and relationship to, a specific context.

Level 3. *Projection*: Perception and comprehension form the basis of projecting future actions of elements in the environment. The highest level of SA allows for anticipating future events and timely decision making.

SA influences decision making, and subsequent actions change the state of the environment, which in turn begins a new cycle of perception starting with Level 1. The whole process is influenced by task/system characteristics as well as individual factors.

4.3.2 Relationship between mental models and Situation Awareness

The relationship between SA and mental models is conceptualised by Endsley [24, p. 61] in her definition, 'Mental models embody stored long-term knowledge about these systems that can be called upon to direct problem solving and interaction with the relevant system when needed'. Mental models represent rather static knowledge about system functionality, which however grow and evolve with experience. In contrast, a situation model is considered '... very dynamic, representing the human's knowledge and understanding of the *present state* of the system' [24, p. 62]. As described in Section 4.2.2, Durso et al. [15] make the same distinction between mental model and situation model. Endsley [24] uses situation models and SA synonymously as the levels of perception, comprehension and projection of future states. Situation models are considered largely influenced by a person's mental models, which guide attention to specific aspects of a situation, influence how this information is interpreted and therefore affect projections about future states.

4.3.3 Situation Awareness as comprehension process

Recent approaches define SA as a comprehension process, analogous to theories of reading comprehension [15, 17, 18]. In these examples, subsequent to Endsley's model, cognitive mechanisms underlying the construction of SA are specified in detail. In particular, the perception of new elements in a situation (bottom-up processes) activates associated knowledge structures in long-term memory (top-down processes), according to the two-stage Construction–Integration Theory by Kintsch [19]. This process forms the actual 'mental situation model', which triggers the activation of coherent actions, which in turn affects change in the situation, and subsequently demands updating the situation model. As the amount of available information exceeds the capacity of working memory, an additional mechanism is postulated: long-term working memory stores the excess information, but provides it immediately when it becomes relevant. The schemata stored in long-term memory encode experiences of different driving situations and therefore also knowledge about the functions of assistance systems. Analogous to findings in reading comprehension research [25], an inappropriate mental model hinders and delays the comprehension process. Moreover, experiences that demand reorganisation of the mental model due to the addition of new information should result in higher cognitive load (the processing-load hypothesis, [26]).

4.3.4 Discussion of existing Situation Awareness measures and requirements

SA measurement techniques have mainly been developed and applied in aviation research. There are several classifications of SA assessment methods, including those by Durso and Gronlund [11], who propose three general approaches:

1. *Subjective measures*: System operators judge their own SA on Likert scales, e.g. Situation Awareness Rating Technique (SART; [27]), Situation Awareness-Subjective Workload Dominance (SA-SWORD; [28]), Situation Awareness Rating Scales (SARS; [29]).
2. *Query methods*: A simulation is stopped, all information is removed and the accuracy of answers to predefined questions for this specific situation indicates the level of SA, e.g. Situation Awareness Global Assessment Technique (SAGAT; [30]).
3. *Implicit performance measures*: Performance-based measures (e.g. lane keeping and responsiveness to critical events) and event-based measures such as the Situation Present Awareness Method (SPAM; [31]).

The main criticism of subjective measures is that operators cannot possibly be aware of their own SA [32]. Operators report what they perceive and know, and this may differ from a real-life situation. Furthermore, subjective SA ratings are highly influenced by self-assessment of performance. Finally, errors and systematic bias in human judgement and memory undermine self-reported results. However, subjective measures provide insights into underlying cognitive processes involved in SA ratings and provide an indication of an operator's confidence level regarding his SA.

Query methods remove the problem of collecting SA data after task completion as they are applied in the situation itself. Furthermore, tools such as SAGAT are subject to numerous validation studies and directly assess SA by objective and unbiased data. However, query methods require simulators and therefore cannot be applied in the field or in real time. Furthermore, freezing a simulation is artificial and intrudes on the primary task; that is, questions posed could affect what aspects of the situation operators look for. Finally, query methods, as well as subjective methods, focus on conscious and explicit knowledge, whereas implicit aspects cannot be assessed. This is critical in the driving domain as many processes are skill-based and therefore automated and unconscious.

The major drawback of implicit performance measures is the indirect measurement approach. In particular, this approach may not reflect actual SA level, as poor performance can result from factors other than low SA and vice versa. The main advantage of performance measures is that they can be collected objectively and non-intrusively. Moreover, implicit knowledge of a situation is brought into the SA evaluation. Furthermore, the transfer of knowledge into observable behaviour can be assessed in terms of timing and appropriateness.

To assess SA in connection with behavioural changes when using ADAS, several requirements were established within ADAPTATION: The measurement technique should not be based only on subjective ratings of operators, because users' awareness of changes in their own behaviour is incomplete. Query methods require a freeze in simulation, which presumably has an impact on behavioural adaptation. In particular, simulation freezing could guide users' attention to certain aspects of a situation and, as noted for subjective ratings, thereby only report explicit and conscious knowledge. Therefore, the SA assessment technique should focus on implicit behavioural measures. This approach seems to best suit the needs of the ADAPTATION project as it is non-intrusive, covers implicit knowledge, assesses changes in behaviour over a longer time period and can be applied in real-time in the dynamic domain of driving.

4.3.5 Development of the Situation Awareness measurement technique

The approach chosen for ADAPTATION is largely based on and extends the work of Rauch, Gradenegger, and Krüger [33], where a secondary task is used to assess SA. The central assumption is that 'Situationally aware drivers are able to decide in accordance with the demands of the actual driving situation if execution of a secondary task is safely possible or not' [33, p. 4]. Thus, prioritisation of the driving task or secondary task is used as an indicator for SA. The decision to execute a secondary task involves processes of perception, comprehension and prediction in a given situation. Rauch [34] differentiates the SA concept from related concepts such as anticipation, hazard perception and workload. Anticipation usually requires stable and unambiguous laws, whereas driving is highly dynamic, circumstances change frequently and situations can be ambiguous. Hazard perception focuses solely on critical situations, whereas SA implies a more general process of action planning and execution, which is also active in non-critical situations. And finally, in contrast with the concept of workload, SA enables drivers to reduce or avoid future workload by prioritising tasks in advance. Tsang and Vidulich [41, p. 262] argue that 'Workload is primarily a result of the limited attentional resources of humans, whereas SA is a cognitive phenomenon emerging from perception, memory, and expertise'.

Rauch et al. [33] used two different secondary tasks in their driving simulator studies: The first was a highly standardised, externally paced task, where drivers had to read a sequence of numbers displayed on a central console [35]. The task in the second study was self-paced, interruptible and consisted of navigating according to instructions, using a realistic menu system [36]. For both tasks, drivers were given the opportunity to perform a task at predetermined points on the route, cued by a question mark appearing in the overhead display. Within a given amount of time, drivers decided whether or not to accept the offer by pushing a button on the steering wheel. Results show that drivers adapted their interaction with secondary tasks depending on the predicted demands of the situation; this was used as an indicator for SA [33].

The measurement approach applied in the ADAPTATION project is based on the same principles, but differs in the type of secondary task used. The secondary task is continuously present and participants are instructed to solve as many of these tasks as possible, while driving safely. The opportunity to complete a secondary task is always present, and thus allows continuous and sensitive measurements for the duration of a simulator trial by calculating the frequency of solved tasks within a certain time period. To obtain a sufficient number of responses, a relatively easy task must be chosen that can be processed continuously and quickly. The Surrogate Reference Task (SURT, [37]) meets these requirements. Details about the setup can be found in Section 4.4.2.4. Based on the definition of SA given at the beginning of this chapter, the frequency of secondary tasks solved should decrease with an increase in the demands of a situation.

4.4 Study 1: Driving simulator study

4.4.1 Aims and research questions

ADAS aim to support drivers by partially automating driving subtasks. Adaptive Cruise Control, ACC, as one such system, maintains a driver-set vehicle speed and constant time headway in relation to a slower leading vehicle, without the need for using the brake or accelerator pedal. However, ACC is not yet operative in all situations due to sensor limitations. Problems in detecting objects ahead can, for instance, occur at narrow bends, in adverse weather conditions, as well as in sensing stationary or small vehicles. Therefore, drivers need to acquire a correct mental model; that is, they must know and understand such limitations in order to use the system in a safe and efficient manner [14, 16].

This driving simulator study systematically investigates the effect of divergent initial mental models of ACC (i.e. varying in correctness), using the same system in identical situations. The main research questions focus on how different initial mental models of ACC affect system trust and acceptance over time, and how a user's mental model evolves with experience. Effects on trust and system acceptance are reported in Beggiato and Krems [38]. The focus of this chapter is the application of the mental model questionnaire and its results.

Regarding the development of an ACC mental model, it is assumed that the provision of different preliminary information induces a correspondingly different mental model about system functionality. With experience, a mental model is expected to approach a realistic view of a system's capabilities. According to Construction–Integration Theory [19], expected but not experienced limitations should become less activated in the mental model network and therefore show a tendency to be forgotten over time. Unexpected but experienced limitations should have the greatest impacts due to a greater need for restructuring the existing mental model. The mental model questionnaire aims to provide a tool for testing these hypotheses.

4.4.2 Method and material

4.4.2.1 Sampling and participants

The study was designed using a matched sample approach, based on the model of Cotter and Mogilka [4]. The model factors 'system' and 'situation' were controlled for by providing the same ACC functionality and the same simulator track for all participants. A two-step sampling procedure was applied to match relevant driver characteristics (details on the matching procedure are reported in Beggiato and Krems [39]). Finally, three experimental groups were created with 17 participants in each, balanced in all confounding variables. The sample consisted of 25 male and 26 female students with a mean age of 24 years (SD 2.37).

4.4.2.2 Research design and procedure

A two-way (3×3) repeated measures mixed design was used with three system descriptions as between-subject factors (Correct, Incomplete, Incorrect) and three consecutive trips as within-subject factors. The average time interval between trips was 13 days, with a minimum of 7 and a maximum of 24 days. As a cover story, test subjects were told that the study aimed to test a driver assistance system, namely DriveFree. The terms 'adaptive cruise control' and 'ACC' were never used in order to prevent information search by participants between simulator trials.

Upon arrival at the driving simulator, participants read a one-page ACC description. Afterwards, all participants completed questionnaires related to mental model, acceptance and trust. Before driving on the simulated highway track, participants drove for 5 km on a city route to practise using the driving simulator and the ACC as well as practise performing secondary tasks. Then, participants drove the 56 km simulated highway track. They were instructed to use the ACC in such a way that they could complete as many secondary tasks as safely as possible. After each driving trip, all participants filled in questionnaires related to mental models, acceptance and trust. The system description was not shown again after Trip 1.

4.4.2.3 Facilities and driving simulator track

Driving simulator and ACC: A fixed-base driving simulator (STISIMDrive 100w) at Chemnitz University of Technology was used for the study. It provided a 135° horizontal field of view and was composed of a BMW 350i driving cab with automatic transmission. The ACC was integrated in the simulation software and allowed specific manipulation of functionality in defined driving situations. In general, the ACC worked at all driving speeds, and included stop-and-go functionality. During the entire simulation, the ACC headway time was set to 2 s and speed to 100 km/h. To maintain constant conditions, participants were not allowed to change these settings while driving or turn off the system. Pressing the accelerator or brake pedal temporarily disabled the ACC; when the pedals were released, the ACC functions were immediately reactivated. No ACC alarms were implemented in order to simulate critical situations without prior warning. The system was programmed to recognise all leading vehicles except motorbikes. In three defined situations (i.e. two narrow bends and one episode of heavy fog), recognition of the lead vehicle was disabled temporarily to simulate system failure.

Simulator track: The simulator track comprised a 56 km long two-lane highway with an average driving time of 35 min and 32 s. With the exception of a 100 km/h speed limit in construction zones, speed limit was set to 120 km/h. All subjects drove the same track, every time. The type and colour of the surrounding vehicles differed in each session, with the exception of white and large vehicles, such as trucks or buses. During driving, all participants experienced cut-in situations from the left (normal car/motorbike/white car), construction zones with different leading vehicles (normal car/white truck), approaching situations involving queuing (normal car/white truck ahead) and different weather conditions (good weather/light fog/heavy fog).

4.4.2.4 Secondary task: SURT

As a secondary task, participants had to continuously solve the Surrogate Reference Task SURT [37]. The software was provided by the German Aerospace Center (DLR; Deutsches Zentrum für Luft- und Raumfahrt) in Braunschweig. The visual search task consisted of 50 randomly arranged circles presented on a 10-inch touch screen on the central console (Figure 4.1). Test subjects had to select one circle with an approximately one-third-greater radius and touch it with their right hand. Therefore, the SURT requires visual perception combined with a manual response. Immediately after touching the screen, a green light appeared if the participant had selected the correct half of the screen and a red light if the wrong half of the screen was chosen. After a delay of 0.5 s, a new picture with randomly arranged circles was presented. The 49 small circles had a diameter of 0.8 cm, and the bigger circle 1.05 cm.

Figure 4.1 Driving simulator setup and SURT

4.4.2.5 System description

To create different initial mental models of ACC, a one-page system description was designed. The first section dealt with general system functionality and was the same for all groups. Further, it included an explanation of longitudinal control automation, speed and time headway settings, temporary deactivation due to braking or accelerating, and the absence of system reaction upon encounters with traffic signs, pedestrians, oncoming traffic and traffic lights. The second part of the description differed between the three experimental groups. Group 1 (correct) was informed about potential system failures at narrow bends, in bad weather conditions (heavy rain, fog or snow) and with small vehicles, such as motorbikes. These problems actually occurred during the simulator run. Group 2 (incomplete) did not receive information about the potential failures mentioned in Group 1. Group 3 (incorrect) had the same realistic information as Group 1, but two supposed potential failures were additionally indicated: The ACC could have detection problems with large vehicles (busses, trucks) and with white or silver cars due to reflection. These problems, however, did not occur during the driving simulation.

4.4.3 Results

4.4.3.1 Mental model

To assess changes in drivers' mental models of ACC, the questionnaire was presented immediately after reading the system description and after each of the three trips. Figure 4.2 shows profiles of the three groups and the nine questions that distinguish the groups.

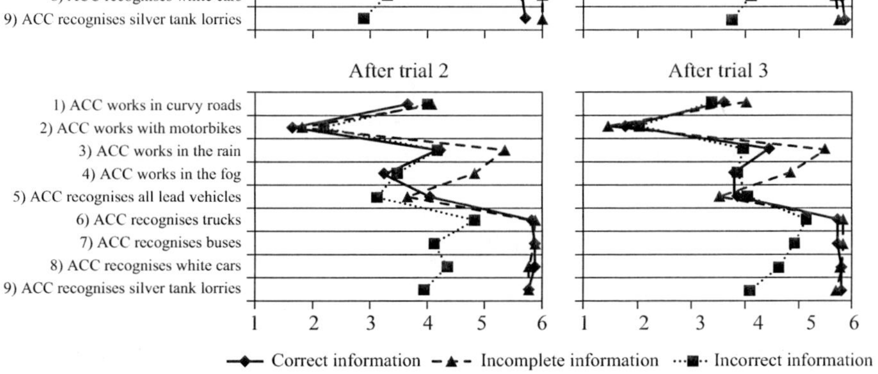

Figure 4.2 Development of ACC mental model (1 = totally disagree, 6 = totally agree) (reproduced with permission from [38])

Items 1–5 focussed on differences between the incomplete and other two groups, whereas Items 6–9 distinguished between the incorrect group and the other two groups. The incomplete group reported almost no anticipated problems in all of the nine question items. The incorrect and correct groups were aware of potential problems at curvy roads, with motorbikes, and in rain and fog. In addition, the incorrect group suspected potential problems with trucks, busses, white cars and silver lorries. As a manipulation check it can therefore be concluded that different preliminary ACC information led to the desired distinctive mental model within the three groups.

The mental model questionnaire enables the transformation of subjective interpretations of real-world statements, such as 'problems at narrow bends can occur', into numbers. As a result, comparisons between the experimental groups can be calculated by using the distance between the profiles (details are reported in Beggiato and Krems [38]). With experience and overall items, mental model profiles shifted towards the correct profile. The incomplete group, in particular, became aware of problems with motorbikes (Item 2), curvy roads (Item 1) and problems due to the ACC's failure to recognise all leading vehicles (Item 5).

With further practice, mental model profiles converged towards the correct group: For the expected, but non-occurring failures questioned in Items 6–9, a continuous shift towards the correct group was observed for the incorrect group. Negative interactions in terms of unexpected but experienced failures (i.e. with motorbikes, curvy roads, fog) led to a faster and stronger shift in mental model than the experience of expected, but non-occurring failures (i.e. with trucks, buses, white cars).

4.4.3.2 Situation Awareness

SA results are presented in detail for one situation to explain the analysis opportunities using the continuously presented secondary task. The situation chosen is the approach to the queue at the beginning of a construction zone and subsequent car following. A stationary white truck was positioned in the right lane at the beginning of the construction zone. As soon as the time headway between the driver and the white truck fell below 4 s, the truck started moving towards a speed of 80 km/h, retained throughout the entire 2.5 km length of the construction zone. Overtaking on the left lane was impossible because of barriers placed on the left lane. The ACC recognised the white truck and maintained a constant speed of 80 km/h, 2 s behind the truck. There were no problems in detecting the white truck, although the incorrect group received information that the ACC could potentially fail when large or white vehicles appeared.

Exploratory analysis: The number of problems solved in the continuously presented secondary task (SURT) was used as an indicator for SA. On average, participants solved 543 SURT items per trip (range: 115–1,082, $SD = 219.14$), corresponding to an average frequency of 0.252 Hz (range: 0.003–1.753 Hz, $SD = 0.214$). As the task can be performed very rapidly (the theoretical maximum being two items per second), SURT completions can be displayed as a frequency over time. This allows for exploratory analysis in combination with the synchronised video file. Figure 4.3 shows the queue situation with the white truck leading in Trip 1 for Driver 1. The spikes in the first graph display SURT reactions and the lines in the second graph show frequency of SURT reactions in Hz. The *x*-axis shows elapsed time in seconds. This exploratory analysis of SURT frequency is a powerful tool for identifying and visualising situations associated with reductions in SURT frequency.

Hypothesis testing: To analyse if the three experimental groups differed in SA in a given situation, the mean relative percentage of solved SURT tasks was calculated for every group and trip from the beginning (30,500 m) to the end (33,000 m) of the situation, as shown in Figure 4.4. This procedure renders the curves of every group comparable and shows differences between groups in SURT frequency over time.

During Trip 1, the incorrect group showed a reduced frequency of SURT reactions at the beginning of the experimental situation. In contrast, both other groups showed a linear and constant completion of secondary tasks. The provision of initial information to the incorrect group that problems could occur with white and large vehicles is evidenced in delays in secondary task completion and indicates that they were aware of this potentially critical situation. In the incorrect group, the recovery in SURT frequency started approximately 650 m after the beginning of the situation, at 31,150 m. At this position, the slope of the percentage curve again became similar to the other groups. In line with results from the mental model questionnaire, this delay effect was reduced in Trips 2 and 3.

In summary, the observed trend in the mental model questionnaire was confirmed by the objective secondary task data: In Trip 1, the incorrect group anticipated potential problems with white/large vehicles and therefore reduced their engagement in the secondary task. After actually experiencing no problems with the ACC in these situations, awareness of potential danger decreased in subsequent trips, as indicated by a higher frequency of solved secondary tasks in the same situation.

4.5 Study 2: On-road study

4.5.1 Aims and research questions

The on-road study was designed as a field test with the main aim of assessing and distinguishing between learning and integration phases of ACC use, for first-time users. Here, the learning phase is defined by the period during which a driver discovers the system, learns how it operates, identifies its limits and discovers the utility of the system. In the integration phase, the driver integrates the system into their management of the overall driving task [41]. In addition to distinguishing between learning and integration, the evolution of trust, acceptance and mental models of the system was investigated. Concerning the mental model, research assessed (1) the appropriateness of the mental model created by reading the user's manual and (2) changes due to on-road experience over time.

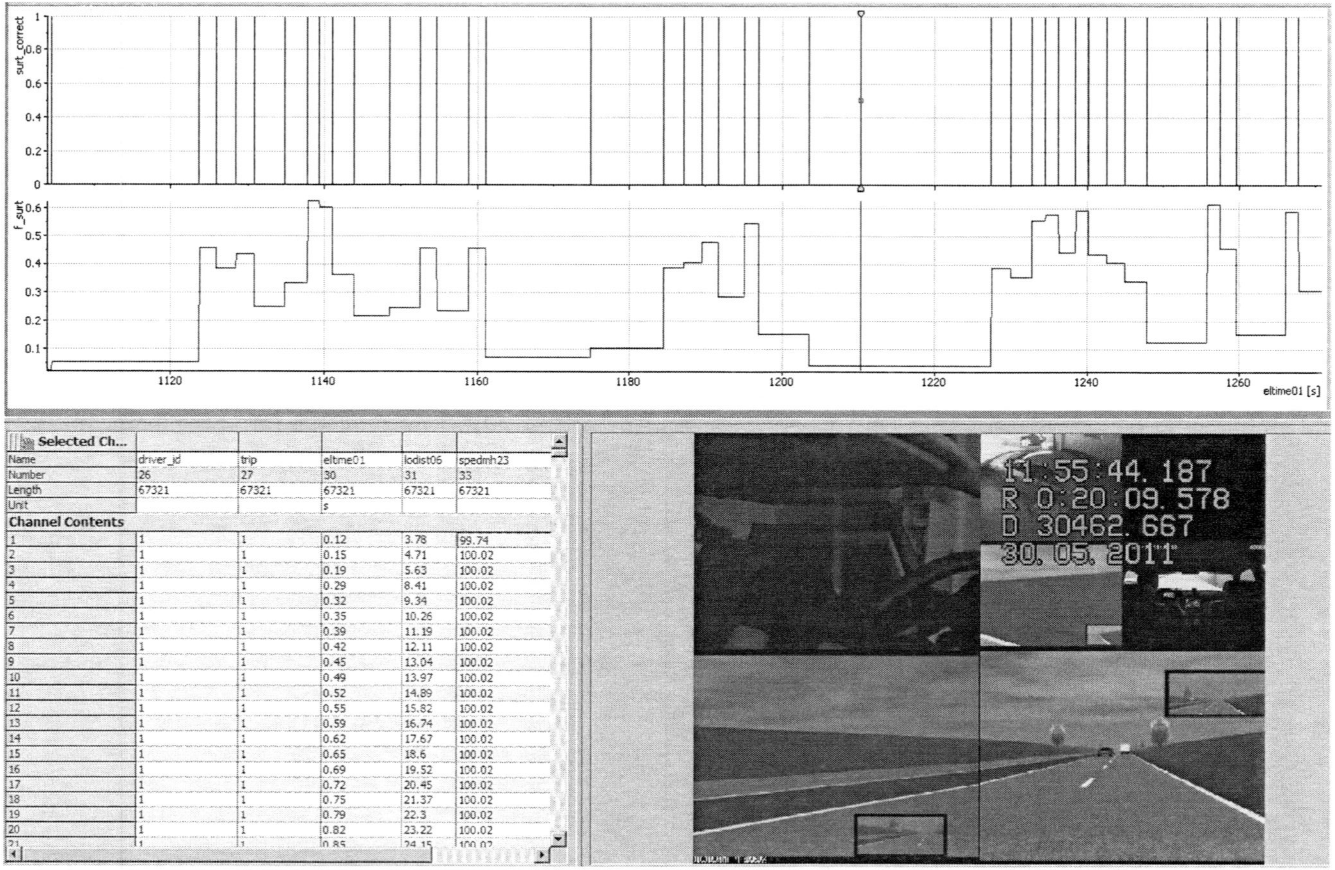

Figure 4.3 Exploratory analysis of secondary task frequency

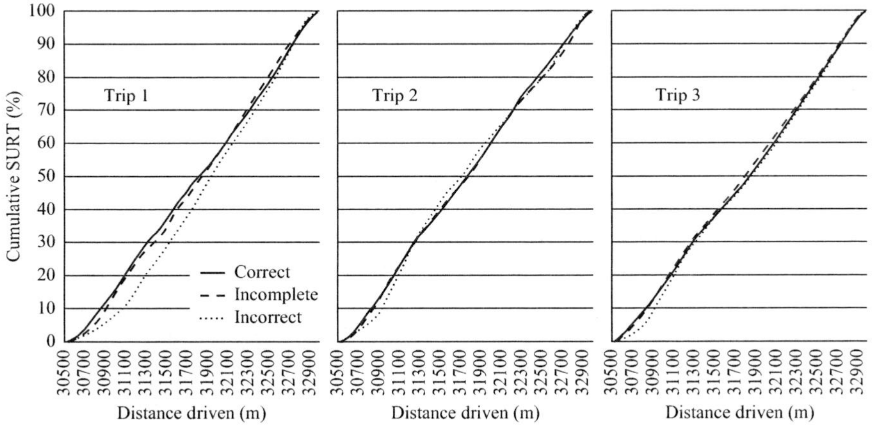

Figure 4.4 Cumulative percentage of solved SURT tasks by trip and experimental group

4.5.2 Method and material

4.5.2.1 Research design and procedure

The repeated measures study design analysed the effects of growing system experience. Fifteen participants drove the experimental vehicle ten consecutive times over a period of two months. The ten sessions had to be completed on different days, and it was not possible to complete more than two sessions per week. In the first session, participants drove without the ACC to familiarise themselves with the car, route and navigation instructions. A researcher accompanied drivers during Trip 1 in case a need for help arose. In subsequent sessions, participants performed the ride alone and were instructed to use the ACC system whenever possible. The decision to use the system was entirely up to the participants; that is, when they felt sufficiently safe and comfortable to use it. Drivers were paid for their participation.

During the first session, participants were informed about aspects of ethical and safety procedures. After familiarisation with the car, they received a copy of the ACC section of the BMW owner's manual and were instructed to read it before the next trip. By the second session researchers confirmed that all participants had read the manual and filled in the mental model questionnaire, before and after driving with ACC. The same mental model questionnaire was applied after sessions three, five and ten.

Route: The 37 km long route was the same in every trial to ensure valid comparisons between sessions and participants. Located in the vicinity of Chemnitz, the route comprised three connected road types: urban, arterial and motorway. An approximately 2.5 km stretch of urban roadway started at the University parking lot and had a speed limit of 50 km/h. This was followed by the arterial road, approximately 5 km in length and with a speed limit of 70 km/h. The motorway road stretch was approximately 11 km long and had no legally imposed speed limit.

To complete the entire route, participants exited the motorway and returned to the starting point via the same route: motorway (11 km), arterial road (5 km) and urban road (2.5 km).

4.5.2.2 Sampling and participants

Specific pre-requisites were established for the selection of participants. They had to possess a valid driver's licence, should have reached a minimum age of 25 years and should have driven at least 50,000 km. Additionally, little or no previous knowledge/experience of ACC was required. Due to contractual obligations regarding insurance provided by the vehicles' rental agency, drivers also had to be employees of Chemnitz University of Technology. Fifteen individuals were selected to obtain a homogeneous group regarding variables that could play a role in the interaction with in-vehicle systems. The final sample consisted of seven women and eight men with a mean age of 28 years (range = 25–32 years, SD = 1.82), with a mean of 144,800 km total driving experience (range = 50,000–460,000, SD = 107,961) and no experience with ACC.

4.5.2.3 Facilities and apparatus

Vehicle and ACC: The experiment was conducted in a BMW 525d equipped with ACC, including stop-and-go functionality. The ACC automatically maintained a set cruising speed, reduced it when approaching a slower leading vehicle travelling in the same lane and always maintained the headway distance set by the driver. When the vehicle ahead reduced its speed and came to a complete stop, the ACC's stop-and-go function simultaneously slowed the participant's car until it came to a complete stop. Once active, the system could be deactivated or put into a standby position by pressing the brake or accelerator pedals, respectively. Visual feedback of system status was given in the control panel of the vehicle, located behind the steering wheel. When manual intervention was called for, visual signs were displayed, and in specific cases, an auditory tone was produced. The following system limitations were described in the owner's manual:

- Detection problems may occur with small vehicles, such as motorbikes or bicycles.
- Automatic braking may not function in short-headway cut-in situations or when deceleration of the leading vehicle occurs too rapidly.
- The system does not react to stationary objects in front (e.g., stationary vehicles at traffic lights). Furthermore, it does not react to oncoming or crossing traffic, pedestrians and traffic lights.
- In situations with swerving vehicles, the other car moving from an adjacent lane into your lane can only be detected when it is fully in your lane ahead.
- Detection can be delayed or even fail in sharp bends, due to limited sensor range. Moreover, vehicles in adjacent lanes may be detected.
- Obstructions on the sensor, such as dirt, snow, ice or heavy rain, can restrict the detection of vehicles.

4.5.3 Results

4.5.3.1 Mental model

The mental model questionnaire was used to determine how users interpreted the limitations stated in the manufacturer's manual and how these interpretations changed with experience. The questionnaire was completed by participants five times during the study: Before Trip 2 (when they had read the ACC manual), and after Trips 2, 3, 5 and 10. Figure 4.5 shows results for all questions over time. The order of questions in Figure 4.5 is presented such that profiles are easy to read; that is, items with statistically significant changes over times are placed at the top of the figure (n.b. the original order of questions is indicated by 'mm' number in brackets).

Figure 4.5 ACC mental model profile by trip

Regarding reaction of the ACC to motorcycles (mm26), participants were most uncertain and sceptical at the beginning. Over time they judged this potential limitation as significantly less critical: $F(4, 56) = 4.69$, $p = .002$, $\eta_p^2 = .251$. However, interview data showed that no one had encountered a motorbike due to winter weather conditions. These results are in line with the simulator study: if a potential limitation has not been experienced, it tends to drop from a participant's mental model.

In questions addressing the ACC's reaction to all leading vehicles (mm20), a clear and statistically significant trend towards agreement could be observed: $F(2.91, 40.79) = 4.14$, $p = .013$, $\eta_p^2 = .228$. Clearly, the system performed better in reality than expected from the description in the manual. The same trend can be observed regarding the ACC's ability to regulate distance in cut-in situations: mm17, $F(4, 56) = 7.77$, $p < .001$, $\eta_p^2 = .357$; after experiencing Trip 1 with the ACC, the evaluation score increased from 3.93 to 4.87. Similar results were obtained for questions concerning the ACC's reaction to trucks (mm11): Although some subjects were doubtful before Trip 1, almost all participants rated this aspect as problem-free after completing Trip 10: $F(2.18, 30.47) = 3.43$, $p = .042$, $\eta_p^2 = .197$.

A rise-and-fall effect can be observed regarding the ability of the ACC to handling all speeds that might occur: mm28, $F(2.72, 38.13) = 6.98$, $p = .001$, $\eta_p^2 = .333$, and to react to stationary objects: mm4, $F(4, 56) = 6.85$, $p < .001$, $\eta_p^2 = .329$. Although most were sceptical after reading the manual, participants became more confident that the system could handle these situations by Trip 3. However, after Trip 5, and especially after Trip 10, confidence decreased again. Interviews showed that with increasing system experience, participants reported more situations where the ACC could not manage all speed differences. Moreover, problems were encountered with stationary cars that were located ahead of the vehicle at traffic lights. Most participants mentioned that they had not really tested if the ACC could handle these situations. Therefore, system limitations were only realised later.

Participants' initially indicated weak agreement that ACC warns when exceeding speed limits: mm30, $F(2.47, 34.63) = 3.81$, $p = .024$, $\eta_p^2 = .214$, and reacts to potholes: mm29; $F(4, 56) = 3.85$, $p = .008$, $\eta_p^2 = .216$. This decreased noticeably over time. Obviously participants expected such functionality after reading the manual. The same decreasing trend was observed regarding the ability of the ACC to adjust speed in anticipation of bends in the road: mm25, $F(4, 56) = 6.83$, $p < .001$, $\eta_p^2 = .328$. The manual stated that the ACC would reduce speed slightly before bends. However, it apparently did not work as participants expected.

All of the remaining items showed no statistically significant changes over time. Most of these were distractor items used to ensure that general system functionality was well understood. As no significant changes in items concerning stop-and-go functionality were found (mm19, mm13, mm32), it can be inferred that the manual's description corresponded well with the participants' on-road experiences. The same can be concluded for system warnings when encountering stationary objects (mm22).

4.6 Discussion and conclusions

4.6.1 Mental model questionnaire

The questionnaire discussed here is based on the concept of mental models as long-term knowledge structures, which represent a user's understanding of situation-specific system functioning [11]. The standardised questionnaire approach offers some advantages: The assessment of profiles permits a quantification of mental model facets. This allows for statistical tests between different groups as well as within subjects over time. The latter is considered useful in the context of understanding processes of behavioural adaptation, as it provides relevant insights by tracking changes in the mental model. The mental model questionnaire allows the transformation of statements into numbers, thereby making it useful for checking: how users perceive and interpret information about a system; how their mental model changes over time (e.g. due to experience); how mental models differ between groups (or persons) and descriptions provided by different manufacturers; which information is unclear; and so on. Due to standardisation, statistics can be calculated, such as profile distances or differences in single items. The questionnaire can be easily administered and the analysis is straightforward.

Results of the driving simulator study show that the questionnaire distinguishes between groups with a specifically induced initial mental model. According to differing initial descriptions of ACC, the incomplete group reported almost no problems immediately after reading the description. The incorrect and correct groups were aware of potential problems at curvy roads, with motorbikes, and in rain and fog. In addition, the incorrect group suspected potential problems with trucks, buses, white cars and silver tank lorries. As hypothesised, based on Construction–Integration Theory [19], mental model profiles converged over time towards the profile of the correct group. Bottom-up information – in terms of concrete system experience – updated the mental model accordingly, and these changes were apparent in mental model profiles.

The questionnaire was also applied in the context of the on-road study. Specifically, we investigated how information from a real ACC manual is understood and interpreted by users and how this interpretation changes due to experience with the system. Results show that non-experienced system limitations (e.g. motorbikes) tend to be judged as unproblematic over time. Participants' positive evaluation of the ACC's ability to handle cut-in situations and identify different vehicle types increased with experience; thus, the system performed better in these situations than participants expected from reading the manual. With on-going practice, users tried out more ACC functions and realised limitations with higher speed differences and stationary objects. The optimistic evaluation in the first sessions decreased to a realistic view after several trips. The description of stop-and-go functionality and warnings for stationary objects matched on-road experience as no statistically significant changes were observed over time.

The studies here show that the mental model questionnaire appears to provide insights into the construction of a mental model and its development over time.

However, there are some limitations of this approach: Designed as a standardised questionnaire, relevant items must be selected in the development phase. In contrast to qualitative methods, such as focus groups or in-depth interviews, users do not include their own topics but only rate existing statements. Moreover, items need to be pretested to make certain that all questions are correctly understood. Another limitation concerns unconscious/implicit aspects of mental models [12]. The questionnaire technique only captures conscious aspects stored in long-term memory. Automatic processes or implicit knowledge cannot be assessed. In summary, the mental model questionnaire is a useful tool for quantifying conscious aspects of participants' mental models, in a standardised and comparable way.

4.6.2 Situation Awareness measurement

The measurement of SA developed in the ADAPTATION project is designed as an implicit performance-based approach, based on participants' continuous completion of a secondary task (SURT). It is based on the principle of Rauch, Gradenegger, and Krüger [33] that situationally aware drivers decide – according to the demands of a present or imminent situation – if the execution of a secondary task is possible without compromising driver safety. In contrast to prior studies, the secondary task is not only implemented in specific situations but is continuously present, whereby users are asked to solve as many secondary tasks as possible without neglecting the primary task of driving.

A major advantage of this implicit measurement approach is the possibility of assessing SA online without the need to interrupt a given situation. Artificial and intrusive freezing of a simulation is therefore not necessary. Moreover, the measure is not biased by human judgement or memory errors in retrospective SA evaluation. As task completion is a behavioural indicator, implicit knowledge also enters into the SA evaluation, and not only conscious and explicit facts mentioned by users. Therefore, the technique is applicable in the dynamic domain of driving, as many processes are automated, unconscious and integrated in a holistic assessment. Furthermore, it is considered especially useful for assessing SA in connection with behavioural adaptation to ADAS, as it does not strongly interfere in the dynamic interactions between the driver, the system and the environment.

Results of the driving simulator study show that the measure is sensitive to user's awareness of potentially critical situations. Only the experimental group who were given incorrect preliminary information received information that the ACC could have problems with large and white vehicles. The stationary white truck located at the beginning of the construction zone was therefore a cue for a potentially critical situation. Thus, in this group, the frequency of SURT completions, when approaching this situation, decreased significantly compared to the other two experimental groups. The incorrect group was aware that this situation could be critical and consequently reduced their engagement in the secondary task. In line with results of the mental model questionnaire, SA diminished in subsequent trips as participants experienced flawless ACC performance in this situation.

However, there are some limitations of the SA measurement technique using a continuously presented visual search task. First, it can hardly be applied in on-road experiments, as the continuous engagement in secondary task completion under real traffic conditions poses safety and ethical problems. Second, the total number of completed secondary tasks depends not only on SA, but also on other factors, such as motivation to engage in a task and task priority, as well. This is shown by the rather wide range of completed tasks (from 115 to 1,082 per trip) in the driving simulator study. Differences in the total number of solved secondary tasks will be subject to further research on, for example, personality factors and age. However, when trip duration is not too long it may be assumed that participants maintain a consistent level of motivation. To solve the problem of differences in SURT completions, the absolute rate could be transformed into relative values, such as percentages (Figure 4.4), rendering the evolution of completion rates comparable. Another limitation of the technique is the duration of a situation: while the technique is useful and appropriate for long-lasting situations, such as car following or approaching, it is not suitable for situations where unexpected events require an immediate reaction, for example, in the case of emergency braking. For these situations, other measures, such as brake reaction time, are better indicators. However, the SA measure can provide insights into whether drivers are aware in advance of potentially critical situations.

Acknowledgement

The research has received funding from the European Community's Seventh Framework Programme (FP7/2007-2013) under grant agreement no. 238833/ ADAPTATION project (www.adaptation-itn.eu).

References

1. Kazi T. A., Stanton N. A., Walker G. H., Young M. S. 'Designer driving: drivers' conceptual models and level of trust in adaptive cruise control'. *International Journal of Vehicle Design*. 2007;45(3):339–360
2. Saad F. 'Dealing with behavioural adaptations to advanced driver support systems' in Cacciabue P. C. (ed.). *Modelling Driver Behaviour in Automotive Environments*. London: Springer; 2007. pp. 147–162
3. Stanton N., Young M. S. 'Driver behaviour with adaptive cruise control'. *Ergonomics*. 2005;48(10):1294–1313
4. Cotter S., Mogilka A. *Methodologies for the Assessment of ITS in Terms of Driver Appropriation Processes Over Time*. HUMANIST Project Deliverable 6 of Task Force E; 2007. Available from http://www.noehumanist.org/documents/Deliverables/TFE/E-6-HUMANIST_TRL_deliverable_VA1.pdf [Accessed 23 January 2013]
5. Rudin-Brown C. M., Noy Y. I. 'Investigation of behavioral adaptation to lane departure warnings'. *Transportation Research Record* (no. 1803); 2002. pp. 30–37

6. Weller G., Schlag B. 'Verhaltensadaptation nach Einführung von Fahrerassistenzsystemen' in Schlag B. (ed.). *Verkehrspsychologie. Mobilität – Verkehrssicherheit – Fahrerassistenz*. Lengerich: Pabst Science Publishers; 2004. pp. 351–370
7. Cahour B., Forzy J. F. 'Does projection into use improve trust and exploration? An example with a cruise control system'. *Safety Science*. 2009;47(9): 1260–1270
8. Rajaonah B., Tricot N., Anceaux F., Millot P. 'The role of intervening variables in driver–ACC cooperation'. *International Journal of Human-Computer Studies*. 2008;66(3):185–197
9. Rudin-Brown C. M., Parker H. A. 'Behavioural adaptation to adaptive cruise control (ACC): implications for preventive strategies'. *Transportation Research Part F: Traffic Psychology and Behaviour*. 2004;7(2):59–76
10. Carroll J. M., Olson J. R. *Mental Models in Human-Computer Interaction: Research Issues about What the User of Software Knows*. Committee on Human Factors, Commission on Behavioral and Social Sciences and Education, National Research Council. Washington, D.C.: National Academy Press; 1987
11. Durso F. T., Gronlund S. D. 'Situation awareness' in Durso F. T., Nickerson R., Schvaneveldt R., Dumais S., Lindsay S., Chi M. (eds.). *Handbook of Applied Cognition*. New York, NY: John Wiley & Sons, Ltd.; 1999. pp. 284–314
12. Bellet T., Bailly-Asuni B., Banet A., Mayenobe P. 'A theoretical and methodological framework for studying and modelling drivers' mental representations'. *Safety Science*. 2009;47(9):1205–1221
13. Makoto I. 'Toward overtrust-free advanced driver assistance systems'. *Cognition, Technology & Work*. 2012;14(1):51–60
14. Rajaonah B., Anceaux F., Vienne F. 'Trust and the use of adaptive cruise control: a study of a cut-in situation'. *Cognition, Technology and Work*. 2006;8(2):146–155
15. Durso F. T., Rawson K. A., Girotto S. 'Comprehension and situation awareness' in Durso F. T., Nickerson R. S., Dumais S. T., Lewandowsky S., Perfect T. J. (eds.). *Handbook of Applied Cognition*. 2nd edn. Chichester, UK: John Wiley & Sons, Ltd.; 2007. pp. 163–193
16. Seppelt B. D., Lee J. D. 'Making adaptive cruise control (ACC) limits visible'. *International Journal of Human–Computer Studies*. 2007;65(3):192–205
17. Baumann M., Krems J. F. 'Situation awareness and driving' in Cacciabue P. C. (ed.). *Modelling Driver Behaviour in Automotive Environments: Critical Issues in Driver Interactions with Intelligent Transport Systems*. London: Springer; 2007. pp. 253–265
18. Krems J. F., Baumann M. 'Driving and Situation Awareness: a cognitive model of memory-update processes' in Greenlee M. W. (ed.). *New Issues in Experimental and Applied Psychology*. Lengerich, Germany: Pabst; 2009. pp. 56–75
19. Kintsch W. *Comprehension: A Paradigm for Cognition*. New York, NY: Cambridge University Press; 1998

20. Cherri C., Nodari E., Toffetti A. *Review of Existing Tools and Methods* (Del. 2.1.1 of AIDE IST-1-507674-IP); 2004. Available from http://www.aide-eu.org/pdf/sp2_deliv_new/aide_d2_1_1.pdf [Accessed 23 January 2013]
21. Durso F. T., Sethumadhavan A. 'Situation awareness: understanding dynamic environments'. *Human Factors*. 2008;50(3):442–448
22. Popken A., Krems J. F. 'Automation and situation awareness' in Boy G. (ed.). *The Handbook of Human-Machine Interaction. A Human-Centered Design Approach*. Farnham: Ashgate; 2011. pp. 75–90
23. Endsley M. R. 'Situation Awareness Global Assessment Technique (SAGAT)'. *Proceedings of the National Aerospace and Electronics Conference (NAECON)*. New York, NY: IEEE; 1988. pp. 789–795
24. Endsley M. R. 'Situation models: An avenue to the modeling of mental models'. *Proceedings of the Human Factors and Ergonomics Society 44th Annual Meeting*; 2000. pp. 61–64
25. Zwaan R. A., Magliano J. P., Graesser A. C. 'Dimensions of situation model construction in narrative comprehension'. *Journal of Experimental Psychology: Learning, Memory, and Cognition*. 1995;21(2):386–397
26. Zwaan R. A., Radvansky G. A., Hilliard A. E., Curiel J. M. 'Constructing multidimensional situation models during reading'. *Scientific Studies of Reading*. 1998;2(3):199–220
27. Taylor R. M. *Situation Awareness Rating Technique (SART): The Development of a Tool for Aircrew Systems Design*. Situational awareness in aerospace operations. AGARD-CP-478, Neuilly sur Seine, France: NATO Advisory Group for Aerospace Research and Development, 1990. pp. 3/1–3/17
28. Vidulich M. A., Hughes E. R. 'Testing a subjective metric of situation awareness'. *Proceedings of the Human Factors & Ergonomics Society 35th Annual Meeting*. Santa Monica, CA: Human Factors & Ergonomics Society; 1991. pp. 1307–1311
29. Waag W. L., Houck M. R. 'Tools for assessing situational awareness in an operational fighter environment'. *Aviation, Space, and Environmental Medicine*. 1994;65(5, Suppl.):13–19
30. Endsley M. R. 'Predictive utility of an objective measure of situation awareness'. *Proceedings of the Human Factors Society 34th Annual Meeting*. Santa Monica, CA: Human Factors & Ergonomics Society; 1990. pp. 41–45
31. Durso F. T., Bleckley M. K., Dattel A. R. 'Does SA add to the predictive validity of cognitive tests?'. *Human Factors*. 2006;48(4):721–733
32. Jones D. G. 'Subjective measures of situation awareness' in Endsley M. R., Garland D. J. (eds.). *Situation Awareness Analysis and Measurement*. Mahwah, NJ: Lawrence Erlbaum Associates; 2000. pp. 101–114
33. Rauch N., Gradenegger B., Krüger H.-P. 'Darf ich oder darf ich nicht?: Situationsbewusstsein im Umgang mit Nebenaufgaben während der Fahrt'. *Zeitschrift für Arbeitswissenschaft*. 2009;63(1):3–15
34. Rauch N. *Ein verhaltensbasiertes Messmodell zur Erfassung von Situationsbewusstsein im Fahrkontext* [dissertation]; 2009. Available from http://opus.bibliothek.uni-wuerzburg.de/volltexte/2009/3722/ [Accessed 22 January 2013]

35. Schömig N., Metz B., Krüger H.-P. 'Anticipatory and control processes in the interaction with secondary tasks while driving'. *Transportation Research Part F: Traffic Psychology and Behaviour*. 2011;14(6):525–538
36. Metz B., Schömig N., Krüger H.-P. 'Attention during visual secondary tasks in driving: adaptation to the demands of the driving task'. *Transportation Research Part F: Traffic Psychology and Behaviour*. 2011;14(5):369–380
37. Mattes S., Hallén A. 'Surrogate distraction measurement techniques: The Lane Change Test' in Regan M., Lee J. D., Young K. L. (eds.). *Driver Distraction: Theory, Effects and Mitigation*. Boca Raton, FL: CRC Press; 2009. pp. 107–122
38. Beggiato M., Krems J. F. 'The evolution of mental model, trust and acceptance of adaptive cruise control in relation to initial information'. *Transportation Research Part F: Traffic Psychology and Behaviour*. 2013;18:47–57
39. Beggiato M., Krems J. F. 'The effects of preliminary information about adaptive cruise control on trust and the mental model of the system: a matched-sample longitudinal driving simulator study' in de Waard D., Merat N., Jamson H., Barnard, Y., Carsten, O. M. J. (eds.). *Human Factors of Systems and Technology*. Maastricht, the Netherlands: Shaker Publishing; 2012. pp. 63–74
40. Saad F., Hjälmdahl M., Cañas J., Alonso M., Garayo P., Macchi L., et al. *Literature Review of Behavioural Effects* (Del. 1.2.1 of AIDE IST-1-507674-IP); 2004. Available from http://www.aide-eu.org/pdf/sp1_deliv/aide_d1-2-1.pdf [Accessed 23 January 2013]
41. Tsang P., Vidulich M. A. 'Mental workload and situation awareness' in Salvendy G. (ed.). *Handbook of Human Factors & Ergonomics*. Hoboken, NJ: Wiley; 2006. pp. 243–268

Chapter 5

Early adopters' mental model of Adaptive Cruise Control (ACC) and its influence on behavioural adaptation to the system

Giulio F. Bianchi Piccinini[1], Anabela Simões[2] and Carlos M. Rodrigues[3]

Abstract

Adaptive Cruise Control (ACC) is a system that, through the automation of the longitudinal driving task, aims to increase drivers' comfort. Previous studies into ACC showed that behavioural adaptations might occur following its use. Those studies were mainly conducted with drivers who had never used the system before and, to date, little information is available on the behaviour of actual ACC users. Hence, this chapter describes the main findings obtained from focus group discussions and a small scale naturalistic Field Operational Test (nFOT) performed with early adopters. The findings illustrate that behavioural adaptations to the system emerged during both studies and that an improper driver's mental model of the system might be among the triggering causes.

5.1 Introduction

5.1.1 Adaptive Cruise Control (ACC) and behavioural adaptations

Intelligent Transportation Systems (ITS) 'integrate telecommunications, electronics and information technologies with transport engineering in order to plan, design, operate, maintain and manage transport systems' [1]. The application of ITS to the transport network (including vehicles) is expected to make a significant contribution to the improvement of environmental performance, efficiency, safety and security. In the wide context of ITS applications available, Advanced Vehicle Control Systems (AVCS) is the functional area that takes care of enhancing vehicle's control in order to allow safer and more

[1] ISEC UNIVERSITAS, Lisbon, Portugal

[2] CIGEST/ISG, Lisbon, Portugal

[3] Faculdade de Engenharia da Universidade do Porto, Porto, Portugal

comfortable travels [2] and that includes the system Adaptive Cruise Control (ACC), which is the device targeted in this chapter. Adaptive Cruise Control, enhancing the functionalities of the conventional cruise control system, maintains the speed adaptively to a forward vehicle according to the settings selected by the driver. The controls, located on the steering wheel, enable the driver to define the speed and the headway. When the system does not detect any vehicle ahead, the ACC brings the equipped vehicle's speed to the value set by the driver. On the other hand, when the ACC identifies a vehicle travelling ahead (based on the information received by the radar), the distance is adjusted according to the value selected by the user. The system can be overridden or deactivated by the driver at any time in order to allow him/her to take back the control of the vehicle. In addition, the system emits a warning when it is not able to handle strong decelerations of the vehicle in front. In this situation, the driver must regain control of the longitudinal vehicle dynamic. The system presents some limitations caused by the restricted field of vision of the radar sensor and by the finite braking capacity of the vehicle. Those limitations are recognised by the car makers and are stated in the owner's manual; for example the ACC might lose the detected vehicle in curves and might not detect small vehicles such as motorbikes or vehicles not driving in the centre of the lane or vehicles at close quarters. In addition, the system cannot react to slow moving or stationary vehicles and might not work in city traffic, at intersections, on slippery surfaces, in poor visibility and heavy rain.

The ACC has, as its objective, the partial automation of the vehicle's longitudinal control and the alleviation of driver's workload in a convenient manner [3] and, for this reason, has been marketed by car makers as a 'comfort and driving pleasure' device. In accordance with the expectations, several studies conducted on the topic demonstrated that, while driving with ACC activated, participants reported a lower workload level [4–8]. However, in terms of driving performance, researchers stressed that some negative effects might occur such as later reaction to safety critical events [4–5, 8, 9], increase in driving speed [5] and deteriorated lane keeping performance [9]. These changes in the drivers' behaviour due to the usage of ACC are regarded as behavioural adaptations, based on the definition issued by the Organization for Economic Co-operation and Development (OECD): 'behavioural adaptations are those behaviours which might occur following the introduction of changes to the road-vehicle-user system and which were not intended by the initiators of the change.... They create a continuum of effects ranging from positive increase in safety to a decrease in safety' [10]. Behavioural adaptations can evolve from the complex interplay of different factors [11]:

(a) Specific features of the assistance systems, e.g. the level of automation and the type of human–machine interface
(b) Characteristics of the driving situations, e.g. more intense or more relaxed driving tasks
(c) Individual driver characteristics, e.g. sensation seeking and locus of control

With regard to the temporal aspects that have an impact on behavioural adaptations, two main phases are considered [12]:

1. 'Learning and appropriation' phase: during this period, the driver discovers the system, learns how it operates and identifies its limits.
2. 'Integration' phase: the driver, through the experience using the system in different road situations, reorganises his/her activity by integrating the system in the management of the overall driving task.

During these two temporal phases and the 'Learning and appropriation' phase in particular, the driver acquires elements that are essential for the construction of the mental model of the system. Based on this mental model, the user decides (consciously or unconsciously) when to drive manually and when to activate the system [13] and, also, how much to trust the system [14]. These decisions might have repercussions on the occurrence of negative behavioural adaptations to the system.

5.1.2 Mental models

A mental model (also called mental representation) is a dynamic representation or simulation of the world [15]. In more detail, mental models are structures, stored in the Working Memory for a specific time and built through a matching process between the Long Term Memory and the information retrieved from the external reality [16]. In the more specific interaction with a system, a mental model can be defined 'as the mechanisms whereby humans are able to generate descriptions of system purpose and form, explanations of system functioning and observed system states, and predictions of future states' [17]. The mental model of the system directly influences the interaction and the cooperation between the user and the system because it reflects users' opinions about the characteristics (functioning principles, usage conditions) of the device. An incorrect mental model might produce an improper usage of the device/system or a misunderstanding of the information provided by the device/system. For example in aviation, if the pilot's mental model of the Flight Guidance System (FGS) is not correct, there is the risk to incur in mode confusion (a phenomenon in which pilots become confused about the status of the FGS and interact with it erroneously) [18]. Alternatively, in industrial processes, if the mental model of a machinery/system is not accurate, the operator might ignore the messages (e.g. warnings, guides) issued by the machinery/system and, therefore, take the wrong action [19].

Research on mental models has been conducted in the air traffic control field [20] and in the domain of industrial processes [21]. In the automotive sector, researchers have underlined that the appropriateness of driver behaviour with advanced technologies depends on the driver's mental model of the in-vehicle systems [22]. Indeed, the proper usage of an in-vehicle system is related to the driver's ability to anticipate and predict the system's behaviour. This competence of the driver derives, in turn, from the driver's mental model of the system, which, if correctly shaped, permits the driver to comprehend what the system is doing at any point in time. On the contrary, if the mental model is not accurate, the expected benefits of in-vehicle technologies (decrease in drivers' errors and mitigation of the consequences) might be weakened or eliminated.

With regard to the use of ACC, the 'mental model' concept is especially important taking into account the system's limitations (as mentioned earlier in this chapter). If the driver's mental model of the system is not accurate, the driver might be unaware of some (or all) of the limitations of ACC and, as a consequence, risky situations might originate. For example, considering that the ACC cannot react to a stationary vehicle, if the driver's mental model of the system does not include this limitation and this situation occurs, the driver might not be able to promptly brake and avoid the vehicle. In addition, if the driver's mental model of the system does not take into account the limited braking capacity of the ACC, the user might think that the system is able to work in every driving condition (including the hard braking situations) and, then, he/she might not react if an emergency braking is required. These examples represent cases of negative behavioural adaptations to the system (the driver relies on the system, even in driving situations not suited for the usage of the system and, as a result, critical situations occur), and they are caused by an improper mental model of the ACC.

The ACC working principle and its limitations are reported in detail in the owner's manual but, unfortunately, the vast majority of drivers do not leaf through the manual and those who do read approximately 50% only [23]. So, the mental model of the ACC that drivers build before and during the usage of the system might often be inaccurate, as confirmed by previous studies on the subject [24, 25]. As a consequence, the ACC users might change their driving behaviour (for instance, activating the ACC in driving environments not suited for its usage and where, before the purchase of the system, they were in firm control of their vehicle) and be confronted with critical situations for safety (as indicated above). This possibility is especially high for the early adopters of the system, considering that, due to their limited experience with the ACC, their mental model might be still under construction and might present some gaps. This assumption is confirmed by the fact that the driver's mental model of ACC becomes more or less precise depending on the failures encountered and on the experience acquired with the system [26].

5.1.3 Scope of the research

Previous studies mentioned earlier in this chapter showed that, in some cases, behavioural adaptations might appear following the usage of ACC. However the relationship between the driver's mental model of the system and the occurrence of behavioural adaptations has not yet been described experimentally. As reported above, an improper mental model of the functional limitations of ACC can lead to critical situations while driving with the system activated; for example the driver might not avoid a standing vehicle that was not detected by the ACC. In order to fill the existing research gap, this study will investigate the driver's mental model of ACC and the corresponding behavioural adaptations.

Differing from previous studies conducted on ACC [4, 5, 8, 9], this research included early users of the system in the sample. The rationale behind this choice is that drivers who use a vehicle equipped with ACC might behave differently from the non-users, as their mental model is more refined following the experience gained during the usage of the system. Hence, studying their mental model of ACC and their behaviour while driving with the system aims to acquire knowledge about how the system is understood and utilised by actual ACC adopters (and not by participants selected exclusively for the experiment but not used to the system). Therefore, if the drivers' mental model of the system is not correct and if any behavioural adaptations occur as a consequence of the improper mental model, it means that ACC drivers are continuously exposed to danger while they are driving with the system activated. Such conclusion should then lead to take appropriate countermeasures for the elimination of the negative consequences linked to the usage of the system.

Some studies have been already performed with users of the ACC. In the United States a survey was carried out with ACC users with the aim of collecting information on users' general perceptions, patterns of use and understanding of the system and its limitations [27]. From the results of the survey, three clusters of ACC users were distinguished, the 'unaware', 'unsure' and 'aware'. Some concern arose for the drivers in the 'unaware' and 'unsure' clusters, due to the combination of the improper mental model of the system (low awareness of system's limitations) and the high level of trust. During the driving with the ACC activated, these drivers might change their behaviour, relying on the system even in situations that the ACC cannot handle. As a result, the drivers might not be able to react in case the ACC fails, given that their mental model of the system is not accurate.

Focus group sessions were held with ACC users in Sweden in order to understand the usage, the driving behaviour and the risks associated with the ACC utilisation [24]. From this study, the users appeared satisfied about the system and also stated that they had already experienced some critical situations with the ACC, e.g. in curvy roads and roundabouts, or during overtaking manoeuvres. Overall, the researchers concluded that, for many participants, the functioning principle of the ACC was still based on a rudimentary mental model of the system.

Again in Sweden, a questionnaire survey was performed to understand drivers' specific experiences and patterns of use with the ACC [25]. The results showed that, overall, the drivers understood the limitations of the system and this awareness was related to the system's ownership, i.e. the longer the drivers had the system, the more they were aware of its limitations. However, the author also underlined that more effort is required to help drivers in developing an appropriate mental model of the ACC. The implementation of appropriate in-vehicle interfaces would be a proper move in this direction, in order to deliver to the driver the information about the general working principle of the system (especially for the earliest utilisations) and about the behaviour of the system in specific situations.

A driving simulator study was performed in the United States in order to better understand the behaviour of ACC users when driving with the system [28]. Through the data collected (subjective and performance measurements), drivers were divided in three groups, conservative, moderately risky and risky. The risky drivers showed high trust in the system and used it more in the simulator compared with the other groups. In addition, they had an improper mental model of the ACC because they expected the system to work even in situations where it could not, e.g. with stationary vehicle. The partial mental model of the system could lead the drivers in the risky group to incur in negative behavioural adaptations (usage of the system in inappropriate situations) and, possibly, to fail to override the system in case of a functional limitation.

Overall, the studies conducted so far revealed that the users of ACC are not completely aware of the limitations of the system (i.e. their mental model of the system is not yet complete) and therefore are confronted with the risk to incur in behavioural adaptation. For example, they use the system in driving situations where it should not be adopted and, as a consequence, critical situations arise, such as the drivers might react late or not react if the system doesn't detect a still vehicle, a motorcycle or vehicles at close quarters. Furthermore, the research performed was mainly based on subjective assessments, e.g. questionnaires, focus groups interviews or driving simulator studies, and, therefore, there is the need to validate previous research with on-road real driving studies. It is important to mention that, up to now, research has been mainly carried out in the United States and Northern Europe. However, driver's behaviour and performance differ across cultures, and South European drivers show more inclination to speeding behaviour and aggressive driving compared to North European drivers [29]. The usage of the ACC (through the setting of speed and headway) might be closely related to the aggressive behaviour of the driver, considering that more aggressive drivers set closer headway and higher speed. Then the usage of the system might vary across different road cultures, motivating the need for research on the ACC in the Southern Europe.

In order to acquire more information about the usage of ACC in Portugal, the research described in this chapter was conducted with early adopters of the system. In detail, the study intended to achieve the following objectives:

1. To collect users' opinions and perceptions of ACC and, in particular, to what extent do drivers appreciate the technical limitations of ACC (what is their mental model of the system)
2. To understand better the usage of the system and, especially, how the usage of ACC changes driver's behaviour with respect to road safety (which are the negative behavioural adaptations caused by the usage of the system)

In order to reach the research target, two methods have been adopted:

1. Focus groups interviews to collect users' opinions on ACC and to determine the patterns of system's use (including possible critical situations experienced)
2. Naturalistic Field Operational Test (FOT) to confirm the data collected during the focus groups discussions, through an on-road real driving experiment

5.2 Focus groups discussions

5.2.1 Procedure

The focus groups discussions were performed with the purpose of collecting users' opinions about the system and getting a deeper knowledge concerning the patterns of use of the system (including possible critical situations experienced).

Table 5.1 Number of participants (percentage in brackets) for each ACC experience category (participants of the focus groups)

Less than 50 km	**51–200 km**	**201–500 km**	**501–1,000 km**	**1,001–3,000 km**	**More than 3,000 km**
0 (0%)	2 (15.4%)	2 (15.4%)	1 (7.7%)	3 (23.7%)	5 (38.4%)

Source: Reproduced with kind permission from Inderscience Enterprises Limited. Inderscience retains copyright of the original article [34].

As reported in previous studies [24, 30–32], this technique is well suited to the scope of the research, considering that focus groups interviews permit gaining the widest range of ideas and feelings about a system and to show differences in perspectives among the participants [33]. The initial criteria defined for participants' selection were:

1. Participants should be experienced drivers (more than 150,000 km driven after getting the driver's licence)
2. Participants should have a minimum experience with the ACC (more than 500 km driven with the system activated)

However, due to the narrow population of ACC users, it was not possible to find enough drivers falling within the second selection criterion and drivers with shorter experience with ACC were also accepted. The complete information about participants' experience in driving with the ACC is reported in Table 5.1.

The focus groups were conducted in Portugal with 13 early adopters of ACC aged 33–61 years (mean = 44.3; SD = 8.0). The focus group was divided in two sessions with, respectively, six and seven participants. Unfortunately only one female participant could be included, indicative of males being greatly over-represented among ACC users.

The focus groups were performed by a team composed of a moderator, an assistant moderator and a note taker. The questioning route followed by the moderator included questions about four topics: (1) the satisfaction with ACC, (2) the critical situations occurring while driving with the system, (3) the type of usage for the system and (4) the suggestions for improvements of the system. Before beginning the discussion, the participants filled in a questionnaire requiring personal information and another questionnaire containing general questions about the ACC. A more detailed description of the procedure, the data analysis and the results of the focus groups discussions can be found in [34].

5.2.2 Results

Analysis of the data was based on the questionnaires presented and on the audio-visual recordings of the discussions. The questionnaires were useful to obtain some demographic information about the panel (age, gender, driving experience, total mileage travelled, mileage driven in the last 12 months, type of vehicle used) and about the usage of the system (mileage driven with the ACC activated, road

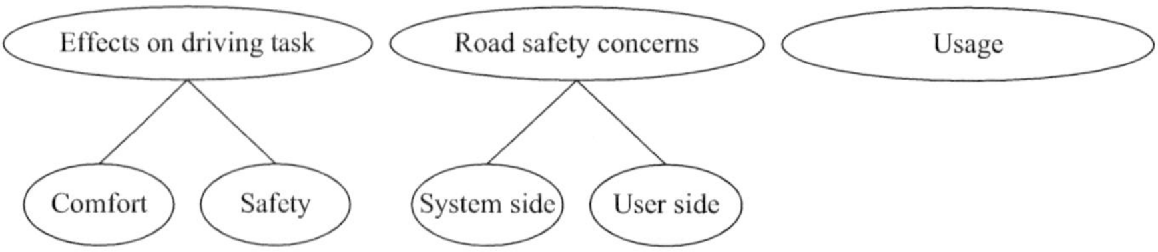

Figure 5.1 Themes emerging from the discussion (reproduced with permission from [34])

context, traffic level and weather conditions in which the system is used, speed and headway set with the system). Concerning the audiovisual information of the focus groups, a thematic analysis approach was adopted, through the identification of patterns emerging from the data [35]. The videos were first watched and transcribed verbatim. Then, the transcribed discussions were read and codes were identified along the text. Through an iterative procedure, a hierarchical coding system was established, with higher level codes representing the themes of the focus groups discussions. Overall, three themes resulted from the discussion with relative sub-themes (Figure 5.1).

Globally, from the focus groups discussions, it was found that participants have a satisfactory experience with the ACC. In general, driving with the ACC is considered by the participants as more comfortable and safer than driving without the system. Concerning comfort, the most positive aspect is that drivers can reach the final destination in a more relaxed condition, thanks to the assistance provided by the system. The only reason for discomfort mentioned during the focus groups is the abrupt braking actions undertaken by the system when detecting vehicles travelling at lower speeds, particularly in conditions of heavy traffic.

With regard to safety, the participants reported lower speeds and safer distances to the vehicle in front. In contrast with the studies conducted in Sweden [24, 25], only one driver experienced a critical situation while driving with the system activated (the ACC accelerates the vehicle in curves because, due to the road path, it loses the target vehicle). Finally, on the negative side, the participants reported high concern for the drivers travelling behind the equipped car. As a result of the continuous braking activity undertaken by the system (cited above), they might judge as odd the behaviour of the vehicle equipped with the ACC, especially when there is a large amount of space in front of it.

The most relevant result of the study is that some examples of behavioural adaptations were reported during the discussion. Drivers admitted to engage more frequently in secondary tasks (using the mobile phone for calling and surfing on the web, reading, working) when driving with the ACC activated. In addition, some participants set short headways while driving with the ACC (probably, with the objective of avoiding the abrupt braking behaviour of the system) and exhibited usages that might be detrimental to road safety (driving using exclusively the ACC buttons, releasing the feet from the brake and accelerator pedals or driving seeking to be in the wake of the vehicle in front).

Finally, concerning the usage of the ACC, participants stated they utilise the system mainly on roads supporting high speed and with stable or low traffic conditions. Those patterns of use are, probably, the consequence of some limitations of the actual system:

1. Low speed roads are distinguished by windy paths and, therefore, in those driving conditions, the system might lose the target vehicle.
2. In dynamic traffic conditions, the ACC might often change the target vehicle and, as a consequence, abrupt braking behaviours might originate.

5.2.3 Discussion

In conclusion, the focus groups discussions performed with early adopters of ACC delivered some positive conclusions concerning the satisfaction and the type of usage of the system. However, some worrisome findings emerged, with respect to the impact of the ACC on road safety:

1. Early adopters of ACC were not fully aware of the critical situations that might occur when driving with the system activated.
2. Early adopters of ACC set short headways and exhibited usages that might be detrimental to road safety (driving using exclusively the ACC buttons and seeking the vehicle in front).
3. Early adopters of ACC admitted to undertaking distracting tasks (calling, surfing on internet, etc.) more frequently when driving with the ACC compared to driving without the system.

With respect to the first two findings, the cause can be framed within the concept of mental models, as described earlier in this chapter. In cases where the driver's mental model of the ACC is not accurate, the driver might not be completely aware of the critical situations occurring with the system or might use the system in inappropriate ways. Both circumstances will engender changes in the driver's behaviour, which has negative impact on road safety (or, in other terms, negative behavioural adaptations to the system).

Regarding the third finding, it can be assumed that the drivers might be tempted to engage more frequently in distracting activities while driving with ACC activated due to the partial automation of the driving task caused by the ACC. Indeed, through the automation of the longitudinal driving task, the drivers perceive a lower level of workload (drivers are released from the need of pressing the pedals and constantly monitoring the distance to the preceding vehicle) and, therefore, they might be more tempted to undertake other activities as explained by the compensatory control model of Hockey [36]. As before, the increased engagement in secondary tasks during the driving with ACC delineates a case of drivers' negative behavioural adaptation to the system (the visual and cognitive distraction associated to the secondary task and attributable to the usage of the ACC represent a treat for road safety).

Considering that focus groups interviews are solely based on participants' opinions, it was not possible to have a complete picture of actual users' behaviour while driving with ACC activated. Hence, it was decided to conduct a small-scale nFOT that aspired to complete the focus groups interviews performed, filling its limitations and shading more light on its main findings.

5.3 Naturalistic Field Operational Test

5.3.1 Procedure

From the combination of Naturalistic Driving Study (NDS) and Field Operational Test (FOT), the definition of naturalistic Field Operational Test (nFOT) was adopted to define studies realised in natural driving conditions and which aim to evaluate the relation between the overall system driver–vehicle environment and the driving behaviour, the accident risk and the efficiency of the countermeasure taken. Then, the nFOT is especially addressed to research the factors that can explain the occurrence of an accident or to evaluate a new technology [37]. In the specific case of this study, the nFOT aimed at the second objective. The system under analysis was the ACC and the evaluation included the following aspects:

1. Drivers' patterns of usage of the system (selected speed and headway, usage in different types of roads and in different traffic levels)
2. Critical situations occurred during the driving with the system activated
3. Utilisations of the system that might be prejudicial for road safety

The detailed targets of the nFOT were defined, based on the results of the focus groups discussions. In addition to the above-mentioned objectives, the nFOT also aimed to quantitatively assess the drivers' mental model of ACC, through a questionnaire, already adopted in a previous study on ACC [26]. The questionnaire was applied at this stage in order to deepen the findings of the focus groups discussions, which assumed that the driver's mental model of ACC was not completely accurate, since the participants were not aware of the critical situations occurring with the system and since they seemed to use the system in unsuitable ways.

The nFOT was performed between June and September 2012 with a total duration of 2 months (taking into account a short interruption) and a luxury vehicle was borrowed from a national dealer and instrumented with a specific platform developed in the frame of the FP7 European project INTERACTION. The acquisition platform included the following elements:

1. Four cameras, which allowed the recording of the driver, the left side of the vehicle, the instrument panel and the road ahead (Figure 5.2).
2. One GPS/GPRS module to localise the vehicle and record the speed.
3. One triaxial accelerometer to record the accelerations on the three axis (x, y and z).
4. Three microswitch sensors to measure the depression of the pedals (accelerator, brake and clutch).
5. One computer, located in the trunk of the vehicle: the computer permitted to run the software that received the signals coming from the sensors and the videos coming from the cameras; in addition, the computer temporarily stored all the information collected.

Figure 5.2 Overall view of the video images collected during the nFOT

Table 5.2 Number of participants (percentage in brackets) for each ACC experience category (participants of the nFOT)

Less than 50 km	**51–200 km**	**201–500 km**	**501–1,000 km**	**1,001–3,000 km**	**More than 3,000 km**
1 ($11.\overline{1}$%)	0 (0%)	1 ($11.\overline{1}$%)	2 ($22.\overline{2}$%)	3 ($33.\overline{3}$%)	2 ($22.\overline{2}$%)

Overall, nine participants took part in the experiment, eight males and one female (as for the focus groups discussions, it was difficult to find female users of ACC). The participants' age ranged from 37 to 65 years old (mean = 49.11; SD = 9.05) and their driving experience ranged from a minimum of 19 to a maximum of 43 years (mean = 30.22; SD = 8.18). All the participants had driven more than 150,000 km and therefore can be considered expert drivers (according, as well, to the criteria set for the focus groups discussions). The yearly mileage was higher than 20,000 km for all participants, except one who drove between 5,001 and 10,000 km. The participants were all users of ACC and the mileage driven with the system is reported in Table 5.2.

The participants picked up the instrumented vehicle at the dealer and drove it for a certain period according to the availability of the vehicle and the participant. Unfortunately, due to organisational reasons, it was not possible to guarantee that all the participants used the car for the same amount of time. Within six months after the end of the test, the participants sent back (by post or email) two questionnaires that they were asked to complete:

1. A 30-items questionnaire to quantitatively assess the mental model of Adaptive Cruise Control [26].
2. A 12-items scale of trust in automation [38] to determine the trust in ACC.

Both questionnaires were translated from English to Portuguese by three researchers and then back-translated in order to check the consistency in meaning between the two versions of the questionnaire.

5.3.2 Results

The video data were analysed through the software ELAN, developed by the Max Planck Institute for Psycholinguistics and freely downloadable from the website http://tla.mpi.nl/tools/tla-tools/elan. (For more information about the coding tool, please see [39].)

The first objective of the nFOT was the understanding of drivers' patterns of usage of ACC, with special interest in three aspects:

1. The settings of ACC (speed and headway) chosen by the participant.
2. The type of roads where the ACC is used.
3. The level of traffic in which users prefer to adopt the system.

Figure 5.3 reports the percentage of time for which each participant, while driving with ACC activated, selected a speed higher than the speed limit or a speed lower or equal to the speed limit. Overall, from the graph, it is evident that most of the participants (seven out of nine) set, for the majority of time, speeds higher than the speed limit. However, from those results, it is not possible to state whether the usage of ACC brings the drivers to increase or reduce their travelling speed, and therefore no conclusion can be drawn about the drivers' behavioural adaptations to the system (affecting the speed). More research is required to compare the driving behaviour while using ACC and while driving without the system.

Figure 5.4 shows the percentage of time for which each participant selected a specific headway while driving with the system. The ACC allowed the drivers to choose a distance from the vehicle ahead ranging from 1 to 5 intervals, with 1 interval being equal to about 1 second and 5 intervals corresponding to about 2.5 seconds. The results presented in Figure 5.4 show that drivers, while using ACC, adopt the smallest headway available almost continuously. This outcome is a confirmation of the findings obtained during the focus groups discussions where the participants admitted to adopting short headways in order to avoid the ACC losing track of the vehicle in front and, as a consequence, inducing an abrupt braking behaviour of the vehicle (especially, in dynamic traffic conditions). The setting of short headways might represent an example of behavioural adaptation to the system because the driver, due to a particular aspect of the external environment (in this case, the highly dynamic traffic) and on account of the usage of the system, is driven to select a short headway to the vehicle in front. However, as in

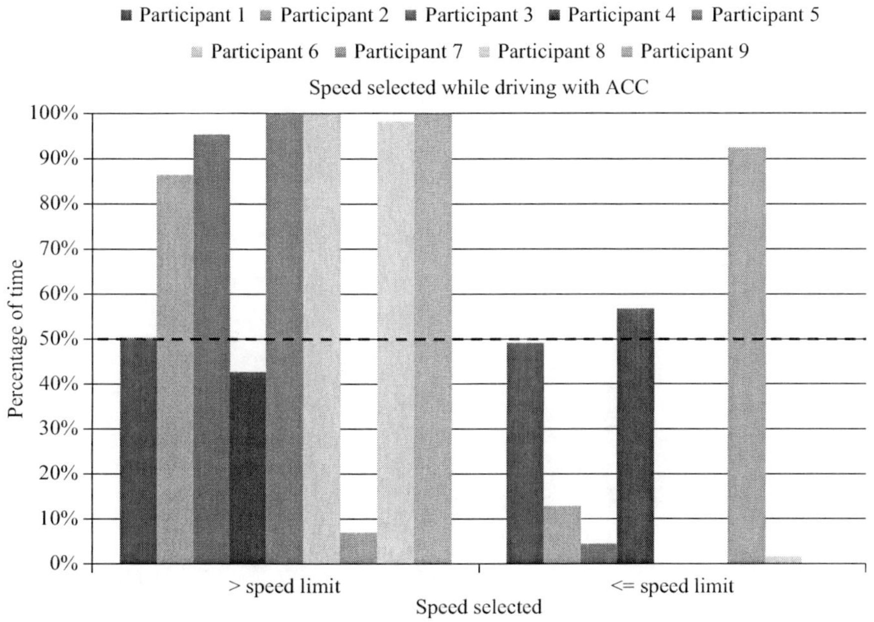

Figure 5.3 Speed selected by the participants while driving with ACC

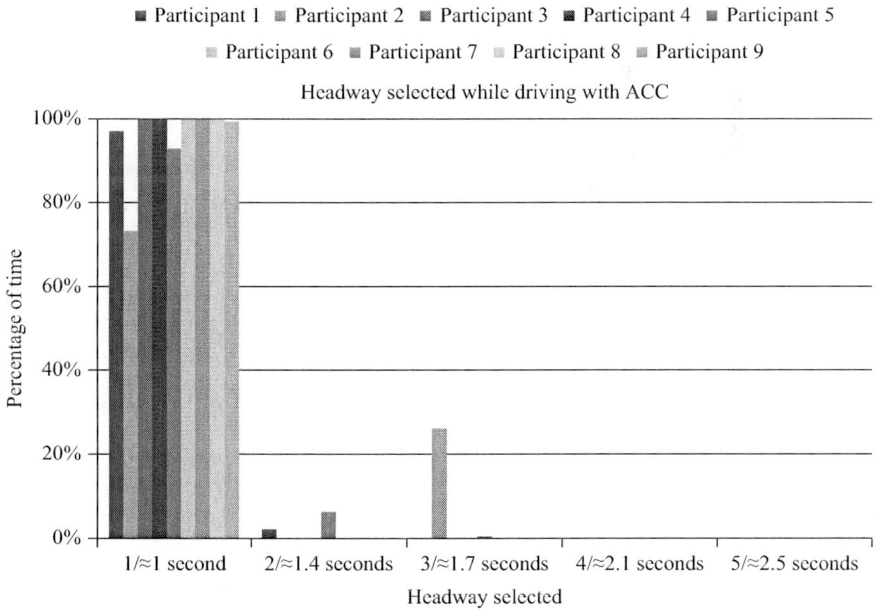

Figure 5.4 Headway selected by the participants while driving with ACC

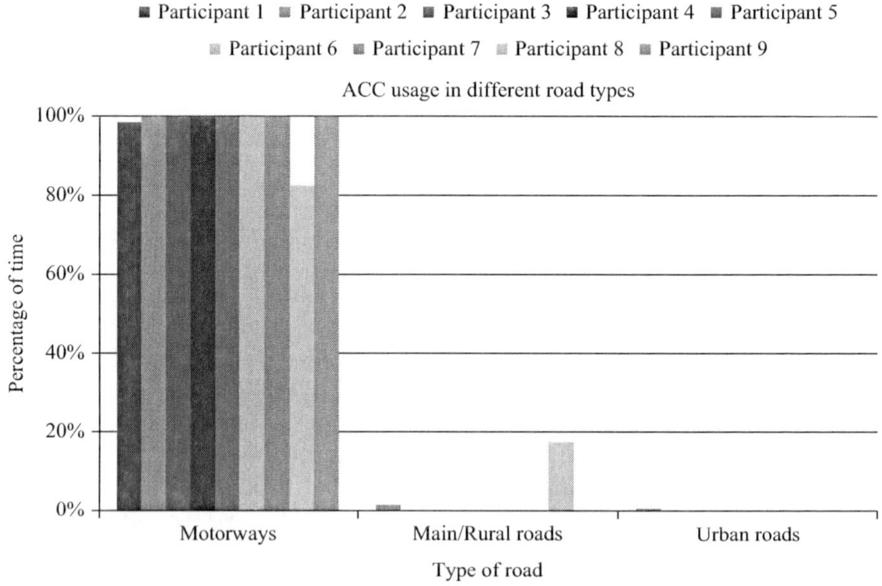

Figure 5.5 Usage of the ACC by type of roads

the case of the speed, there is no information available about the behaviour of the participants when driving without the system. Thus, more research is required to compare the headway selected while using ACC and while driving without the system.

Concerning the type of roads where the participants used the system, the results are reported in Figure 5.5 (always in percentage of time). For the classification of the roads, three categories were considered:

1. *Motorways*: roads, generally located outside the urban areas, with physical separation between the two carriageways, with at least two lanes for each direction and speed limit equal to 120 km/h (unless differently specified)
2. *Main/Rural roads*: roads, generally located outside the urban areas, with/without physical separation between the two carriageways and speed limit higher than 50 km/h and lower or equal to 110 km/h
3. *Urban roads*: roads, inside populated areas, with/without physical separation between the two carriageways and speed limit equal or lower than 50 km/h

Overall, it appears clear that drivers adopt the ACC almost exclusively on motorways, probably because, in such environment, the usage of the system is smoother, due to the fact that the way is distinguished by the scarcity of curves. This usage of the system is actually in accordance with the advice provided in the owner's manual.

Figure 5.6 concerns the usage of ACC in different traffic conditions. For the measurement of the traffic level the definition of Level of Service (LOS) from the

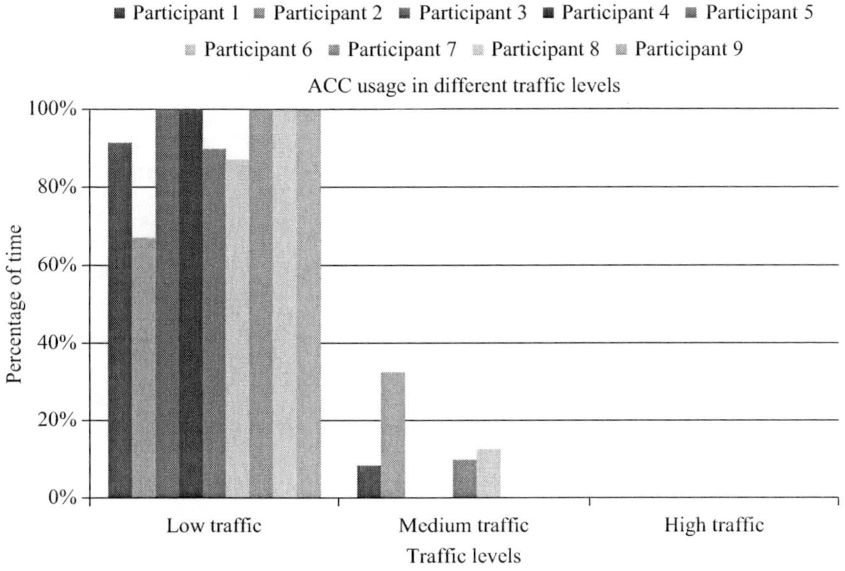

Figure 5.6 Usage of the ACC by level of traffic

Highway Capacity Manual was used [40]. According to this definition, six LOS are identified, with the following characteristics:

- *LOS A*: free-flows operation, vehicles are almost completely unimpeded in their ability to manoeuvre within the traffic stream (density $\leq$ 11 passenger car/mile/lane)
- *LOS B*: reasonably free-flows operation, vehicles' manoeuvre within the traffic stream is only slightly restricted (11 $\leq$ density $\leq$ 18 passenger car/mile/lane)
- *LOS C*: freedom to manoeuvre within the traffic stream is noticeably restricted (18 $\leq$ density $\leq$ 26 passenger car/mile/lane)
- *LOS D*: freedom to manoeuvre within the traffic stream is more noticeably limited and the driver experiences reduced physical and psychological comfort level (26 $\leq$ density $\leq$ 35 passenger car/mile/lane)
- *LOS E*: vehicles are closely spaced, leaving little room to manoeuvre within the traffic stream at speed that still exceeds 49 mph (35 $\leq$ density $\leq$ 45 passenger car/mile/lane)
- *LOS F*: breakdowns in vehicular flow (density >45 passenger car/mile/lane)

Unfortunately, through the frontal camera, it was not possible to have a perfect classification of the LOS. Hence, the traffic conditions were coded in three categories: low traffic (LOS = A or LOS = B), medium traffic (LOS = C or LOS = D) and high traffic (LOS = E or LOS = F).

From Figure 5.6. it is evident that drivers adopt the ACC almost exclusively in low traffic conditions, as is advised in the owner's manual. Indeed, in unstable traffic, the behaviour of the system is not completely reliable due to the continuous change in the target vehicle. Only few participants used the ACC in medium traffic conditions and none of them in high traffic conditions.

Further objectives of the nFOT were the detection of critical situations and improper utilisations of the system during the driving with the ACC activated. Regarding the first objective, three 'cut-in' situations were spotted. The 'cut-in' occurs when a vehicle (travelling on the left or on the right lane relative to the equipped vehicle) suddenly enters the space between the equipped vehicle and the vehicle in front of it. In such situations, the radar of the ACC cannot detect the 'cut-in' vehicle and, therefore, the intervention of the user is required. It is relevant to notice that, in all the three critical situations mentioned, the ACC had been used by the participants in environments where its utilisation is not advised (with demanding traffic or in winding roads). This inappropriate usage of the ACC can be deemed as a considerable factor in the origin of the risky driving situation. Fortunately, in all the mentioned circumstances, the driver was able to react quickly enough to avoid an accident.

With regard to the improper utilisations of the system, none of the behaviours mentioned during the focus groups was found during the procedure of video coding. However, some participants used the ACC in situations in which, according to the instructions reported in the owner's manual, it should not be, such as in winding roads, at motorway entrances/exits, in demanding traffic or during overtaking on double carriage roads. The usage of the ACC in those situations should be avoided because the radar of the system might easily lose the target vehicle and, therefore, provoke undesired fast accelerations of the equipped vehicle (and the associated necessary intervention of the driver).

The last aspect that was investigated during the nFOT was the driver's mental model of the system. As mentioned earlier, a 30-item questionnaire developed by Beggiato and Krems [26] was used to quantitatively assess the driver's mental model of ACC. The questionnaire presents statements about the system and for each assertion the participants can choose a response between 1 and 6 (1 corresponding to 'Completely disagree' and 6 corresponding to 'Completely agree'). Some examples of statements included in the questionnaire are 'The system works in motorways' or 'The system warns when the intervention of the driver is required'. Some interesting findings can be obtained from the questionnaire with regard to the drivers' understanding of the ACC working principle.

Regarding the 'cut-in' situation previously described, only one participant was aware that the system cannot always detect the vehicle that enters in the gap between the equipped vehicle and the vehicle ahead (Figure 5.7).

Another aspect that was examined through the mental model questionnaire is the possibility of ACC to work in all weather conditions. As reported in the owner's manual, the system might not work properly in poor weather conditions (such as snow or hard rain) due to the possible obstruction of the ACC radar. However three participants were not aware of such limitation of the ACC, considering that they totally agreed that the system could work in any weather condition (Figure 5.8).

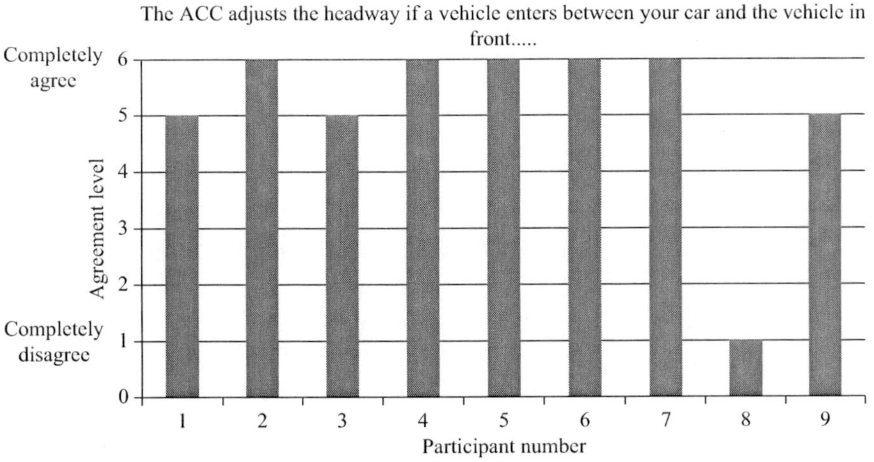

Figure 5.7 Driver's mental model of the cut-in 'situation'

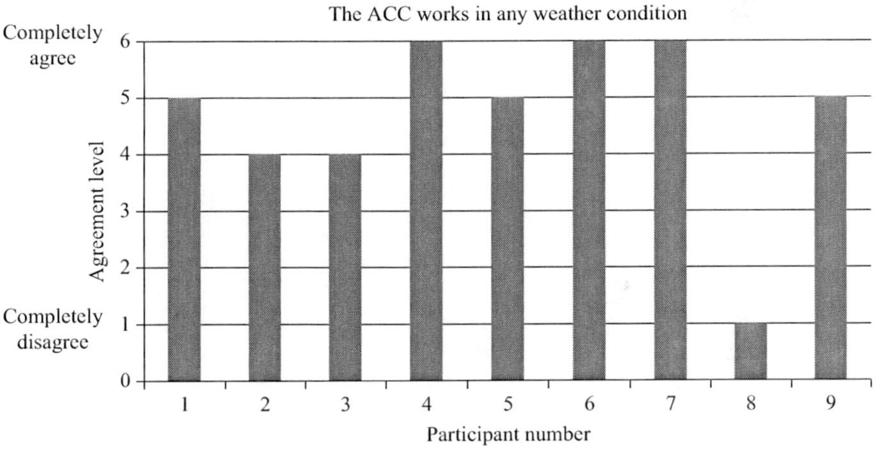

Figure 5.8 Driver's mental model about the usage of ACC in any weather condition

5.3.3 Discussion

The nFOT, conducted with a small sample of early adopters, increased awareness about ACC usage in Portugal. In addition, through the questionnaire applied after the nFOT, some aspects of the drivers' mental model of ACC were clarified. The main findings are summarised as:

1. The participants mostly chose speeds higher than the speed limit when driving with the ACC activated.
2. The participants selected short headways (about 1 second) to the vehicle in front when driving with the system activated.

3. The participants used the system almost exclusively in motorways and in low traffic conditions.
4. In some cases, the participants showed some usages that are not in accordance with the instructions reported in the owner's manual, e.g. usage on winding roads or demanding traffic.
5. The users' mental model of the system is not complete even after the initial usage of the system.

The last finding is especially interesting because it shows that early adopters of ACC, despite their initial usage of the system, are not yet completely aware of the situations in which the functioning of the system might be limited. However, during the real driving, it was not possible to obtain any confirmation of the findings gathered through the mental model questionnaire because, luckily, critical situations never occurred. There are methodological limitations of sample size and driver demographics in this study. Therefore, to know more about the ACC users' ability to react to an imminent critical situation and about behavioural adaptations to the system (comparing the behaviour while driving with and without the system), the usage of driving simulators is suggested.

5.4 Discussion and conclusions

This chapter described two studies conducted with early users of Adaptive Cruise Control (ACC). In the first, focus groups discussions were performed in order to determine users' opinions about the system and patterns of usage. From this study, behavioural adaptations to the system emerged, especially concerning some inappropriate usage of ACC and an increased engagement in secondary driving activities. With regard to the first behavioural adaptation, the cause was supposed to be the inaccurate driver's mental model of the system. However, considering that the focus groups discussions are exclusively based on subjective opinions, a real driving study was conducted with a small sample of users.

The nFOT demonstrated that, when driving with ACC activated, drivers usually select speeds higher than the speed limit and the shortest headway possible (around 1 second). Regarding the second result, taking into account the opinions collected during the focus groups, it is possible that drivers prefer to set the shortest headway in order to avoid that the radar of the system might easily lose the target vehicle, producing an abrupt braking behaviour of the vehicle. This tailgating behaviour might be due to the fact that the study was conducted in South of Europe, where it is common to observe an aggressive driving style, as mentioned in the first part of this chapter. Besides, the ACC is used almost exclusively in motorways and, in general, drivers prefer to adopt it in low traffic conditions. Both types of usage are in accordance with the specifications reported in the owner's manual.

Finally, applying the mental model questionnaire, it was clear that some users are not yet completely aware of the functional limitations of the ACC. However, from the nFOT, it is not possible to estimate if those partial or total omissions in the drivers' mental model can have an impact on road safety during the usage of the system.

Overall, through this study, it is possible to have a clearer indication of how early adopters of ACC use the system. However, some doubts still remain especially regarding two aspects:

1. Does the introduction of ACC provoke any behavioural adaptations regarding the speed and the headway chosen by the users?
2. Which impact might be produced on road safety by the omissions in the drivers' mental model of ACC?

In particular, further research should address the following aspects:

1. Are there any differences between the speeds and headways adopted with and without the system for the early adopters of ACC?
2. Are early adopters of ACC able to promptly react to a critical situation that might rise while driving with the system activated?

The overall study presents some limitations. First, the sample was very limited in both the focus groups discussions and in the nFOT. Also, in view of the fact that in Portugal early adopters of the system are principally businessmen, there was not a balance between men and women in the sample. However, with respect to those limitations, it is important to mention that it was not in the scope of this research to draw a statistical inference about the reference population.

A further limitation is that, during the nFOT, it was not possible to acquire a baseline measurement of participants' behaviour while driving without the ACC activated, due to the limited availability of the experimental vehicle. However, the comparison between the driving task with and without the ACC will be the subject of future assessment to be performed in the driving simulator.

Acknowledgements

This research received funding from the European Commission Seventh Framework Programme (FP7/2007-2013) under grant agreement no. 238833 (Marie Curie Initial Training Network ADAPTATION: 'Drivers' behavioural adaptation over the time in response to ADAS use'). The authors would like to thank the Volvo dealer 'Auto Sueco Minho' for the help provided during the study and colleagues Susana Rôla and Ana Ferreira for assistance during the focus groups. The authors would like to thank Inderscience for the authorisation given to summarise part of the article [34] 'Effects on driving task and road safety impact induced by the usage of adaptive cruise control (ACC): a focus groups study' on which they retain the full copyright.

References

1. Official Journal of the European Union. *Directive 2010/40/EU of the European Parliament and of the Council of 7 July 2010*. Brussels; 2010
2. Sussman J. M. 'ITS: a short history and perspective on the future'. *Transportation Quarterly*. 1996;50(4):115–125

3. International Organization for Standardization (ISO). *Intelligent Transport Systems – Adaptive Cruise Control Systems – Performance Requirements and Test Procedures*. ISO 15622; 2010
4. Stanton N. A., Young M., McCaulder B. 'Drive-by-wire: the case of driver workload and reclaiming control with adaptive cruise control'. *Safety Science*. 1997;27(2/3):149–159
5. Hoedemaeker M., Brookhuis K. A. 'Behavioural adaptation to driving with an adaptive cruise control (ACC)'. *Transportation Research Part F: Traffic Psychology and Behaviour*. 1998;1(2):95–106
6. Törnros J., Nilsson L., Östlund J., Kircher A. 'Effects of ACC on driver behaviour, workload and acceptance in relation to minimum time headway'. *Proceedings of the 9th World Congress on Intelligent Transport Systems*. Chicago, IL, USA, 14–17 October 2002
7. Young M. S., Stanton N. A. 'Taking the load off: investigations of how adaptive cruise control affects mental workload'. *Ergonomics*. 2004;47(9): 1014–1035
8. Vollrath M., Schleicher S. and Gelau C. 'The influence of cruise control and adaptive cruise control on driving behaviour – a driving simulator study'. *Accident Analysis and Prevention*. 2011;43(3):1134–1139
9. Rudin-Brown C. M., Parker H. A. 'Behavioural adaptation to adaptive cruise control (ACC): implications for preventive strategies'. *Transportation Research Part F Traffic Psychology and Behaviour*. 2004;7(2):59–76
10. OECD. *Behavioural Adaptations to Changes in the Road Transport Systems*. Organization for Economic and Co-operation Development Publications; 1990
11. Saad F. 'Some critical issues when studying behavioural adaptations to new driver support systems'. *Cognitive, Technology & Work*. 2006;8(3):175–181
12. Cacciabue P. C., Saad F. 'Behavioural adaptations to driver support systems: a modelling and road safety perspective'. *Cognitive, Technology & Work*. 2008;10(1):31–39
13. Boer E. R., Hoedemaeker M. 'Modeling driver behavior with different degrees of automation: a hierarchical decision framework of interacting mental models'. *Proceedings of the 17th European Annual Conference on Human Decision Making and Manual Control*. Valenciennes, France; 1998
14. Rajaonah B., Anceaux F., Vienne F. 'Trust and the use of adaptive cruise control: a study of a cut-in situation'. *Cognitive, Technology & Work*. 2006;8:146–155
15. Craik K. *The Nature of Explanation*. Cambridge: Cambridge University Press; 1943
16. Bellet T., Bailly-Asuni B., Mayenobe P., Banet A. 'A theoretical and methodological framework for studying and modelling driver's mental representations'. *Safety Science*. 2009;47(9):1205–1221
17. Rouse W. B., Morris N. M. 'On looking into the black box: prospects and limits in the search for mental models'. *Psychological Bulletin*. 1986;100(3): 349–363

18. Butler R. W., Miller S. P., Potts J. N., Carreno V. A. 'A formal methods approach to the analysis of mode confusion'. *Proceedings of the 17th AIAA/IEEE Digital Avionics Systems Conference*. Bellevue, USA; 1998
19. Wickens C. D., Kessel C. 'Failure detection in dynamic systems'. *Human Detection and Diagnosis of System Failures. NATO Conference Series*. 1981;15:155–169
20. Mogford R. H. 'Mental models and situation awareness in air traffic control'. *International Journal of Aviation Psychology*. 1997;7(4):331–341
21. Bainbridge L. 'Mental models in cognitive skill: the example of industrial process operation' in Rogers Y., Rutherford A., Bibby P. A. (eds.). *Models in the Mind: Theory, Perspective and Application*. New York, NY: Academic Press; 1992. pp. 119–143
22. Stanton N. A., Salmon P. M. 'Human error taxonomies applied to driving: a generic driver error taxonomy and its implications for intelligent transport systems'. *Safety Science*. 2009;47(2):227–237
23. Mehlenbacher B., Wogalter M. S., Laughery K. R. 'On the reading of product owner's manuals: perceptions and product complexity'. *Proceedings of the Human Factors and Ergonomics Society Annual Meeting,* Baltimore, MD, September 2002. CA: Sage Publications; 2002;46(6):730–734
24. Strand N., Nilsson J., Karlsson I. C. M., Nilsson L. 'Exploring end-user experiences: self-perceived notions on use of adaptive cruise control systems'. *Intelligent Transport Systems (IET)*. 2011;5(2):134–140
25. Larsson A. F. L. 'Driver usage and understanding of adaptive cruise control'. *Applied Ergonomics*. 2012;43(3):501–506
26. Beggiato M., Krems J. F. 'The evolution of mental model, trust and acceptance of adaptive cruise control in relation to initial information'. *Transportation Research Part F Traffic Psychology and Behaviour*. 2013;18:47–57
27. Dickie D. A., Boyle L. N. 'Drivers' understanding of adaptive cruise control limitations'. *Proceedings of the Human Factors and Ergonomics Society Annual Meeting*, San Antonio, USA, October 2009. CA: Sage Publications; 2009;53(23):1806–1810
28. Xiong H., Boyle L. N., Moeckli J., Dow B. R., Brown T. L. 'Use patterns among early adopters of adaptive cruise control'. *Human Factors: The Journal of the Human Factors and Ergonomics Society*. 2012;54(5):722–733
29. Özkan T. *The Regional Differences between Countries in Traffic Safety: A Cross-Cultural Study and Turkish.* University of Helsinki, Finland: Doctoral thesis; 2006. Available from University of Helsinki E-thesis website http://ethesis.helsinki.fi/en/ [Accessed 20 July 2013]
30. Young K. L., Regan M. A. 'Use of manual speed alerting and cruise control devices by car drivers'. *Safety Science*. 2007;45(4):473–485
31. Pereira M., Simões A., Lancelle V., Bruyas M. P., Britschgi V., Diez J. L., et al. 'A focus group approach towards an understanding of drivers' interaction with in-vehicle technologies'. *Proceedings of European Conference on Human Centred Design for Intelligent Transport Systems*, Berlin, Germany, April 2010. Lyon: HUMANIST Publications; 2010:389–399

32. Shams M., Shojaeizadeh D., Majdzadeh R., Rashidian A., Montazeri A. 'Taxi drivers' views on risky driving behaviour in Tehran: a qualitative study using a social marketing approach'. *Accident Analysis and Prevention*. 2011;43(3): 646–651
33. Krueger R. A., Casey M. A. *Focus Groups: A Practical Guide for Applied Research*, 4th edn. CA: Sage Publications; 2009
34. Bianchi Piccinini G. F., Simões A., Rodrigues C. M. 'Effects on driving task and road safety impact induced by the usage of adaptive cruise control (ACC): a focus groups study'. *International Journal of Human Factors and Ergonomics*. 2012;1(3):234–253
35. Braun V., Clarke V. 'Using thematic analysis in psychology'. *Qualitative Research in Psychology*. 2008;3(2):77–101
36. Hockey G. R. J. 'Compensatory control in the regulation of human performance under stress and high workload: a cognitive-energetical framework'. *Biological Psychology*. 1997;45(1–3):73–93
37. Victor T., Bärgman J., Gellerman H., Hjälmdahl M., Hurtig, S., Kircher K., et al. *Sweden-Michigan Naturalistic Field Operational Test (SeMiFOT) Phase 1: WP 5 Evaluation of Methodology. Final Report* (SAFER Rep. 2010:02). Chalmers Univ. of Technology, Gothenburg, Sweden; 2010
38. Jian J. Y., Bisantz A. M., Drury C. G. 'Foundations for an empirically determined scale of trust in automated systems'. *International Journal of Cognitive Ergonomics*. 2000;4(1):53–71
39. Wittenburg P., Brugman H., Russel A., Klassmann A., Sloetjes H. 'ELAN: a professional framework for multimodality research'. *Proceedings of the 5th International Conference on Language Resources and Evaluation,* Genoa, Italy, May, 2006. Available from http://www.lrec-conf.org/proceedings/lrec2006/ [Accessed 20 July 2013]
40. Highway Capacity Manual. *Highway Capacity Manual*. Washington, D.C.: Transportation Research Board of the National Academies; 2010

Chapter 6

Intelligent driver support systems: Effects of learning and aging

Vera Berthon-Donk[1], Marc Grosjean[1] and Gerhard Rinkenauer[1]

Abstract

Driving is a complex task, which relies on multiple cognitive and sensory-motor processes to ensure a safe maneuvering of the vehicle. To assist drivers with these processes, a variety of Intelligent Driver Support Systems (IDSS) have been developed. Although most IDSS become very useful over time, they initially often lead to an increase in overall task complexity, which can become excessive, especially for older drivers. Indeed, aging is associated with general changes in cognitive (e.g., slower information processing) as well as sensory-motor (e.g., decrease in sensory-motor sensitivity) functioning. These changes generally cause older adults to experience more difficulties while driving, particularly when additional tasks (e.g., paying attention to traffic signs in an unfamiliar environment or processing information from an unfamiliar IDSS) need to be performed. In this chapter, we review recent research and present a novel empirical study aimed at understanding how younger and older adults learn to drive with IDSS in multitask driving situations. Taking into account the cognitive models of multitask performance and learning, the focus is on what kind of behavioral changes occur with increasing practice with these systems, whether younger and older adults learn at the same rate, and whether they rely on different strategies to cope with increases in task complexity.

6.1 Introduction

Driving relies on multiple cognitive and sensory-motor processes to ensure safe maneuvering of the vehicle. To assist drivers in their cognitive and motor abilities, a variety of Intelligent Driver Support Systems (IDSS) have been developed, such as navigation devices, lane-keeping, and collision-avoidance assistants. Although most IDSS become very useful over time, they often initially lead to an increase in overall task complexity [1, 2], which can become excessive, especially for older drivers.

[1] Leibniz Research Centre for Working Environment and Human Factors (IfADo), Dortmund, Germany

Indeed, aging is associated with general changes in cognitive as well as sensory-motor functions, which may affect safe driving. Cognitive changes include significant declines in divided attention [3–5], selective attention [6], working memory [7], and executive functions [8], such as planning [9, 10], problem solving [11], and decision making [12]. Changes in sensory-motor functions include delayed motor-response speed, leading to increased reaction times [13–16], decreases in flexibility [e.g., 17], and coordination [precision of movement; 18]. An additional difficulty when defining the aging process is that age-related changes vary widely in their onset and amount [19]. An additional difficulty when defining the aging process is that age-related changes vary widely among individuals in their onset, amount, and rate. As a result, large variations in driving performance between aging individuals exist and chronological age, alone, can therefore not be used as a reliable index to define age-related changes in basic driving skills [19].

Changes generally cause older adults (over 50 years old) to experience more difficulties while driving, particularly when additional tasks need to be performed. This has been observed both in laboratory settings, for example, when people are asked to pay attention to (abstract) secondary tasks [e.g., visual search tasks; 20, 21], and in real-life settings, when conversing with passengers or someone on the phone [22], or driving with IDSS [23]. Under such dual-task or multitask driving situations, most studies find that older adults suffer more than younger adults, both in terms of primary- (driving) and secondary-task performance.

In light of these findings, the aim of this chapter is to further our understanding of how younger and older adults learn to drive with IDSS in multitask driving situations. To do so, we begin by focusing on dual-task performance; how people behave when trying to perform two tasks at the same time. After considering what dual-task paradigms involve and discussing a number of models that have been developed to explain the interference that can arise in such situations, we review a variety of studies that have examined how aging affects dual-task performance. We subsequently explore how learning takes place in dual-task situations. In particular, we discuss what learning actually is, how it can be quantified, and then present a model of skill acquisition that can account for learning effects. This is followed by a review of studies that examined the effect of aging on learning in dual-task (driving) situations. Finally, we conclude this chapter by presenting a simulated driving experiment that sought to further address the following issues: (1) how does learning affect dual-task performance? (2) how does aging affect learning in dual-task situations? and (3) if age-related effects are observed, is there a qualitative difference in how younger and older adults learn?

6.2 Dual-task performance

A dual-task paradigm is a situation in which participants are instructed to perform two tasks simultaneously [24]. To assess the potential costs this may incur, dual-task performance is compared to the respective single-task performances. If a person is able to perform two tasks simultaneously without a drop in performance in either task, then attention is assumed to have been successfully divided. However, if performance drops in either task, this is referred to as *dual-task interference*. When interference occurs, it is assumed that both tasks compete for the same processes or processing resources, thereby potentially limiting attention to one task at a time. For examples, see [25].

6.2.1 Models of dual-task interference

Several theoretical frameworks have been developed to explain dual-task interference and the assumed failures of divided attention. For example, *central information bottleneck theory* [25, 26] posits that structural constraints lead to a central bottleneck that only allows us to fully process one task at a time. In particular, certain aspects of the second task, such as selecting an appropriate response, cannot start until the same process has been completed for the first task. As a result, one of the two tasks (generally the secondary task) will be performed more poorly (e.g., slower) than in single-task conditions. Along different lines, the *attentional resource theory* proposed by Kahneman [27] states that dual-task interference is caused by competing demands for attentional resources. Thus, the more tasks that have to be handled simultaneously, the less attentional resources will be available for each of them. Kahneman refers to this interference as *capacitive interference*.

Of particular interest for this current chapter, Wickens [28, 29] proposed a *four-dimensional multiple-resource model* as an extension to Kahneman's resource theory that can account for distraction (interference) effects in dual-task or multitask driving situations. According to this model there are four categorical dimensions, each containing different discrete "levels." The four dimensions of the model are shown in Figure 6.1 and described below. The basic assumption is that interference will be greater when two (or more) tasks require the same level of a given dimension (e.g., two tasks demanding auditory perception).

1. *Processing stages*: This dimension indicates that resources used for perceptual and cognitive activities are different from those needed for the selection and execution of responses. This stage trichotomy is supported by both experimental as well as neuro-anatomical evidence [29, 30]. According to this dimension, there should be more interference between perceptual and cognitive tasks, as they are thought to rely on common resources. For example, visual search on a navigation display coupled with mental rotation (e.g., a map) will lead to interference in a driving situation.
2. *Perceptual modalities*: This dimension indicates that auditory perception uses different resources than visual perception. Wickens' model suggests that cross-modal time-sharing (combining a visual with an auditory task) will create less interference than intra-modal time-sharing, [e.g., combining a visual task with another visual task; 29]. In a driving environment, an IDSS with a high visual load can easily lead to interference as driving relies heavily on vision as well.
3. *Processing codes*: This dimension indicates that analogue/spatial activity uses different resources from categorical/symbolic activity (which is usually verbal

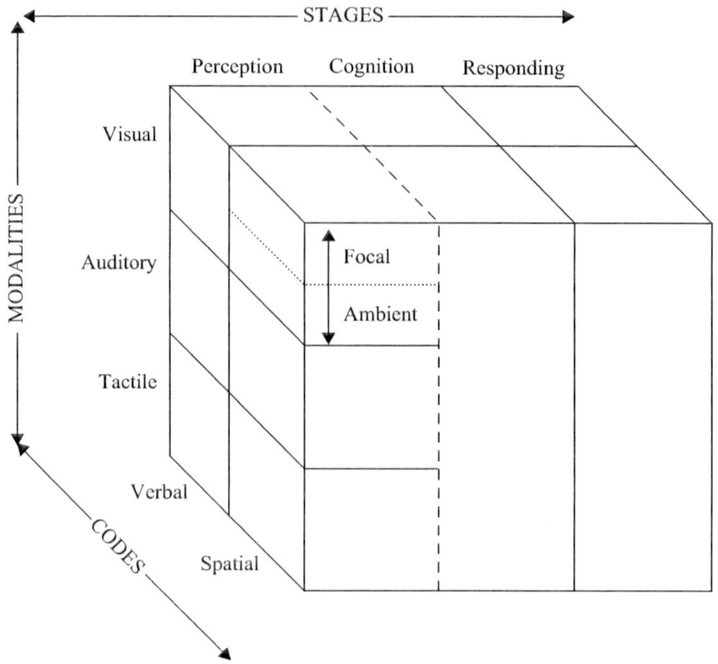

Figure 6.1 *Wickens' four-dimensional multiple resource model (figure reproduced with permission from Wickens [29, page 163, figure 1])*

or linguistic). Evidence for this separation comes from studies [29] showing a relatively efficient time-sharing between manual and vocal responses. Brain research has also shown that this separation can be associated with the functioning of the two cerebral hemispheres; the left hemisphere is more specialized in processing nonverbal auditory information, whereas the right hemisphere is more specialized in processing verbal information presented visually. This is done by using resources that are qualitatively different from those of the other hemisphere and that cannot be made available to the other hemisphere [31]. In driving, which relies greatly on spatial working memory, adding a detailed manual task [e.g., 13], might cause more interference than adding a verbal task [e.g., 32].

4. *Channels of visual information*: This dimension is nested within visual spatial resources and refers to two aspects of visual processing, focal and ambient vision. Focal vision is required for object recognition, visual search, and other tasks requiring high visual acuity, for example, searching for information on a navigation display or looking for a specific road sign among others. Ambient vision is used for sensing orientation and ego movement [33], such as for lane-keeping or adjusting distance to a leading vehicle. According to Wickens' framework [29], focal and ambient vision show efficient time-sharing because they use separate resources.

As Wickens notes himself [29, 34], although multitasking can theoretically be achieved when the tasks rely on separate resources, certain tasks can demand or attract so much attention that concurrent tasks are ignored altogether. One example he mentions is a study by Strayer and Johnston [32] who found that drivers became so preoccupied with a cellular phone task, that they would completely neglect aspects of the concurrent driving task, even though the two tasks were not really structurally related. According to Wickens [29], two types of interference can thus occur. First, *structural interference* when two or more tasks require the same processing stages, perceptual modalities, processing codes, or visual channels, and second, *capacitive interference* when multitasking leads to conflicts in terms of the overall allocation of attentional resources.

6.2.2 Effects of aging on dual-task interference

A multitude of studies have shown that age-related cognitive and sensory-motor changes have an effect on performance in dual-task settings. Interestingly, a meta-analysis on aging and dual-task performance by Riby et al. [35] revealed that age effects are task-dependent; tasks that rely on controlled processing (i.e., involving considerable mental resources) showed greater age-related dual-task impairments than tasks that rely on automatic processing (i.e., involving little or no mental resources). For example, Wilschut et al. [21] found differences in how younger (20–22 years old) and older adults (50–70 years old) switch their attention between driving and a secondary task. This consisted of a visual search task that included two difficulty levels, an easier and a more difficult task. Results showed that performance on the driving task as well as the secondary task decreased when compared with (single-task) baseline performance for both age groups. However, the effect was stronger for older adults; in addition to a significant drop in driving performance (i.e. less accurate lane-keeping), performance on the secondary task decreased to a point at which it was not attended to by the older adults (i.e., they failed to perform the secondary task at all), especially when this task required the more difficult conjunction visual search.

Riby et al. [35] suggest that task-dependency plays a role in age-dependent dual-task interference, as driving while performing a visual conjunction search requires more controlled processing, which results in a stronger decrease in driving performance for older as compared to younger adults. Wilschut and colleagues [21] furthermore propose that older adults' allocation of attentional resources is insufficient, resulting in overall performance decreases that go beyond what would be expected from general cognitive slowing due to aging only. Unfortunately the authors do not provide an explanation for this insufficient allocation of resources. A possible explanation comes from a study by Hahn et al. [36] who conducted an experiment aimed at examining age-related differences in stimulus processing in a dual-task driving situation. In accordance with Wilschut and colleagues [21, 37], age effects were observed both in terms of an increase in reaction time on a visual attention task as well as in a decrease in tracking performance on a driving-like task. The observed deficit in the secondary visual-attention task was explained by the greater difficulty for older (57–70 years old) adults to differentiate between relevant and irrelevant stimuli. According to the authors, this results in a disproportionate amount of attentional resources being allocated to irrelevant stimuli, impeding an efficient allocation of resources across tasks.

In light of the studies reviewed above, there is good evidence to suggest that aging negatively affects dual-task performance. However, the effects of aging are actually task-dependent, in that they mainly arise when tasks involve controlled as opposed to automatic processing [35]. Moreover, drops in performance seem not only due to age-inherent cognitive and sensory-motor changes, but also due to changes in the allocation of resources as well, that may rely on people's (decreasing) ability to distinguish relevant from irrelevant information.

6.3 Learning in dual-task situations

One question that has not received a large amount of attention, so far, is whether age-related effects in dual-task driving situations diminish or even disappear with practice. In this section we take a closer look at learning in dual-task situations. We will first define what learning is, how it can be quantified, and then review a model relevant for learning in dual-task situations. Lastly, we will review some studies which examined the effect of aging on learning, in general, and on learning in dual-task situations, in particular.

6.3.1 Models of learning

The Oxford Dictionary defines learning as "the acquisition of knowledge or skills through experience, practice, or study, or by being taught" [38]. Learning is often a result of rehearsal or practice. It should be noted that changes in knowledge or skills that are a result of maturation (e.g, a baby being able to track a movement at about one month of age), the intake of medication or drugs (e.g, leading to changes in perception), structural changes in the brain (e.g., in the case of Parkinson's disease), or due to fatigue (e.g., inducing changes in action sequences) cannot be considered a result of learning [39, cited by 40]. Skill acquisition is a specific form of learning referring to a type of prolonged learning about a family of events that occurs with extended practice. Practice within this framework refers to the effect of repeated task performance and is operationalized by the number of practice trials or the amount of time practicing a specific task [41]. Through many pairings of similar stimuli with particular responses, a person can begin to develop knowledge representations of how to respond in certain situations. These representations can be retrieved more easily and reliably than memories of single events and, as such, skilled behaviors can become routines and even automatic [42].

One of the methods used to quantify learning has been to find a function that describes how a performance changes with practice. A well-known example is the *power law of practice* [43], which can be expressed with the following equation:

$$T = NP^{-c},$$

where T is the time to complete a task, P is the number of practice trials, and N is performance time on the first trial of the task. The parameter c in this equation is the learning rate. The value of c is usually between 0 and 1, and is preceded by a minus sign to capture the negatively accelerated nature of the learning curve. The closer the value of c is to 1, the faster the learning rate.

Although the power law of practice can be widely applied to all forms of knowledge acquisition, it is based on the assumption that something will slow down the learning process at some point. That is, after a certain amount of practice a so-called learning plateau is often reached, where performance essentially ceases to improve. Such a plateau can be due to a decrease in motivation, a lack of automatisation of (partial) actions, a lack of transfer from already existing knowledge to a new task, or an over fixation on parts of the task only, thereby forgetting other parts of the task [39, cited by 40].

In order for a skill to become "automatic," that is, efficient, unintentional, and unconscious [44], it is generally assumed that people go through a number of stages. For example, the *Adaptive Control of Thought–Rational* (ACT–R) theory of skill acquisition by Anderson [45] explains the acquisition of skills as a process comprising three main stages: A *cognitive*, an *associative*, and finally an *autonomous* stage. In the initial stage of learning, the *cognitive* stage, people solve problems in new domains by applying unspecific problem-solving productions to explicit (i.e., coded and stored) knowledge they have about this domain. Productions are condition–action pairs that specify that if a certain state occurs in working memory, a particular action should take place. What is more, they require attention and working memory, which makes their execution slow, effortful, and relatively inefficient. The second stage of learning, the *associative* stage, corresponds to the acquisition of problem-solving routines through knowledge compilation. With practice, productions are progressively adapted to the task at hand, making problem solving more efficient and more reliable. Separate components are combined into one routine. Finally, in the third stage, the *autonomous* stage, routines are triggered in an autonomous manner and do not demand any cognitive resources, thereby freeing up resources to do other activities in parallel.

For skill acquisition to take place according to ACT-R, several memory types exist: working memory, declarative memory, procedural memory, among others. Working memory is the part of memory used for interaction with the outside world. It encodes information from the outside world and stores this information in declarative knowledge or performs pattern matching with production rules in procedural knowledge. It also retrieves information from declarative memory or retrieves production rules from procedural memory. Declarative memory is where factual information is stored (e.g., Berlin is the capital of Germany). Procedural memory is where production rules are stored. The interaction between these different types of memory enables the development of cognitive strategies resulting in skill acquisition. Examples of such strategies include *proceduralisation* (combining different chunks of explicit knowledge into procedures), *composition* (a sequence of individual productions is progressively collapsed into a larger task-specific production), and *generalization* (the application of similar production rules to different situations). All these methods allow freeing up resources in working memory and attention. According to this theory, mental load should decrease with practice as actions become more automated. In a dual-task setting, this would imply that more resources should become available for other tasks, hence diminishing the amount of dual-task interference over time.

6.3.2 Effects of aging on learning in dual-task situations

If we consider that skill acquisition and dual-tasking rely heavily on memory and attention [46], which decline at older ages [3–5, 7], learning in dual-task situations should differ between older and younger adults.

At a general level, aging has an effect on cognitive functioning as a whole, resulting in deficient problem-solving skills, poor sustained attention, and an impairment in the generation of reliable goal structures needed for skill acquisition [47]. One reason for a decrease in cognitive functioning at older age is a decline in memory capacities, which has an effect on learning both sensory-motor as well as cognitive tasks. Verwey et al. [48] found that older adults (55–62 years old) make less use of motor chunks (i.e. integrated motor sequence representations) when learning movement patterns in a dual-task setting due to age-related declines in memory functions (especially limited explicit sequence knowledge). This resulted in poorer sensory-motor performance when compared with younger adults.

Memory capacity is also needed to manage and coordinate multiple tasks at the same time. Voelcker-Rehage and Alberts [49] found that this ability was impaired for older adults (65–75 years old), even after extended practice on a task including both a sensory-motor as well as a cognitive task. Göthe et al. [50] found that this decrease in capacity to manage and coordinate multiple tasks at the same time resulted in a qualitative difference in the way younger (16–19 years old) and older (64–77 years old) adults deal with dual-task requirements. Specifically, most younger adults made a transition from serial to parallel processing with practice. Older adults, however, apparently did not make this transition. The authors explain these findings by age-related changes in the executive system of older adults, which heavily relies on memory processes. This system is needed to assign processing resources or processing time to concurrent tasks, and for scheduling processing steps in a way that minimizes interference [51], for example, by holding only one task set in operative mode at any time (serial processing). With practice, the executive system might overrule the serial-processing constraint, allowing parallel processing of two task sets. Older adults' executive systems seem to be more conservative than those of younger adults, in that they cannot overrule the serial-processing constraint and thereby will not achieve perfect time-sharing.

Taken together, these studies suggest that age-related declines in memory functions have an effect on learning in dual-task situations, resulting in poorer performance on sensory-motor tasks and difficulties in handling dual-task situations. As we are interested in dual-task situations involving driving, we will now consider studies that attempted to address these situations. Surprisingly few studies have been done in this domain, despite the recent rise of IDSS and entertainment systems in cars.

Shinar et al. [52] used a driving simulator to perform a dual-task driving study in which participants simultaneously performed a mathematical operations task or an emotionally involving phone conversation. Their experiment included three driver groups: (1) young/novice drivers (all, but one, were 18 years old), (2) experienced drivers (30–33 years old), and (3) older drivers (60–71 years old). Practice effects were measured as performance changes over 5 sessions that extended over 14 days (with 1–4 days in between each session). Although the authors found an effect of age, with older adults showing poorer performance on all tasks compared to younger adults, they also found that practice had a positive effect on performance. In particular, age-related differences in driving performance decreased over sessions and by the fifth session all three age groups had nearly identical levels of driving performance. The authors concluded that with sufficient practice, dual-tasking in a driving environment is possible for both younger and older drivers. However, these results were challenged by later studies from Cooper and Strayer [53] and Chisholm et al. [54]. Both studies looked at the effects of practice on secondary-task related driver distraction over multiple practice sessions in a driving simulator. They included only one (younger) comparison age group and they both found that, although practice increased dual-task performance, older participants did not reach younger participants' baseline performance levels (i.e., single-task performance levels).

A general conclusion from these driving-related dual-task studies is that dual-task performance becomes better with practice. However, care should be taken when interpreting the results of these studies because the experimental settings might not have allowed for proper experimental control (e.g., as is the case with naturalistic phone calls). Moreover, not all studies included different age groups, making it difficult to draw any general conclusions regarding age-related effects on learning in dual-task driving situations. As a consequence we conducted an experiment aimed at examining the effect of practice over multiple sessions on driving and visual secondary task performance for younger and older adults in a controlled laboratory setting.

6.4 An empirical study

6.4.1 Hypotheses

Based on our review of the literature (see above), the following hypotheses motivated this study: (1) learning will improve dual-task performance, (2) older adults will learn more slowly than younger adults, and (3) younger and older adults will rely on different strategies to learn in dual-task driving situations. In what follows, we will discuss the different hypotheses in more detail, present the experimental setup and provide a summary of major findings.

As to our first hypothesis, we expected on the basis of Anderson's ACT-R theory [45] that practice would lead to improved performance in both the driving task and a secondary task. According to Anderson [45], tasks become (partially) automated with practice, thereby freeing up resources. According to Wickens' model, this leads to a decrease in the resources needed for performing each task, which results in improved dual-task performance with practice for both age groups. However, in line with Wickens' four-dimensional multiple-resource model [29, 34], we expected some interference along the perceptual modality dimension. This interference was expected for both younger and older drivers. Taking potential age-related changes in executive control into account, we expected the amount of interference to be greater for the older drivers, leading to less improvement in dual-task performance as compared to younger adults.

Our second hypothesis pertained to how aging affects learning in a dual-task driving situation. As already shown by previous studies, [e.g., 48, 50, 53, 55], age-related differences are likely to persist, even after extended practice. Although according to Andersons' model of skill acquisition [45], the driving task should become (partially) automated and hence overall dual-task performance should increase, we expected skill acquisition to be slower for older adults than for younger ones. This is because the cognitive processes that underlie skill acquisition (attention, memory and executive functions) are less efficient in older than in younger adults.

Finally, our third hypothesis was based on the prediction that capacitive interference would lead to differences in learning between age groups. As observed by different researchers [49, 50], aging is associated with changes in the executive functions that are needed to manage dual-task situations. Therefore, older adults should engage in more serial processing of the dual-task situation, which, depending on how people prioritize the tasks, might lead to performance declines on the driving task and/or on the secondary task [21, 37].

6.4.2 Experimental setup

A group of 10 older adults (over 65 years old) was compared with a group of 10 younger adults (25–35 years old). Standardized tests were used to assess visual acuity, color blindness, ocular dominance, and cognitive abilities/performance (e.g., manual tracking capacities, speed of cognitive processing). Subjective ratings of mental effort were assessed with the Rating Scale of Mental Effort (RSME) [56] and with the NASA Task Load Index, (NASA-TLX) [57].

To account for effects of practice and learning, the experiment consisted of four sessions of approximately three hours each. The first three sessions took place within a timeframe of two weeks; the fourth session took place three weeks after the third session. As illustrated in Figure 6.2, the driving task consisted of the Lane-Change Test (LCT) [58]. The LCT is an automotive tracking task that has the goal to assess changes in performance on lane-change maneuvers in a simplified driving environment while performing additional non-driving related activities (i.e., secondary tasks). It consists of a straight three-lane road without other traffic and with lane-change signs at both sides of the road which indicate lane-changes to the left or the right of 1 or 2 lanes. Each track consists of 18 signs. At a constant speed of 60 km/h, participants are presented with a traffic sign approximately every 10 seconds. There are six possible lane-change types which each occur exactly

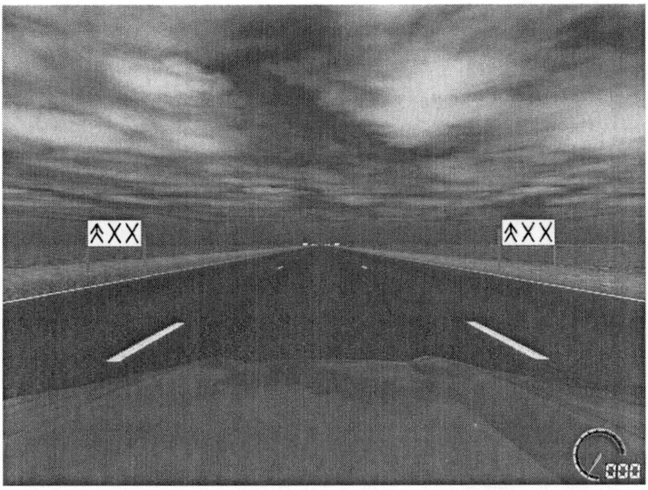

Figure 6.2 Screenshot of the Lane-Change Test (LCT) showing the straight 3-lane road without other traffic and with lane-change signs at both sides of the road which, in this case, indicate a lane-change to the left of 1 lane

Figure 6.3 Examples of the visual search task used as a surrogate Intelligent Driver Support System (IDSS). In the left panel, 16 items are presented, including 1 target (dark grey square) and 15 distractors (light grey squares and dark grey triangles). In the right panel, 25 distractor items are presented without a target

three times on each track. All participants received end-of-block feedback about their driving performance [59].

We used a visual search task (VST) [60] as a surrogate IDSS and employed conjunction search [i.e., target and distracters share similarities in more than one single visual property such as colour and shape; 60] to have a sufficiently challenging task [21]. Figure 6.3 presents two examples of the visual search task. Participants were instructed to look for target stimuli (light grey triangle or dark grey square) that were presented among distractors (dark grey triangles and light grey squares). Their task was to press as fast and as accurately as possible one of two buttons on the back of the steering wheel to indicate whether a target was present or absent. Task difficulty was manipulated by presenting different sizes of symmetric matrices consisting of 3×3 (9) items, 4×4 (16) items, or 5×5 (25) items. The dual-task condition consisted of a combination of the LCT and the VST, both as described above. An additional manipulation of task difficulty was added by either presenting the VST while performing a lane-change (more difficult) or while lane-keeping (easier). Each block of 18 signs in the LCT was thus associated with 36 VST trials.

It was emphasized in the written instructions and orally by the experimenter that in the dual-task situations the LCT was always more important than the VST. Participants were instructed to prioritize performance on the LCT over performance on the VST if they were not able to perform the LCT and VST at the same time. Electrooculogram (EOG) recording electrodes allowed for the registration of eye movements. EOG recordings are not precise enough to extract information on specific elements of the visual scene a driver is looking at, but the frequency and duration of vertical eye glances on and off the virtual road can be measured. The driving data of relevance for this chapter were lane-change speed and accuracy. Lane-change speed was expressed as the time it took participants to initiate a lane-change after the appearance of the relevant road sign (lane-change initiation time), as well as the total movement time needed to perform a lane-change (lane-change duration). Lane-change accuracy was defined as the mean lateral deviation of the vehicle (MDEV) with respect to a reference trajectory and the standard deviation of the lateral deviations (SDDEV). For the visual search task, reaction time and the proportion of correct responses were used as dependent variables. Dual-task data were analyzed using the measures described above. Additionally, the Inter-Response Interval (IRI) was calculated which is a variable designed to specifically measure dual-task interference. IRI provides information about potential interactions between the LCT and the VST in dual-task conditions by capturing the time it takes a participant to respond to a VST after initiation of the lane-change maneuver. The larger the IRI value, the higher the degree of interference from the VST on the LCT.

6.4.3 Summary of major findings

We replicated the basic findings of Wilschut and colleagues [21]: adding a secondary task (VST) had a negative effect on some, but not all, driving measures. Lane-change initiation times decreased and lane-change duration increased. Secondary task performance also deteriorated when compared with the single-task baseline condition. This finding is congruent with Wickens' [29] four-dimensional multiple resource model, according to which structural interference will occur if two tasks use the same channels of visual information. Although lane-keeping and lane-changing requires more ambient vision whilst searching for a target on a visual display required more focal vision, it is possible that this dichotomy was not sufficiently clear in the present experiment. Scanning traffic signs in order to change lanes probably required focal vision as well, which may have led to structural interference.

Interference effects were age-related, in that older adults "suffered" more than younger adults in dual-task conditions, resulting in overall poorer performance. In other words, dual-task performance for younger adults was closer to their single-task performance than it was for older adults. This finding might be explained by what Kahneman [27] and Wickens [29, 34] call capacitive interference. The executive control function, which is responsible for the distribution of attentional resources to each task at hand, becomes overloaded. This effect is stronger for older adults, as their executive control is less flexible [50].

Almost all driving measures improved with practice in both single- and dual-task conditions for both age groups. In particular, lane-change initiation became faster and lane-change duration decreased. The learning effect for lane-change accuracy, however, was strongest in single-task conditions. This effect is not in accordance with findings by Oberauer and Kliegl [61] who found that younger adults, on simple sensory-motor tasks, can reach the same level of performance in single- and dual-task situations. One explanation for the differences between their and our study is that our driving task is not a simple sensory-motor task, but consists of a more complex combination of motor actions.

Learning effects were also observed on secondary (VST) task performance, with a decrease in reaction time and errors for both age groups. Analyses of the subjective measures produced patterns that were consistent with the practice effects discussed above. Both younger and older adults progressively rated the tasks as less demanding, requiring less effort and generating less frustration. IRI, a variable measuring dual-task interference and which, in other words, defines how capable a participant is at performing two tasks at the same time, also decreased with practice for both age groups.

If we now consider our first hypothesis, which pertained to how learning affects dual-task performance, the present data show that learning took place for both younger and older adults, resulting in less interference in a dual-task driving situation over time. This is in accordance with the ACT-R theory of skill acquisition [45], according to which skills become more automated with practice. This, in turn, frees up resources allowing for better time-sharing [29] and resulting in less capacitive dual-task interference.

As to our second hypothesis, which focused on how aging affects learning in dual-task situations, our data show that learning effects differ between age groups. With practice, performance on both the primary as well as the secondary task increased for both age groups. However, learning rates differed between age groups, especially on some of the driving performance measures where learning rates were slower for older adults. IRI was significantly higher for older adults when compare with younger adults. This means that when older adults had to change lanes and perform a secondary task at the same time, they needed significantly more time to respond to the secondary task as compared to younger adults. At the final session, even with prolonged practice, dual-task performance of younger adults always remained better than dual-task performance of older adults (as witnessed by significantly higher IRI values for older adults at the last session). This is in accordance with findings by Göthe et al. [50] and Verwey et al. [48], who found that differences in dual-task performance between age groups persisted even after extended practice. One explanation could be that the less flexible executive functions in older adults do not allow for effective time-sharing [50], leading to capacitive interference that cannot be overcome with practice. Another explanation comes from Anderson's ACT-R theory, according to which practice results in automation, leading to less capacitive interference. However, for automation to occur, cognitive functions such as attention, memory, and executive control are needed, but those functions are known to decline with age [e.g., 7, 8, 36, 48]. Thus, skill acquisition, including automation, will not be as effective for older adults as for younger adults.

Finally, our third hypothesis pertained to potential differences in learning between younger and older adults. The present findings might indeed point to such differences. One explanation could come from the gaze behavior we observed in this experiment. With practice, older adults looked at the VST display (i.e., away from the road) more often and for a longer period of time. Although this behavior has a positive effect on secondary task performance, it might also impede performance on the primary driving task. This is what may have prevented older drivers from improving their dual-task performance by the same amount as young drivers. This explanation is consistent with the views of Göthe et al. [50], according to which older adults have less flexible executive functions that do not allow for perfect time-sharing.

6.5 Conclusion

The present review has shown that learning to tackle novel dual-task driving situations should still be possible at older ages. However, based on studies examining the effects of aging on dual-task performance, it can be expected that age-related differences will remain despite extended practice. Our empirical study, which explored the effects of practice and aging in a simulated dual-task driving situations, revealed that learning had a positive effect on dual-task performance. However, although learning took place for both age groups, older drivers learned at a slower rate. One possible explanation for this discrepancy comes from preliminary findings in eye-movement behavior, but future studies examining qualitative differences in learning between younger and older adults are needed to improve our understanding of dual-tasking in a driving environment, in general, and the development and use of IDSS, in particular.

Acknowledgment

This research received funding from the European Community's Seventh Framework Program (FP7/2007–2013) under grant agreement no. 238833/ADAPTATION project (www.adaptation-itn.eu).

References

1. Hancock, P.A. & Parasuraman, R., "Human factors and safety in the design of intelligent vehicle-highway systems (IVHS)," *Journal of Safety Research*, vol. 23, pp. 181–198, //Winter 1992.
2. Vollrath, M., Schleicher, S., & Gelau, C., "The influence of Cruise Control and Adaptive Cruise Control on driving behaviour – A driving simulator study," *Accident Analysis & Prevention*, vol. 43, pp. 1134–1139, 2011.
3. Brouwer, W.H., Waterink, W., Van Wolffelaar, P.C., & Rothengatter, T., "Divided attention in experienced young and older drivers: Lane tracking and visual analysis in a dynamic driving simulator," *Human Factors: The Journal of the Human Factors and Ergonomics Society*, vol. 33, pp. 573–582, 1991.
4. Ponds, W.H.M., Brouwer, W.H., & Wolffelaar, P.C., "Age differences in divided attention in a simulated driving task," *Journal of Gerontology: Psychological Sciences*, vol. 43, pp. 151–156, 1988.
5. Wild-Wall, N. & Falkenstein, M., "Age-dependent impairment of auditory processing under spatially focused and divided attention: An electrophysiological study," *Biological Psychology*, vol. 83, pp. 27–36, 2010.
6. Sweeney, J.A., Rosano, C., Berman, R.A., & Luna, B., "Inhibitory control of attention declines more than working memory during normal aging," *Neurobiology of Aging*, vol. 22, pp. 39–47, 2001.
7. Borella, E., Carretti, B., & De Beni, R., "Working memory and inhibition across the adult life-span," *Acta Psychologica*, vol. 128, pp. 33–44, 2008.
8. Adrian, J., Postal, V., Moessinger, M., Rascle, N., & Charles, A., "Personality traits and executive functions related to on-road driving performance among older drivers," *Accident Analysis & Prevention*, vol. 43, pp. 1652–1659, 2011.
9. Allain, P., Nicoleau, S., Pinon, K., Etcharry-Bouyx, F., Barré, J., Berrut, G., Dubas, F., & Gall, D.L., "Executive functioning in normal aging: A study of action planning using the Zoo Map Test," *Brain and Cognition*, vol. 57, pp. 4–7, 2005.
10. Sorel, O. & Pennequin, V., "Aging of the planning process: The role of executive functioning," *Brain and Cognition*, vol. 66, pp. 196–201, 2008.
11. Diehl, M., Willis, S.L., & Schaie, K.W., "Everyday problem solving in older adults: Observational assessment and cognitive correlates," *Psychology and Aging*, vol. 10, pp. 478–491, 1995.
12. Henninger, D.E., Madden, D.J., & Huettel, S.A., "Processing speed and memory mediate age-related differences in decision making," *Psychol Aging*, vol. 25, pp. 262–270, June 2010.
13. Alm, H. & Nilsson, L., "The effects of a mobile telephone task on driver behaviour in a car following situation," *Accident Analysis & Prevention*, vol. 27, pp. 707–715, 1995.
14. Rubichi, S., Neri, M., & Nicoletti, R., "Age-related slowing of control processes: Evidence from a response coordination task," *Cortex*, vol. 35, pp. 573–582, 1999.

15. Shanmugaratnam, S., Kass, S.J., & Arruda, J.E., "Age differences in cognitive and psychomotor abilities and simulated driving," *Accident Analysis & Prevention*, vol. 42, pp. 802–808, 2010.
16. Cerella, J., "Information processing rates in the elderly," *Psychol Bull*, vol. 98, pp. 67–83, July 1985.
17. Davidse, R.J., "Ouderen en ITS: Samen sterk(er)?," Stichting Wetenschappelijk Onderzoek Verkeersveiligheid SWOV2004.
18. Poston, B., Van Gemmert, A.W.A., Barduson, B., & Stelmach, G.E., "Movement structure in young and elderly adults during goal-directed movements of the left and right arm," *Brain and Cognition*, vol. 69, pp. 30–38, 2009.
19. Llaneras, R.E., Swezey, R.W., Brock, J.F., Rogers, W.C., & Van Cott, H.P., "Enhancing the safe driving performance of older commercial vehicle drivers," *International Journal of Industrial Ergonomics*, vol. 22, pp. 217–245, 1998.
20. Merat, N., Anttila, V., & Luoma, J., "Comparing the driving performance of average and older drivers: The effect of surrogate in-vehicle information systems," *Transportation Research Part F: Traffic Psychology and Behaviour*, vol. 8, pp. 147–166, 2005.
21. Wilschut, E.S., Rinkenauer, G., Brookhuis, K.A., & Falkenstein, M., "Effects of Visual Search Task Complexity on Lane Change Task Performance," presented at the European Conference on Human Centred Design for Intelligent Transport Systems (pp. 23–33), Lyon, France, 2008.
22. McCarley, J.S., Vais, M.J., Pringle, H., Kramer, A.F., Irwin, D.E., & Strayer, D.L., "Conversation disrupts change detection in complex traffic scenes," *Human Factors*, vol. 46, pp. 424–436, Fall 2004.
23. Mahr, A. & Müller, C., "A Schema of Possible Negative Effects of Advanced Driver Assistant Systems," presented at the International Driving Symposium on Human Factors in Driver Assessment, Training and Vehicle Design, Lake Tahoe, CA, United States, 2011.
24. Pashler, H., "Dual-task interference in simple tasks: Data and theory," *Psychol Bull*, vol. 116, pp. 220–244, September 1994.
25. Pashler, H., "Processing stages in overlapping tasks: Evidence for a central bottleneck," *Journal of Experimental Psychology: Human Perception and Performance*, vol. 10, pp. 358–377, 1984.
26. Ruthruff, E., Hazeltine, E., & Remington, R., "What causes residual dual-task interference after practice?," *Psychological Research/Psychologische Forschung*, vol. 70, pp. 494–503, 2006.
27. Kahneman, D., "Attention and effort," ed Englewood Cliffs, NJ: Prentice-Hall, 1973.
28. Wickens, C.D., "Processing resources in attention," in *Varieties of Attention*, R. Parasuraman and D. R. Davies, Eds., ed Hillsdale, NJ: Erlbaum, 1984, pp. 239–257.
29. Wickens, C.D., "Multiple resources and performance prediction," *Theoretical Issues in Ergonomical Science*, vol. 3, pp. 159–177, 2002.

30. Wickens, C.D., "Multiple resources and mental workload," *Human Factors: The Journal of the Human Factors and Ergonomics Society*, vol. 50, pp. 449–455, June 1, 2008.
31. Polson, M.C. & Friedman, A., "Task-sharing within and between hemispheres: A multiple-resources approach," *Human Factors*, vol. 30, pp. 633–643, October 1988.
32. Strayer, D.L. & Johnston, W.A., "Driven to distraction: Dual-task studies of simulated driving and conversing on a cellular telephone," *Psychological Science*, vol. 12, pp. 462–466, November 1, 2001.
33. Horrey, W.J., Wickens, C.D., & Consalus, K.P., "Modeling drivers' visual attention allocation while interacting with in-vehicle technologies," *Journal of Experimental Psychology: Applied*, vol. 12, pp. 67–78, 2006.
34. Wickens, C.D., "Time sharing and multiple resource theory," in *Applied Attention Theory*, C. D. Wickens and J. S. McCarley, Eds., ed New York: CRC Press, 2007.
35. Riby, L.M., Perfect, T.J., & Stollery, B.T., "The effects of age and task domain on dual task performance: A meta-analysis," *Journal of Cognitive Psychology*, vol. 16, pp. 863–891, 2004.
36. Hahn, M., Wild-Wall, N., & Falkenstein, M., "Age-related differences in performance and stimulus processing in dual task situation," *Brain Research*, vol. 1414, pp. 66–76, 2011.
37. Wilschut, E.S., "The impact of in-vehicle information systems on simulated driving performance," PhD, Psychology, Rijksuniversiteit Groningen, Groningen, 2009.
38. (2013). *Oxford Dictionary of English* [Online]. Available: http://oxforddictionaries.com/definition/english/learning?q=learning
39. Schermer, F.J., *Lernen und Gedächtnis (2. überarbeitete und erweiterte Auflage). Grundriss der Psychologie, Band 10.* Stuttgart: Verlag W. Kohlhammer, 1998.
40. Totzke, I., Hofmann, M., Meilinger, T., Rauch, N., & Schmidt, G., *Kompetenzerwerb für Fahrerinformationssysteme: Einfluss des Lernprozesse auf die Interaktion mit Fahrerassistenzssystemen*. Offenbach: Berthold Druck, 2004.
41. Anglim, J.K., "Strategies in skill acquisition: Reconciling continuous models of the learning curve with abrupt strategy shifts," PhD, Psychological Sciences, The University of Melboure, Melbourne, 2011.
42. Speelman, C. & Kirsner, K., "Skill acquisition: History, questions, and theories," in *Beyond the Learning Curve: The Construction of Mind*, C. Speelman and K. Kirsner, Eds., Oxford: Oxford University Press, 2005.
43. Newell, A. & Rosenbloom, P.S., "Mechanisms of skill acquisition and the law of practice," in *Cognitive Skills and Their Acquisition*, J. R. Anderson, Ed., ed Hillsdale, NJ: Erlbaum, 1981, pp. 1–55.
44. Charlton, S.G. & Starkey, N.J., "Driving without awareness: The effects of practice and automaticity on attention and driving," *Transportation Research Part F: Traffic Psychology and Behaviour*, vol. 14, pp. 456–471, 2011.

45. Anderson, J.R., "Acquisition of cognitive skill," *Psychological Review*, vol. 89, pp. 369–403, 1982.
46. Repovs, G. & Baddeley, A., "The multi-component model of working memory: Explorations in experimental cognitive psychology," *Neuroscience*, vol. 139, pp. 5–21, Apr 28, 2006.
47. Peretti, C.-S., Danion, J.-M., Gierski, F., & Grangé, D., "Cognitive skill learning and aging: A component process analysis," *Archives of Clinical Neuropsychology*, vol. 17, pp. 445–459, 2002.
48. Verwey, W.B., Abrahamse, E.L., Ruitenberg, M.F.L., Jiménez, L., & Kleine, d.E., "Motor skill learning in the middle-aged: limited development of motor chunks and explicit sequence knowledge," *Psychological Research*, vol. 75, pp. 406–422, 2011.
49. Voelcker-Rehage, C. & Alberts, J.L., "Effect of motor practice on dual-task performance in older adults," *Journal of Gerontology: Psychological Sciences*, vol. 62B, pp. 141–148, 2007.
50. Göthe, K., Oberauer, K., & Kliegl, R., "Age differences in dual-task performance after practice," *Psychology and Aging*, vol. 22, pp. 596–606, 2007.
51. Baddeley, A., "Exploring the central executive," *The Quarterly Journal of Experimental Psychology A*, vol. 49, pp. 5–28, 1996.
52. Shinar, D., Tractinsky, N., & Compton, R., "Effects of practice, age, and task demands, on interference from a phone task while driving," *Accident Analysis & Prevention*, vol. 37, pp. 315–326, 2005.
53. Cooper, J.M. & Strayer, D.L., "Effects of simulator practice and real-world experience on cell-phone - Related driver distraction," *Human Factors*, vol. 50, pp. 893–902, 2008.
54. Chisholm, S.L., Caird, J.K., & Lockhart, J., "The effects of practice with MP3 players on driving performance," *Accident Analysis & Prevention*, vol. 40, pp. 704–713, 2008.
55. Caird, J.K., Chisholm, S.L., & Lockhart, J., "Do in-vehicle advanced signs enhance older and younger drivers' intersection performance? Driving simulation and eye movement results," *International Journal of Human-Computer Studies*, vol. 66, pp. 132–144, 2008.
56. Zijlstra, F. & Van Doorn, T., "The construction of a scale to measure perceived effort," PhD, Department of Philosophy and Social Sciences, Delft University of Technology, Delft, The Netherlands, 1985.
57. Hart, S. & Staveland, L., "Development of NASA-TLX (Task Load Index): Results of empirical and theoretical research," in *Human Mental Workload*, P. Hancock and N. Meshkati, Eds., ed Amsterdam: North Holland, 1988, pp. 139–183.
58. Mattes, S., "The lane-change task as a tool for driver distraction evaluation," in *Quality of Work Products in Enterprises of the Future*, H. Rausch and H. Bubb, Eds., ed: Ergonomia Verlag, 2003.
59. Berthon-Donk, V., Grosjean, M., & Rinkenauer, G., "Effect of feedback on performance in the Lane-Change Test," in *Sixth International Driving*

Symposium on Human Factors in Driver Assessment, Training and Vehicle Design, Olympic Valley – Lake Tahoe, CA, 2011, pp. 277–283.

60. Treisman, A.M. & Gelade, G., "A feature-integration theory of attention," *Cognitive Psychology*, vol. 12, pp. 97–136, 1980.
61. Oberauer, K. & Kliegl, R., "Simultaneous cognitive operations in working memory after dual-task practice," *Journal of Experimental Psychology: Human Perception and Performance*, vol. 30, pp. 689–707, August 2004.

Chapter 7

Researching safety issues with intersection assistance systems for the older driver

Mandy Dotzauer

Abstract

Even though road traffic safety strategies and policies have resulted in a significant decrease in crash fatalities, special attention needs to be paid to implementing strategies that will protect older drivers. Over the next few decades, the number of older persons will increase noticeably; and therefore the number of active older drivers will rise as well. Mobility and driving is an important aspect of life, giving people a sense of independence whether being young or old. Older drivers make up a unique age group that faces different difficulties in traffic from younger drivers and thus require special support. Intersection assistance might be a promising approach for supporting older drivers. This has previously been investigated, but only sporadically and in short-term studies. In order to investigate the safety benefits of an intersection assistant for older drivers, and because their learning phase is longer than that of younger drivers, such assessments should involve multiple sessions spread over a prolonged period of time.

7.1 Introduction

As revealed through demographic studies, the population is becoming older due to increasing life expectancy and low levels of fertility. The number of persons age 65 and above will increase rapidly over the next few decades and, in particular, this concerns 'older old' persons over 75 [1]. Driving is probably going to be a more preferred and a more frequently used mode of transportation for future older persons than it is for the present ones. We see, for example, an increasing proportion of this population possessing driver licences, especially women, than previous generations. Therefore, the number of older persons holding a valid driver's licence and being active drivers will probably rise substantially [2]. Because of age-related visual, cognitive, and motor impairments, older drivers are over-represented in

University Medical Centre Groningen, Hanzeplein 1, Department of Neuropsychology, AB 60, 9700 RB Groningen, Netherlands at-fault crashes at intersections [3, 4]. Due to their age-related physical vulnerability, those crashes lead to severe injuries that may prove fatal [5]. Through driving, older persons can maintain a high sense of independence, enabling them to participate in various activities ranging from shopping for groceries to attending social events [6]. Refraining from driving or having a driver's licence revoked might lead to faster ageing, depression, and solitude [4]. Restricting and compensating are not the most favourable solutions to the problem because driving contributes to independent living and subjective well-being, particularly if accessible public transport is not available.

Even though road traffic fatalities have decreased substantially over the last few decades due to the implementation of road traffic safety strategies and policies throughout Europe and other industrialised countries [7], older people need special attention within the road traffic network because of their unique characteristics and needs. Especially for older drivers, it is important not only to prevent crashes from occurring, but also to minimise the negative effects of crashes. For example, an 80-year-old female driver is seven times more likely to be killed compared to a 45-year-old female driving when experiencing the same severity of the crash due to the increased fragility with increasing age [8]. Because of their physical vulnerability, extensive measures for bodily protection would be required. Bodily protection in the form of airbags is provided to drivers and passengers in the event of a crash, but they do not fully compensate for the vulnerability of older persons. Safety at intersections has been improved over the last few decades but mostly through infrastructural means (e.g. reduced speed, roundabouts), but in order to fit the specific needs of the older driver, additional methods need to be developed and implemented. One promising approach is that of an intersection assistant.

This chapter shows how the implementation of various road traffic safety strategies and policies has resulted in a significant decrease in fatalities, but it will also be argued that such developments are not sufficient in terms of improving older drivers' safety. Older drivers make up a unique group with unique characteristics that requires special attention and more tailored support.

7.2 Road traffic safety

Road traffic safety refers to methods and measures used to reduce the risk of road users being involved in a fatal accident or suffering from serious injuries as a result of an accident within the road network [7]. Road users include pedestrians, cyclists, motorists as well as passengers of vehicles and public transport. Even though road traffic safety is concerned with all road users, in this chapter, the focus lies on car-related accidents.

Over the past few decades, the number of fatalities has decreased significantly. Figure 7.1 shows the number of fatalities, accidents, and injuries of 27 European countries from 2000 to 2009. According to the CARE glossary, an accident 'occurs on a public road or on a private road to which the public has right of access [. . .], involves at least one moving vehicle [. . .], involves at least one injured or killed

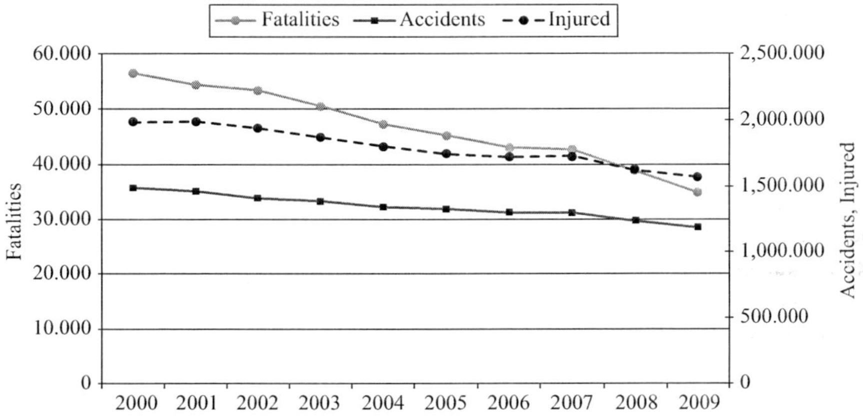

Figure 7.1 Change in fatalities, accidents and injuries in EU-27 from 2000 to 2009 (Source: CARE; United Nations (statistics of road accidents); national reports)

person' [9, p. 8]. On the right *y*-axis, the total number of accidents and the total number of injuries resulting from an accident are displayed. The left *y*-axis shows the total number of fatalities resulting from the accidents. This decreasing trend is not only true for European countries, but also for the Asia-Pacific region and North America. Many countries have incorporated road safety strategies and policies that have shown fruitful effects over the past decade. The increase in road traffic safety can be attributed to changes/improvements in vehicle safety, road infrastructure and vehicle technology [7].

Changes and improvements are based on thorough safety analyses and the two approaches to this are the Haddon matrix and the Systems approach. The Haddon matrix, developed in 1980, is a commonly used approach in the field of safety analysis [10]. It is a two-dimensional model that applies basic principles of public health to motor vehicle-related injuries. The matrix identifies risks before the crash, during the crash and after the crash in relation to the human, the vehicle and the environment. It serves as an analytical tool to help identifying all crash-related factors. After the identification of multiple factors, countermeasures can be developed and implemented. For the pre-crash phase, the goal is to implement countermeasures that prevent the occurrence of a crash. During the crash, preventing injuries and/or reducing the severity of injuries has the highest priority. Post-crash countermeasures are implemented with the goal to reduce adverse outcomes.

The systems approach builds upon the Haddon matrix. It has been realised that often a single factor cannot be identified and that the interaction of different components must be taken into account. With the systems approach, major sources of errors or design weaknesses that contribute to fatalities and severe injuries are identified and rectified. The approach also aims to mitigate the severity and consequences of injuries. The model represents the highly complex interplay of elements of the system (motor vehicle, roads, road users and physical, social and economic environment). Significantly, it is realised that humans are vulnerable and that humans make mistakes; therefore, a safe road traffic system is the one that accommodates and compensates for human vulnerability and fallibility [11].

Overall, the aim of road traffic safety programs is to prevent crashes, prevent/ reduce injuries and save lives in the event of a crash. Initiatives have been taken throughout Europe. Implementation of countermeasures can be categorised in accordance with the Haddon matrix. For example, passive safety is aimed at preventing/reducing injuries in the event of a crash (e.g. vehicle structure crumple zones). Changes in the infrastructure such as implemented low speed zones are aimed at preventing crashes from occurring. Systems such as Anti-lock Braking System (ABS) or Electronic Stability Program (ESP) are advances in vehicle technology that are implemented in order to prevent crashes from occurring.

Vehicle technology has improved steadily over the years. Technology that has been invented and improved is, among other things, aimed at avoiding collision and/or minimising impacts to the body in case of a collision. Vehicle technologies are part of Intelligent Transport Systems (ITS). ITS may be defined as 'systems in which information and communication technologies are applied in the field of road transport, including infrastructure, vehicles and users, and in traffic management and mobility management, as well as for interfaces with other modes of transport' [11, p. 4]. ITS covers a wide range of services, and one of these services is Advanced Driver Assistance Systems (ADAS). ADAS are systems that interact with the driver with the main purpose of supporting the driving task at the tactical and the operational level [12]. Definitions and distinctions between the different levels of driving tasks are explained in detail in Section 7.3.3. In a broader sense, ADAS date back to 1929 when Anti-Lock Braking Systems (ABS) were already used in the field of aviation [13]. ABS is a safety system that prevents the wheels from locking when strong brake force is applied, and therefore uncontrolled skidding can be avoided [14]. In the 1960s, ABS was introduced into cars, but because of costs and low reliability, it was taken off the market again. A decade later, ABS became popular [13]. In 2004, car manufacturers in Europe, the United States and Japan agreed on equipping all new vehicles with ABS [15]. A related autonomous system that has been developed is Electronic Stability Control (ESC). ESC stabilises the vehicle in case of loss of traction. It counters over-steering and under-steering by systematically braking one or more of the wheels [16]. In 1995, the first version of ESC was implemented by Mercedes and, in the following years, other major car manufacturers followed. The European Union passed a law that states that all vehicles entering the market after 1 November 2011 need to be equipped with ESC and that after November 2013 new cars without ESC cannot be sold [17]. Several studies have investigated the effectiveness of ESC on the reduction of crash risk. It has been found that single crash risk was reduced by 33–35% for cars and by 56–67% for SUVs [18].

Besides ABS and ESC, several other ADAS have been introduced into the market over the past decade, usually in the higher price range vehicles. Systems include Adaptive Cruise Control (ACC), Lane Departure Warning (LDW), night vision systems and parking assistants. Cars may be equipped with sensors and cameras that make it possible to keep a safe distance from the vehicle in front (ACC), warn drivers when they unintentionally leave their travelling lane (LDW), enhance vision in the dark, and either assist with parking or park the car parallel in a fully automated manner. Research and development is now taking innovations a step further whereby communication between vehicles (C2C) and communication with the infrastructure (C2I) are being investigated. At the moment, the aim is to design intersection assistance on the basis of either C2C or C2I in order to support drivers at junctions.

Overall, road traffic safety is a major concern throughout Europe and many other industrialised countries. The most important developments in regard to passive safety are seatbelts and airbags, which reduce the severity of injuries in the event of a crash. In regard to road infrastructure it has been found that speed is a major cause of accidents and also for the severity of injuries of people involved in a crash. Therefore, strategies such as low speed zones and roundabouts have been implemented.

The most common support given to the driver is ESC, although different age groups benefit to a different degree from the electronic stability control; young drivers, for example, might benefit the most. When they are involved in fatal crashes, it is often a single vehicle crash on a rural road because they lose control of their vehicle. Older drivers, on the other hand, might benefit the most from intersection assistance. It is only recently that car manufacturers and research institutes have been investigating C2C and C2I communication. The following sections will discuss why these technologies and the development of intersection assistants could provide additional safety, especially for older drivers and contribute to lower crash and fatality rates.

7.3 Intersections and the older driver

Taking a closer look at crash statistics, especially the statistics for urban areas, sheds light on motivations to design and implement intersection assistants. Figure 7.2 shows the trend of fatalities in urban areas in Europe from 2000 to 2009. According to the European Commission, in 2009, 38% of all fatalities in EU-19 occurred in urban areas. Even though the number of urban road fatalities declined between 2000 and 2009, the percentage of all fatalities that occurred in urban areas has increased slightly [19]. Of those fatalities occurring in urban areas, fatalities at junctions accounted for 27% and even though the total number of fatalities has been reduced (as shown in Figure 7.3), the proportion of fatalities occurring at junctions did not decrease and remained slightly over 20% throughout the decade [20].

7.3.1 Demographic change and its consequences

Because of the demographic change over the next decades, intersections and crashes are important issues that need attention. Crash statistics are somewhat alarming: according to the literature, older drivers are over-represented in at-fault

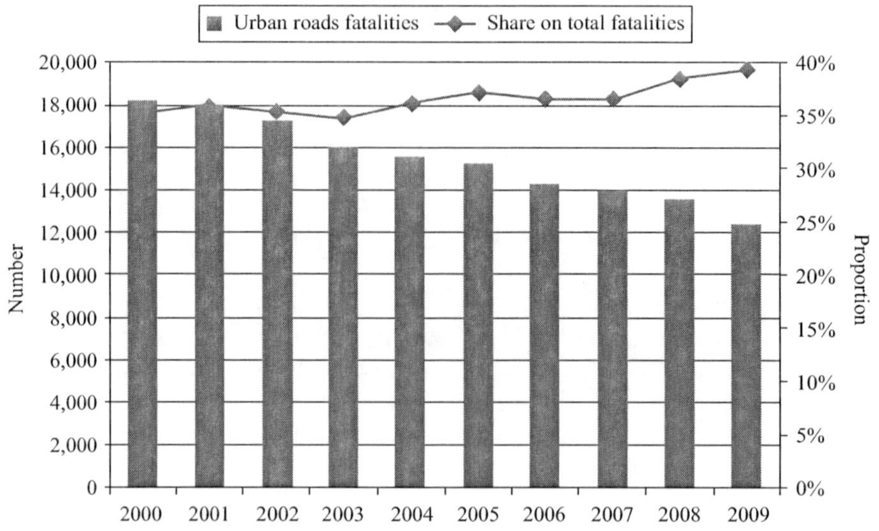

Figure 7.2 Number and proportion of fatalities in urban areas in Europe, 2000–2009 (Source: DaCoTA [19])

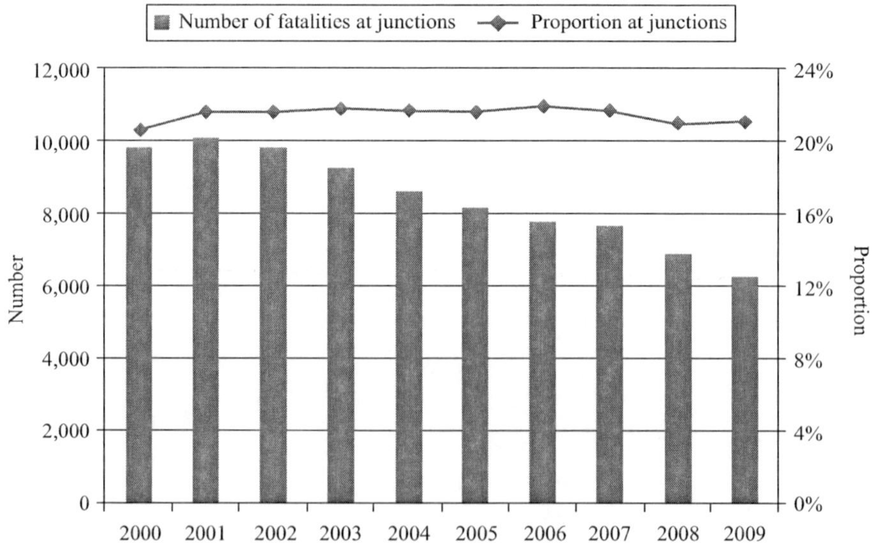

Figure 7.3 Number and proportion of fatalities at junctions in Europe, 2000–2009 (Source: DaCoTA [20])

crashes on intersections [4, 21, 22]. The age structure of the population is already changing and this trend will continue over the next few decades. In Europe, the population of persons aged 65 and older accounted for 17.4% of the total population in 2010. It is projected that the percentage of person 65+ years old will increase to about 30% by the year 2060. Among the old age group, the group of the oldest-old (80 years and older) will treble from 4.7% in 2010 to 12.1% in 2060 [1]. The change in age structure affects all areas of society and requires changes in thinking. The field of transport and mobility is affected as well. As the age structure changes, so does the number of persons holding a valid driver's licence as well as the number of persons operating a vehicle, the percentage of older drivers (65+ years) actively participating in driving, and the travel pattern of the older driver. The OECD [2] has projected driver's licence possessions for the year 2030. For selected European countries, the percentage of persons aged 65 and older holding a valid driver's licence and owning a car is expected to increase significantly over the next few decades. In today's middle-aged age group, most persons hold a valid driver's licence, both males and females. In today's older generation the possession of driver's licences and cars is lower, especially in females. In a few decades, we will see a more equal distribution among older male and female drivers, which is another reason why the percentage of older drivers in the driver population will increase substantially over the next few decades [2].

7.3.2 Age-related impairments, diseases/disorders and intersection performance

Even when persons age in a healthy manner, e.g. not suffering from a neurodegenerative disease, age-related impairments that affect safe driving can be observed. As persons become older, deterioration in vision, cognition and motor functions is observable: three domains relevant for driving. Vision is the most critical sense with regard to driving, and deterioration in many visual functions has been described with ageing, including small changes in basic functions such as visual acuity and contrast sensitivity. Also, limitations of dynamic visual acuity, movement detection, night-time vision and sensitivity to glare have been reported [4].

Higher order visual processes have been investigated when the information reaching the senses has to be distinguished, categorised and attended to; for example, an increase in field dependence, a decrease in visual working memory capacity and a loss of functional visual field [23, 24] have been reported. This area borders on the domain of attention and executive functions. The most distinguished general limitation of attention in old age is slowing of information processing [25]. An important central executive limitation that increases with age is a decreased ability to divide and switch attention, making it more difficult to perform multiple tasks simultaneously or in rapid succession [3, 26]. These effects are much larger in new and complex situations than in complex situations which have become routine [27, 28]. Processing speed and divided attention are important in the constantly changing traffic environment. However, the structure of the driving task often allows compensation to a certain degree, particularly while making decisions at the tactical and strategic level.

Other relevant bodily changes concern motor functions, proprioception and balance [29, 30]. Besides the fact that getting in and out of the car can become more difficult, more concerning may be that older persons experience restrictions in head and body rotation making it more difficult to check for other users in the blind spot and to compensate for their restricted field of view. Proprioceptive limitations in the legs and feet can make it more difficult to locate the pedals and make precise and smooth speed adjustments. Deterioration in motor functions also has an effect on the severity of injuries in the event of a crash, adding to the general finding that older persons are physically more vulnerable; which is a leading cause of severe injuries and fatalities when involved in a crash [4].

Besides age-related impairments and reductions often described as normal or primary ageing, more severe impairments associated with ageing can occur, such as age-related macular degeneration (AMD), Alzheimer's disease (AD) and Parkinson's disease (PD). In AMD, visual acuity declines and a central field defect may occur. In AD, memory and other cognitive functions severely decline as the disease progresses, eventually also including severe impairment in attention and executive functions. PD begins with motor impairments that might affect steering and pedal skills. As PD progresses visual and cognitive impairments can also become symptoms of the disorder. Often older persons with these disorders continue to drive, at least initially. While in more severe form, these disorders are incompatible with driving, many persons with milder forms still drive. Often, however, they manage by severely restricting the situations in which they drive, for example, under good weather conditions and avoiding complex road and traffic situations. ADAS might offer the possibility to help them in compensating and extending the situations in which they feel comfortable and safe to drive in [31, 32].

As already noted, older drivers can often compensate for age-related impairments and, actually, their crash involvement is less concerning than that of young inexperienced drivers. When looking at specific road and traffic situations, it appears that both young and older drivers have more problems than the average driver. Characteristic of young drivers are fatal single vehicle crashes on rural roads [33]; whereas, older drivers are over-represented in at-fault crashes at intersections [4, 21, 22].

In general, older persons have been observed judging distance and speed of other vehicles inaccurately [34], which explains their difficulties in joining and exiting traffic streams [4, 35]. The underlying causes for the inaccurate judgement of speed and distance might be explained through older persons' limitations of dynamic visual acuity and movement detection. Limitations of head and neck rotation could also serve as an explanation for difficulties with joining and exiting traffic streams. Older drivers also tend to react late to traffic conflicts [34, 35, 36]. This might be due to their difficulties in decision making under time pressure. Older drivers also experience difficulties with divided attention, which makes it more challenging for them to see, identify and seek out relevant traffic signs [37]. Consequently, older drivers often fail to yield to the right-of-way [21, 22] and misjudge safe crossing gaps [38], which results in an over-representation of older drivers in crashes when turning left [39, 40]. Passing straight through an intersection is also a problematic undertaking due to inaccurately estimating the distance to other cars [41]. Overall, more crashes occur at stop sign-controlled intersections compared to signal-controlled intersections [38, 41].

Research has identified the most problematic situations for older drivers. When trying to explain why older drivers experience those problems, it becomes obvious that often a single cause cannot be identified. Considering older drivers' difficulties joining and exiting traffic streams, it is more likely that an interaction of visual, cognitive and motor impairment cause these problems. Moreover, no one ages the same: one person might experience a drastic decline in vision, whereas another person struggles with memory problems. Ageing does not have a linear and easily predicted course. Everyone ages at a different speed: a little faster in one domain, a little slower in another. Older persons are a heterogeneous age group; nevertheless, it has been agreed that approaching and crossing an intersection involves several processes resulting in a complex task. Crossing an intersection requires divided attention among several items of information [42]: perceiving and processing changes in the traffic situation, perceiving and processing signals and traffic signs, determining and executing a course of action [43] and making decisions under time pressure [3].

7.3.3 Crossing an intersection: A multi-level task

In order to better understand how age affects intersection performance, we first look at Michon's hierarchical task analysis of driving [44] as applied to the domain of driver impairments by Brouwer and colleagues [3, 45, 46]. In this analysis, three task levels are distinguished: the strategic level, the tactical level and the operational level. The strategic level (navigation) is the highest level. On this level, decisions with regard to route, navigation and time of driving are made. Decisions are usually made before the trip has begun, and also, occasionally, during the trip; for example, when deciding to choose an alternative route because of an expected traffic jam. On the tactical level, which takes place while driving, safety margins are set and adjusted for the trip. This not only includes choosing cruising speed, time-headway and lane position, but also involves considering various manoeuvres such as overtaking and passing. Decisions on the tactical level are only performed occasionally; for example, setting smaller time headway than normal if one is in a hurry, or choosing the middle of three parallel lanes in an unfamiliar town. On the operational level (control), the driver performs second to second lateral and longitudinal control tasks to avoid acute danger and to stay within the margins set on the tactical level. The key difference between tactical and operational level decisions and actions is that the latter are reactive and the former are proactive (anticipatory), not a reaction to immediate danger but a setting of safety margins for the case that actual danger (e.g. vehicle on collision course) manifests itself in the near future.

On the strategic and tactical level, drivers can make adjustments and compensate for their challenges on the operational level. On the strategic level this might include not driving during rush hours or avoiding highly complex intersections. On the tactical level, drivers can choose lower travelling speeds or decide on taking a greater gap, giving them more time to seek the necessary information and make decisions. This compensation for challenges is not infinite: when the driving task becomes too complex and/or impairments are too severe, limitations of attentional capacity cannot be compensated for and other means, e.g. Advanced Driver Assistance Systems (ADAS), may offer a means to support the older driver. However, currently available ADAS are not necessarily designed to fit the needs of the older driver.

7.3.4 New developments, potential benefits and problems

Older drivers make adjustments on the tactical level in order to extract traffic-relevant information from their surroundings. For example, in order to read traffic signs, older drivers travel at lower speed, often under the speed limit [37]. Considering this behaviour outside of the general traffic context, we could conclude that older drivers engage in safe driving behaviour. Taking into account all other road users, especially other drivers, this behaviour might be perceived as traffic blocking, especially when other drivers do not see an obvious reason (e.g. a traffic jam) for travelling under the speed limit; therefore, the older driver becomes a hazard. Hence, older drivers need support that will prevent them from applying compensation strategies at the tactical level.

One promising approach is an intersection assistant providing relevant traffic information, including traffic signs, speed limits and gap sizes, for the upcoming intersection in advance. Receiving information in advance serves two purposes. It takes away uncertainty because the driver knows what to expect and can anticipate; it can thus compensate for decision making under time pressure. It also counters difficulties with divided attention because important information, for example priority regulation at the upcoming intersection, is fed to the driver before reaching the intersection. In theory, giving the older driver information about speed limits, priority regulations and approaching traffic in advance can offset attentional capacity challenges and spare enough resources to fulfil the primary driving task.

Technology needed to realise intersection assistants to provide relevant information about the upcoming intersection in advance involve C2C and C2I communication. Innovations for vehicle-to-infrastructure communication have come from major car manufacturers and one form of intersection assistance involves communication between vehicles and the traffic lights at signalised junctions. In such systems, cars communicate with traffic lights and, after appropriate filtering, the driver is informed about the status of the upcoming traffic light and receives speed recommendations that will ensure a ‘green wave’. Drivers are also warned when they are about to run a red light [47, 48] (some systems are even designed to intervene in such cases). Warnings are also given when a driver makes a hazardous turning manoeuvre [48].

C2C communication is also under investigation in order to bring the intersection assistant forward and Ford is one car manufacturer that investigates this approach. They use cellular mobile telephony and WiFi in order to transmit data such as speed, status of brakes and indicators. If the transmitted data indicates a dangerous situation (e.g. approaching speed of another vehicle is too high) it will warn the driver [49].

Realising intersection assistance as described above involves very complex technology and an effective and efficient interplay of the components making up an assistance system. If one of the components involved does not work properly, the system may fail. Considering Ford's approach to an intersection assistant that uses internet communications, the information would not be provided when internet is not available. This is probably a more pronounced problem in rural areas than in urban areas and needs to be considered.

Besides technological problems that can occur, a major concern is the human–machine interaction. Even though ADAS are intended to improve safety, comfort, traffic efficiency and the environment, hence facilitating driver performance [50], research has also shown that with the introduction of automation, performance deteriorates [51]. One factor contributing to performance deterioration is mental workload. Additional information that is presented to the driver needs to be processed and this requires mental and/or visual resources. Because older drivers already experience difficulties in more complex driving tasks, this additional information might increase mental workload because the driver also needs to pay additional attention to what the system is doing. De Waard and colleagues [52] as well as Davidse and colleagues [53] found that with the introduction of additional information, workload increased.

Another factor that influences ADAS use is trust in automation. In general, it has been observed that drivers start over-trusting ADAS, resulting in reliance on and compliance with the advice given without confirming the accuracy of the information [54–59]. Little is known about older drivers and their trust in automation. Development of trust can go two ways. Older drivers are more reluctant to accept new technologies [60]; therefore, it might be possible that they will reject the automation and not trust it, which in turn leads to an underutilisation. On the other hand, older drivers are aware of their difficulties and impairments [37] as illustrated, for example, by restricted driving [31]. Thus, they might trust the system more than their own abilities and start over-utilising the system, relying on and complying with the advice given.

Automation in the vehicle can also change how the driving task is perceived and what is demanded from the driver. This may lead to a change in driving behaviour. Several studies [58, 59, 61] have investigated the effect of different ADAS on driving behaviour. When ADAS is added to the vehicle, drivers often display a behaviour termed 'behavioural adaptation'. 'Behavioural adaptations are those behaviours which may occur following the introduction of changes to the road–vehicle–user system and which were not intended by the initiators of the change. Behavioural adaptations occur as road users respond to changes in the road transport system such that their personal needs are achieved as a result. They create a continuum of effects ranging from positive increase in safety to a decrease in safety' [62, p. 23]. Of concern is the behaviour that compromises the intended safety benefit. For example, in Rajaonah and colleagues' study [59], participants reacted later to a hazardous condition when driving with ACC and time headway dropped below the safety margin. Therefore we also have to be alert for behavioural adaptation in the use of intersection assistance.

Overall, intersection assistance seems to be a promising technology to support the older driver in negotiating intersections and countering difficulties with decision making under time pressure and divided attention. As recent development shows, the technology has progressed and intersection assistance is now realisable. When designing such systems, thorough research is needed so that intended safety benefits are not compromised by unforeseen human behaviour, and that flaws in design and function are identified before systems are introduced into the market.

7.3.5 Previous research relevant to intersection assistance

Even though the challenges of older drivers have been known for decades, little has been done so far to develop appropriate support for the older driver, but some effort has been made [34, 53, 63–67]. As early as 1991, Staplin and Fisk [63] investigated the effect of advanced left turn information on the decision making performance in young (mean = 37 years) and older (mean = 71 years) drivers in simulator studies. Traffic approached intersections either at a speed of 30 mph (48 km/h) or at 60 mph (96 km/h) and according to the signal presentation, participants needed to decide whether to go or stop. The signal presentation was either redundant or non-redundant. Younger and older drivers made more accurate decisions when advanced information was available. They also found that younger drivers understood the information presented better than older drivers and that redundant information was interpreted more accurately than non-redundant information.

In a driving simulator study, Lee and colleagues [64] examined in-vehicle messaging, presented either alone or in combination with traffic signs. The information was presented on the dashboard. They found that older drivers (mean = 74 years) rated their performance as the same as young drivers (mean = 22 years). Nevertheless, data revealed that older drivers' performance deteriorated in terms of crashes per hour, lane variability and speed variability while interacting with the in-vehicle and roadway information. Overall, their performance was significantly poorer compared to young drivers.

Hanowski and colleagues [65] tested signing, navigation and warning in-vehicle information systems in an instrumented vehicle on a test track. In this study the information was also presented on the dashboard of the vehicle. A signal informed the driver about an event such as a car entering the road ahead. The information was presented for 5 seconds. Even though young drivers (18–25 years) reacted faster to events than older drivers (65–75 years) when the information was present, both age groups benefited from advanced information.

Other studies have been performed that have focused on workload and acceptance rather than driving performance as a result of ADAS use. Louma and Rämä [66] investigated driver acceptance of traffic signs that were presented in the vehicle to young (mean = 20 years) and older (mean = 68 years) drivers. Four different message conditions were tested: (1) visual sign, (2) visual sign and auditory message, (3) visual sign and auditory feedback based on driver behaviour and (4) visual sign in combination with complete instructions. Young and older drivers favoured message type (1) the most. Complete instructions were rated the least desirable by both age groups. Participants thought that the given information increased the effects of traffic signs and traffic safety.

Caird and colleagues [34] investigated whether in-vehicle advanced warning signs improve young (mean = 21 years) and older (mean = 69 years) drivers' intersection performance, especially stopping performance at signal-controlled intersections in a driving simulator study. Information was presented in a head up display (HUD). They found that, overall, young and older drivers were more likely to stop at the intersection when in-vehicle warnings were given, especially at traffic lights with a relatively short amber-onset. As a result of the presence of the warning, drivers adopted a slower intersection approach speed and removed their foot from the accelerator earlier. Older drivers took more time to perceive and process the information given compared with young drivers, but when the decision was made to stop, they compensated by faster reactions and higher deceleration rates.

Ziefle and colleagues [67] investigated driving performance and acceptance of an intersection assistant in a driving simulator experiment. Information about the priority regulation and the traffic density at the upcoming intersection was conveyed to the driver. Results showed that, overall, older drivers (mean = 62 years) drove slower through intersections, but their performance in lane tracking did not differ from young drivers (mean = 27 years). The driving performance (i.e. speed control and lane tracking) of the group that drove without ADAS was superior to the group that drove with ADAS, indicating that ADAS induces additional workload. Young drivers were indifferent as to whether the intersection assistant was helpful, whereas the majority of older drivers rated the intersection assistant as helpful. Despite a driving performance deterioration when the information was presented in the auditory mode (in comparison with the visual mode), older drivers preferred the auditory over the visual interface. Young drivers did not have a preference for one interface over the other.

Davidse and colleagues [53] investigated the effect of advanced in-vehicle information on driver performance and workload of older drivers (mean = 75 years) in a driving simulator study. Information about priority regulation, obstructed view on upcoming intersections, safe gaps and one-way streets was presented to the driver. They concluded that the information conveyed was beneficial for intersection performance. Other traffic approaching the intersection needed less sharp decelerations to prevent collisions so less hindrance was given to other drivers. Driving speed was lower when participants had to yield to the other traffic and also when they received information about an obstructed view at the upcoming intersection. Participants also made less route errors when they received information about a one-way street that they were not allowed to enter. The beneficial effects of messages on driving performance were similar for younger and older drivers. It was also found that ADAS use did not result in workload reduction. It was concluded that longer experience with ADAS needs to be studied in order to draw conclusion about workload over time.

Overall, the studies summarised above show that an intersection assistant has the potential to benefit the older driver while crossing intersections. Information given in advance helps the driver anticipate upcoming events and act upon them. An assistant also helps them to react earlier to unexpected events. Acceptance ratings indicate that such functions are appreciated by the driver. Nevertheless, conclusions are based on findings from short-term studies, so no conclusions concerning what happens to driver performance and driver behaviour over time can be drawn.

7.3.6 Drawbacks, unanswered questions and the need for further research

The studies above show that intersection assistance improves driver performance and also that it is potentially beneficial for the older driver, but all studies have one major drawback: the effects of the system on driver performance and driver behaviour were assessed in short-term studies and the overall consensus is that even though older drivers benefit from the system, they perform worse than younger drivers. As the initially added safety benefit of an intersection assistant might be compromised by changes in driver behaviour, short-term studies cannot answer several important questions regarding the overall benefit to road traffic safety.

Little is known about how ADAS use changes driving behaviour over time and whether negative behavioural adaption occurs. Testing participants in a long-term study and giving them more trials to get acquainted with the system may reveal whether the particular ADAS leads to further positive or negative effects. It will also be interesting to study what happens when the ADAS is taken away from the driver after a longer period of usage, i.e. whether carry-over effects can be observed. Trust and acceptance play an important role when it comes to appropriate ADAS use. Utilizing long-term studies allow for detecting changes in those ratings, which, in turn might lead to changes in ADAS use.

Often, conclusions are drawn that older drivers perform less well than younger drivers. However, short-term studies do not provide an opportunity for the older driver to get acquainted with the system and the new environment. Older drivers still learn, but at a slower pace than younger drivers. Assessing older drivers over a longer period will shed light on learning and how well they learn when being exposed to the system and the environment over a longer period. Knowing that it takes an older person longer to learn something new, drawing conclusion on performance based on a single assessment might be erroneous as well. Longer exposure to ADAS might result in more positive effects for the older driver. Depending on the degree of impairment, ADAS might be beneficial for drivers with impaired executive functions. Performance of older drivers is poorer than that of younger drivers, especially in complex new situations, while in routine situations the difference between younger and older drivers is much smaller.

Another question that short-term studies cannot answer is whether intersection assistance is beneficial for all age groups. Often it is argued that whatever helps the older driver will also support the young driver. Such conclusions are also drawn on the basis of a single assessment and might only show that young drivers learn faster than older drivers and do not allow conclusions to be drawn about added safety benefits. Young drivers experience different problems in traffic from older drivers. Helping them cross an intersection might be inappropriate because they are already able to cross an intersection safely and, therefore, they receive support that is not needed. This may lead to unsafe attention allocation if crossing an intersection becomes so easy that young drivers allocate attention to other tasks not related to driving.

Because so many questions cannot be answered from short-term studies, more extensive research is needed to assess the magnitude of ADAS use, the effectiveness and efficiency of an intersection assistant, and in how far this kind of support is beneficial or detrimental for various age groups.

In a recent study by the author [68] the effects of, and behavioural adaptation to, an intersection assistant on the driving performance and driving behaviour of older drivers (65–82 years) were investigated. A long-term driving simulator study was realised in order to acquaint drivers with the support system and to examine changes in driving performance and behaviour due to ADAS use over time. The implemented intersection assistant was designed to support drivers crossing an intersection safely and it gave advice on whether it is safe to cross an intersection. The advice was based on the driver's time-to-intersection (TTI) as well as the time-to-collision (TTC) with other cars approaching the intersection. The information was presented in a head-up display (HUD). Participants completed 14 sessions in the driving simulator; the first twelve sessions within a four-week time period and the last two after a four-week retention interval. During each session, participants drove through a virtual city and encountered various traffic situations. One of them was crossing uncontrolled intersections at which participants had to yield the right-of-way. Bushes placed near the intersection obstructed the view to the intersection and made the crossing a safety-critical task, forcing the driver to slow down before crossing. Older drivers assigned to the treatment group (driving a car equipped with the assistance) crossed intersections quicker over time, had a higher maximum speed, stopped fewer times, crossed more often with a critical TTC and had a lower minimum TTC than older drivers of the control group. Analyses of gaze showed that when equipped with ADAS, drivers spent more time looking to the road centre and neglecting the left and right.

It was unclear whether these changes in performance and behaviour reflected a safer and more efficient way of driving or negative behavioural adaptation. Therefore, follow-up work [69] has studied the performance of older drivers in relation to the performance of young drivers, to test for long-term effects on performance due to ADAS use, and to gain more insight into the development of trust and acceptance over time. The focus was on investigation and evaluation of the intersection assistant but effects of three other functions (traffic sign recognition, speed advice and collision warning) on driving performance and behaviour were investigated as well. In terms of the intersection assistant, safety critical intersections (bushes along the road) as well as intersections without any view obstructions were investigated. Young inexperienced drivers (18–25 years) completed the same drives as older drivers and followed the same protocol as older participants, but instead of completing 14 sessions, they only completed eight sessions. One of the main findings was that older drivers equipped with ADAS were as quick as young drivers in crossing intersections, but they did not have as high a maximum speed at intersections as young drivers. Also, their driving resulted in larger TTC and less frequent critical crossings compared with young drivers. Older drivers also caused fewer crashes than young drivers.

7.4 Summary and conclusion

In conclusion, road traffic safety programs have led to a significant decrease in fatalities over the past few decades. Efforts have been put not only into preventing crashes from happening but also into minimising the consequences in the event of a crash. The most prominent inventions in terms of protecting the body (passive safety) are the seatbelt and the airbag. The most noticeable road infrastructural changes are roundabouts. In many European countries, traditional intersections are being reconstructed resulting in significant speed reductions while approaching a roundabout compared with an intersection, reducing the probability of a crash. Vehicle technology, such as ESC, is being deployed to help prevent crashes from occurring.

Even though road traffic safety ambitions have been effective in terms of crash and fatality reduction, more attention needs to be paid to older drivers. The number of older drivers will increase significantly over the next few decades and, as recent crash statistic show, older drivers are over-represented in at-fault crashes in intersection. The crash involvement is explained with age-related impairments in vision, cognition and motor functions. In view of older drivers' crash involvement and impairments, it can be concluded that currently available ADAS are not ideally tailored to the need of the older driver. One promising ADAS is intersection assistance which provides relevant traffic information in advance, such as priority regulation.

Intersection assistance has only been sporadically investigated but studies show that an intersection assistant has the potential to benefit the older driver while crossing intersections. A major limitation of previous studies is their duration as results and conclusion drawn are based on single assessments. In terms of the human–machine interaction, it is important to investigate what happens over time. Questions that cannot be answered in short duration studies are how behaviour and performance changes when being exposed to ADAS over a longer period of time; whether, and to what extent, workload, trust and acceptance change; whether older drivers learn and become better over time; and whether effects are the same for young and older drivers. In order to answer these questions, more research is needed in which drivers are studied over an extended period of time.

Acknowledgement

The research received funding from the European Community's Seventh Framework Program (FP7/2007-2013) under grant agreement no. 238833/ADAPTATION project (www.adaptation-itn.eu).

References

1. Demography Report. *Older, More Numerous and Diverse Europeans*. Luxembourg: Publications Office of the European Union; 2011
2. OECD. *Ageing and Transport: Mobility Needs and Safety Issues*. Paris: Organisation for Economic Co-operation and Development OECD; 2001
3. Brouwer W. H., Ponds R. W. H. M. 'Driving competence in older persons'. *Disability and Rehabilitation*. 1994;16(3):149–161
4. Davidse R. 'Assisting the older driver: intersection design and in-car devices to improve the safety of the older driver'. SWOV Dissertatiereeks, Leidschendam, Netherlands; 2007
5. Hewson P. 'Ageing drivers or ageing casualties?' Presented at the Older People in Rural Area (OPeRA) Research Conference, Taunton, UK, 2006. Available from http://www.plymouth.ac.uk/files/extranet/docs/HSW/Hewson Taunton0407206.pdf [Accessed 12 December 2012]
6. Houser A. *Older Drivers and Automobile Safety*. Fact Sheet: AARP Public Policy Institute, Washington, DC; 2005
7. IRTAD. *Road Safety Annual Report* [online]. 2011. Available from http://internationaltransportforum.org/irtadpublic/pdf/11IrtadReport.pdf [Accessed 30 January 2013]
8. Evans L. *Traffic Safety*. Science Serving Society, Bloomfield Hills, MI, 2004
9. *Care Glossary* [online]. 2006. Available from http://ec.europa.eu/transport/roadsafety/road_safety_observatory/doc/care_glossary.pdf [Accessed 30 January 2013]
10. Haddon Jr. W. 'Advances in the epidemiology as a basis for public policy'. *Public Health Report*, 1980;95:411–421
11. *Directive 2010/40/EU of the European Parliament and of the Council on the framework for the deployment of Intelligent Transport Systems in the field of road transport and for interfaces with other modes of transport* [online]. 2010. Available from http://guides.library.vu.edu.au/content.php?pid=270421&sid=2230821 [Accessed 26 July 2013]
12. Östlund J., Peters B., Thorslund B., Engström J., Markkula G., Keinath A., et al. *AIDE Deliverable No. D2.2.5: Driving performance assessment – methods and metrics* [online]. 2005. Available from http://www.aide-eu.org/pdf/sp2_deliv_new/aide_d2_2_5.pdf [Accessed 10 December 2012]
13. Hoffman G. *The History of Car Safety* [online]. 2007. Available from http://autos.aol.com/article/car-safety-history/ [Accessed 10 December 2012]
14. Bester Beifahrer. *ABS- Anti-Lock Braking System* [online]. 2012. Available from http://www.bester-beifahrer.de/de/startseite/lexikon/#a [Accessed 10 December 2012]
15. Auto Motor und Sport. *Der Weg des ABS vom Flugzeug in Auto* [online]. 2010. Available from http://www.auto-motor-und-sport.de/news/abs-die-geschichte-des-anti-blockier-systems-1790991.html [Accessed 30 January 2013]

16. Bester Beifahrer. *ESC- Electronic Stability Control* [online]. 2012b. Available from: http://www.bester-beifahrer.de/de/startseite/lexikon/#e [Accessed 10 December 2012]
17. CarScoop. *EU: Electronic Stability Control to be Standard on All Vehicles from 2014* [online]. 2008. Available from: http://carscoop.blogspot.com/2008/05/eu-electronic-stability-control-to-be.html [Accessed 10 December 2012]
18. Ferguson S. A. 'The effectiveness of Electronic Stability Control in reducing real-world crashes: a literature review'. *Traffic Injury Prevention.* 2007;8(4):329–338
19. DaCoTA. *Traffic Safety Basic Facts 2011: Urban areas. European Road Safety Observatory* [online]. 2011a. Available from http://ec.europa.eu/transport/road_safety/pdf/statistics/dacota/bfs2011_dacota-intras-urbanareas.pdf [Accessed 10 December 2012]
20. DaCoTA. *Traffic Safety Basic Facts 2011: Junctions. European Road Safety Observatory* [online]. 2011b. Available from http://ec.europa.eu/transport/road_safety/pdf/statistics/dacota/bfs2011_dacota-trl-junctions.pdf [Accessed 10 December 2012]
21. Aizenberg R., McKenzie D. M. *Teen and Senior Drivers*. Report No. CAL-DMV-RSS-97-168; 1997
22. McGwin G., Brown D. B. 'Characteristics of traffic crashes among young, middle-aged, and older drivers'. *Accident Analysis and Prevention*. 1999;31:181–198
23. Ball K., Beard B. L., Roenker D. L., Miller R. L., & Griggs D. S. 'Age and visual search: expanding the useful field of view'. *Journal of the Optical Society of America*. 1988;5:2210–2219
24. Ball K., Owsley C. 'Identifying correlates of accident involvement for older driver'. *Human Factors*. 1991;33:583–595
25. Salthouse T. A. 'The processing speed theory of adult age differences in cognition'. *Psychological Review*. 1996;103:403–428
26. Ponds R. W. H. M., Brouwer W. H., & van Wolffelaar P. C. 'Age differences in divided attention in a simulated driving task'. *Journal of Gerontology: Psychological Sciences*. 1988;43:151–156
27. Lowe C., Rabbitt P. 'Cognitive models of ageing and frontal lobe deficits' in Rabbitt P. (ed.). *Methodology of Frontal and Executive Function*, Hove: Psychology Press; 1997. pp. 39–57
28. Hakamiis-Blomqvist L. (ed.). 'Age-related differences in driving: are older drivers more serial?' *Proceedings of Human Factors and Ergonomics Society Annual Meeting*, San Francisco, USA, 1999. pp. 190–194
29. Dietz V. 'Proprioception and locomotor disorders'. *Nature Review Neuroscience*. 2002;3:781–790
30. Ghez C., Gordon J., & Ghilardi M. F. 'Impairments of reaching movements in patients without proprioception: II. Effects of visual information on accuracy'. *Journal of Neurophysiology*. 1995;73:361–372
31. Singh R., Pentland B., Hunter J., Provan F. 'Parkinson's disease and driving ability'. *Journal of Neurology, Neurosurgery & Psychiatry*. 2007;78:363–366

32. Brouwer W. H. 'Autorijden bij dementia en cognitieve functiebeperkingen'. In Mast R. van der Heeren T., Kat M., Stek M., Vandenbulcke M., Verhey F. (eds.). *Handboek Ouderenpsychatrie*. Utrecht: De Tijdstroom; 2010. pp. 183–194
33. Wundersitz L. 'Identifying young drivers at a higher risk of crashing'. University of Adelaide, Dissertation, Australia; 2007
34. Caird J. K., Chisholm S. L., Lockhart J. 'Do in-vehicle advanced signs enhance older and younger drivers' intersection performance? Driving simulation and eye movement results'. *International Journal of Human-Computer Studies*. 2008;66:132–144
35. Horswill M. S., Marrington S. A., McCullough C. M., Wood, J., Pachana N. A., McWilliam J., Raikos M. K. 'The hazard perception ability of older drivers'. *Journal of Gerontology*. 2008;63B(4):212–218
36. Charlton J. L., Fildes B., Koppel S., Muir C., Scully J. (eds.) 'Visual search patterns in older and younger drivers'. Proceedings of Australasian Road Safety Research Policing Education Conference, Wellington, New Zealand, 2005
37. Musselwhite C., Haddad, H. 'Exploring older drivers' perceptions of driving'. *European Journal of Ageing*. 2010;7(3):181–188
38. Oxley J., Fildes B., Corben B., Langford J. 'Intersection design for older drivers'. *Transportation Research Part F: Traffic Psychology*. 2006;9(5):335–346
39. Griffin L. I. *Older Driver Involvement in Injury Crashes in Texas 1975–1999*. Washington, DC: AAA Foundation for Traffic Safety; 2004
40. Mayhew D. R., Simpson H. M., Ferguson S. 'Collisions involving senior drivers: high-risk conditions and locations'. *Traffic Injury Prevention*. 2006;7:117–124
41. Preusser D. F., Williams A. F., Ferguson S. A., Ulmer R. G., Weinstein H. B. 'Fatal crash risk of older drivers at intersections'. *Accident Analysis and Prevention*. 1998;30(2):151–159
42. Owsley C., Ball K., McGwin G., Sloane M. E., Roenker D. L., White M. F., Overley E. T. 'Visual processing impairment and risk of motor vehicle crash among older adults'. *Journal of the American Medical Association*. 1998;279(14):1083–1088
43. Braitman K. A., Bevan K. B., Ferguson S. 'Factors leading to older drivers' intersection crashes'. *Traffic Injury Prevention*. 2007;8(3):267–274
44. Michon J. A. 'A critical view of driver behavior models: what do we know, what should we do' in Evans L., Schwing R. C. (eds.). *Human Behavior and Traffic Safety*. New York, NY: Plenum Press; 1985. pp. 485–520
45. Van Zomeren A. H., Brouwer W. H., Deelman B. G. 'Attentional deficits: riddles of selectivity, speed and alertness' in Brooks D. N. (ed.). *Psychological, Social, and Family Outcome of Severe Blunt Head Injuries*. London: Oxford University Press; 1984. pp. 74–107
46. Brouwer W. H. 'Attention and driving: a cognitive neuropsychological approach' in Leclercq M., Zimmermann P. (eds.). *Applied Neuropsychology of Attention: Theory, Diagnosis and Rehabilitation*. London: Psychology Press; 2002. pp. 230–245

47. VW. *Intersection Assistant* [online]. 2012. Available from: http://www.volkswagenag.com/content/vwcorp/content/en/innovation/driver_assistance/intersection_assistant.html [Accessed 10 December 2012]
48. Bickerstaffe S. *Toyota Launches Car-to-Infrastructure System* [online]. 2012. Available from http://ae-plus.com/technology/toyota-launches-car-to-infrastructure-system [Accessed 10 December 2012]
49. Ford. *Car-to-Car Communication: Preventing Accidents in the Future* [online]. 2012. Available from: http://social.ford.com/our-articles/cars/taurus/car-to-car-communication-preventing-accidents-in-the-future/ [Accessed 10 December 2012]
50. IHRA. *Design principles for Advanced Driver Assistance Systems: Keeping Drivers in-the-Loop*. Informal Document No. ITS-19-07, 19th ITS; 2010
51. Young M. S., Stanton N. A. 'Automotive automation: investigating the impact on drivers' mental workload'. *International Journal of Cognitive Ergonomics.* 1997;1(4):325–336
52. De Waard D., Van Der Hulst M., Brookhuis K. A. 'Elderly and young drivers' reaction to an in-car enforcement and tutoring system'. *Applied Ergonomics.* 1999;30:147–157
53. Davidse R. J., Hagenzieker M. P., van Wolffelaar P. C., Brouwer W. H. 'Effects of in-car support on mental workload and driving performance of older drivers'. *Human Factors*. 2009;51(4):463–476
54. Nilsson L. (ed.). 'Safety effect of adaptive cruise controls in critical traffic situations'. *Steps Forward Proceedings of the Second World Congress on Intelligent Transport Systems;* Yokohama, 1995
55. Stanton N. A., Young M., McCaulder B. 'Drive-by-wire: the case of driver workload and reclaiming control with adaptive cruise control'. *Safety Science.* 1997;27:149–159
56. De Waard D., Van der Hulst M., Hoedemaeker M., Brookhuis K. A. 'Driver behavior in an emergency situation in the automated highway system'. *Transportation Human Factors.* 1999;1:67–82
57. Buld S., Krüger H. 'Die Auswirkung von Teilautomation auf das Fahrverhalten' in Grandt M. (ed.). *Entscheidungsunterstützung für die Fahrzeug- und Prozessführung, DGLR-Bericht 2003-04*. Bonn: DGLR; 2003. pp. 241–253
58. Rudin-Brown C. M., Noy I. 'Investigation of behavioral adaptation to lane departure warning'. *Transportation Research Record Journal of the Transportation Research Board.* 2003;1803:30–37
59. Rajaonah B., Francoise A., Vienne F. 'Trust and the use of adaptive cruise control: a study of a cut-in situation'. *Cognition, Technology and Work.* 2006;8(2):146–155
60. Hancock P. A., Parasuraman R. 'Human factors and safety in the design of intelligent vehicle-highway systems (IVHS)'. *Journal of Safety Research.* 1992;23:181–198
61. Rudin-Brown C. M., Parker H. A. 'Behavioral adaptation to adaptive cruise control (ACC): implications for preventive strategies'. *Transportation Research Part F: Traffic Psychology and Behaviour*. 2004;8(2):59–76

62. OECD. *Behavioural Adaptation to Changes in the Road Transport System*. OECD, Paris; 1990
63. Staplin L., Fisk A. D. 'A cognitive engineering approach to improving signalized left-turn intersections'. *Human Factors*. 1991;33;559–572
64. Lee J. D., Gore B. F., Campbell J. L. 'Display alternatives for in-vehicle warning and sign information: message style, location and modality'. *Transportation Human Factors*. 1999;1(4):347–375
65. Hanowski R. J., Dingus T. A., Gallagher J. P., Kieliszewski C. A., Neale V. L. 'Driver response to in-vehicle warnings'. *Transportation Human Factors*. 1999;1(1):91–106
66. Louma J., Rämä P. 'Acceptance of traffic sign information provided by an in-vehicle terminal'. *Proceedings of the Ninth World Congress on Intelligent Transportation Systems*; Chicago, Illinois, 2002. ITS America 2002
67. Ziefle M., Pappachan P., Jakobs E., Wallentowitz H. 'Visual and auditory interfaces of advanced driver assistant systems for older drivers'. *Proceedings of Computers Helping People with Special Needs*: 11th International Conference; ICCHP, Linz, Austria, 2008. pp. 62–69
68. Dotzauer M., Caljouw S. R., de Waard D., Brouwer W. H. 'Intersection assistance: a safe solution for older drivers?'. *Accident Analysis and Prevention*. 2013;59:522–528
69. Dotzauer M., de Waard D., Caljouw S. R., Poehler G., Brouwer W. H. 'Behavioral adaptation of young and older drivers to an intersection crossing advisory system'. Manuscript submitted for publication, 2013

Chapter 8

Motivational factors when investigating ADAS impacts on driver behaviour

Juliane Haupt[1] *and Ralf Risser*[1]

Abstract

Previous researchers have developed hierarchical approaches to describe driver behaviour and have postulated models in which changes at higher levels in a hierarchy affect characteristics at the lower levels. Thus, in research on behavioural adaptation it is important to investigate not only driving performance outcomes (that are mainly represented on lower levels) but also the characteristics of motivational factors that are represented at higher levels. The two motivational factors that have already been considered in past research dealing with the effects of driver assistance systems are trust and acceptance. According to motivational theories, another factor that should now be considered here is perceived risk as it is, theoretically, highly related to potential changes in driving behaviour. In this chapter, the role of motivational factors including perceived risk, perceived behavioural control, norms, attitudes and intentions when drivers use ADAS is discussed. Reference will be made to an extended version of the theory of planned behaviour and to motivational driver behaviour theories.

8.1 Understanding driver behaviour: Models and approaches that attempt to explain driving behaviour

In order to investigate the effects of ADAS on driver behaviour, it is necessary to first understand driver behaviour and associated internal and external processes and aspects. The driving task itself is a complex, dynamic control-task [14]. According to Nilsson [15] the driving task includes the continuous monitoring of the environment, knowledge of where and when to look, identification of relevant and important, out of all available information, appropriate responses to unexpected events as well as ability to revise and change planned actions. In substance, the driving task contains two basic modules: having the knowledge and applying it

[1] FACTUM OG, Vienna, Austria

(top-down processes) as well as reacting appropriately to (sudden) events (bottom-up processes).

From this perspective it can be noted that not only the driver characteristics are relevant with respect to the observed driving behaviour, but also external factors are important. Fastenmeier and Gstalter [16] drew an analogy to working behaviour by describing both working and driving behaviour as behaviour distinctly related to special situational circumstances. The 'Diamond' [17] (Figure 8.1) illustrates five areas that influence road user behaviour and thus determine problems in traffic.

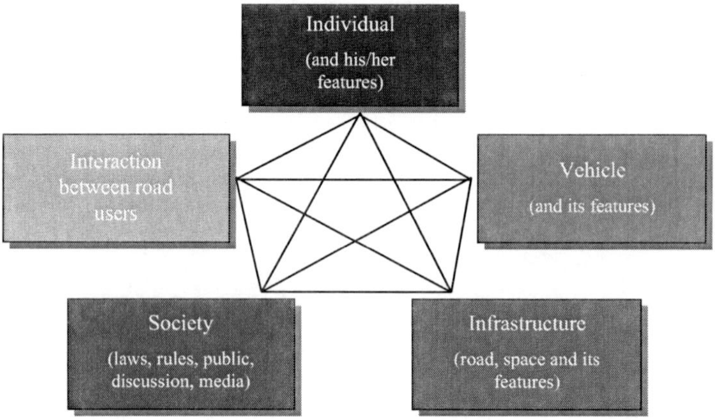

Figure 8.1 The diamond interaction model of relevant areas of the traffic system [17]

Since traffic represents a complex interplay between several factors, no simple deterministic cause-effect relationship can be established. In order to give a useful picture, the areas shown in Figure 8.1 should be considered; (a) the individual itself including his/her motives, attitudes, experiences, mental state, etc.; (b) the vehicle including its physical and technical features and restrictions, and assets like assistance systems, etc.; (c) infrastructure including road and space characteristics like width, surface, space provided for different road users, etc.; (d) society including traffic laws and rules, media reporting and public discussion (for instance conflicts between cyclists and car drivers and the question 'who is right'), etc.; (e) the interaction between road users reflecting power relations and social climate as ad hoc motivating factors (such as extending hand to say 'thank you'; expressing anger by visibly articulating swear words; 'polite' or 'rude' car movements).

The idea behind the diamond model is that any characteristics and changes in one area may influence the factors in other areas and vice versa.

Even when focusing only on the 'individual' area, a lot of other aspects that potentially influence driver behaviour can be identified. Studies of driver behaviour have shown that appropriate driving performance skills (handling the vehicle) are not sufficient for safe driving behaviour on the road (e.g. [18]). Research and

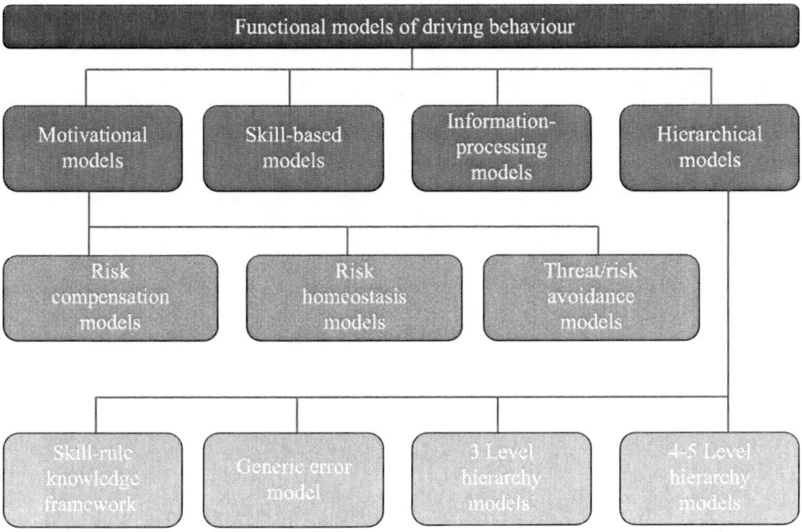

Figure 8.2 Functional models of driving behaviour (adapted from Ranney [20], Keskinen [24], Hatakka [2, 25], Keskinen, Peraaho & Laapotti [3])

established theories stressed that not only performance factors but also motivational aspects are important for the conduct of safe behaviour [19]. A driver may have the skills to handle the vehicle safely, but if he/she is not willing to drive safely, a risky driving behaviour will finally set in.

Ranney [20] gave an overview of several functional models of driver behaviour (see Figure 8.2). In this chapter, the hierarchical approaches will be looked at in detail; additionally, the motivational models of driver behaviour will be introduced.

Hierarchical approaches that aim at explaining driver behaviour usually follow the consideration that any characteristic on higher levels may influence aspects of the lower levels. Frequently quoted hierarchical models moving from strategic aspects of driving behaviour on the highest level to operational aspects on the lowest level are those of Michon [1] or Moe, Engen and Jenssen [21]. Another hierarchical approach that differentiates between a highly automated behaviour on the lowest level and knowledge-based behaviour where decisions require more elaborated thinking before acting is the behaviour taxonomy of Rasmussen [22]. Considerations by other authors have shown that some psychological variables (e.g. level of sensation seeking, personality variables like the BIG FIVE: persons' level of openness, conscientiousness, extraversion, agreeableness and neuroticism) that are not directly related to the driving task may also have an influence on driving performance [23, 69]. On the other hand, driving task models most often do not consider these potential influence factors but just explain what skills are necessary to drive. Thus, relevant aspects that may have a significant influence on the driving performance are usually neglected in these models.

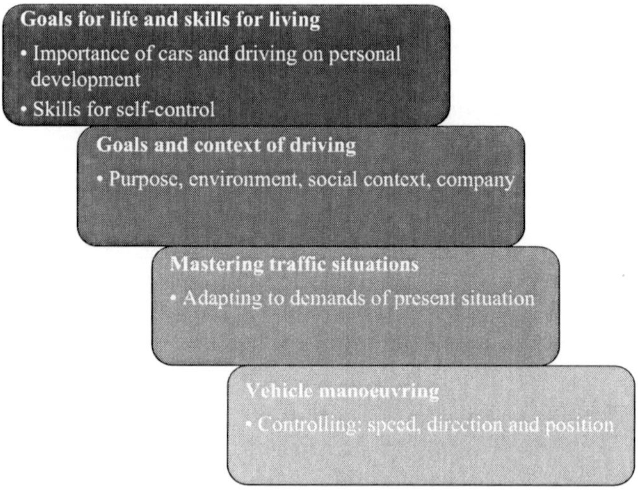

Figure 8.3 Hierarchical model of driver behaviour [25]

One hierarchical model that takes the above arguments into account is the model of Hatakka [2] that considers four behaviour levels of driving (see Figure 8.3).

8.1.1 The Hatakka hierarchical model

8.1.1.1 Vehicle manoeuvring

This first level includes the skills of handling the vehicle: to know how the vehicle works; how to start it, how to steer it; how to operate clutch, gas and brake pedal and switches (e.g. the indicator); how to control speed; and how to determine direction and position of the vehicle. The majority of studies that investigate driver behaviour look at variables on this level such as lane deviation, speed, acceleration or deceleration. Appropriate behaviour on this level is essential. In case that the driver does not have the skills that are necessary on this level, the driving task cannot be accomplished appropriately.

This basic level of driving behaviour is almost always considered in models that explain the driving task and is seen as the 'base of driving'. So, for instance in Moe's driving process model, vehicle manoeuvring is reflected in 'acting'. This level is also represented in Michon's hierarchical driving task approach [1], and as in the other models, it is also the lowest one, labelled as the 'operational' level. Some of the tasks that have to be completed on this manoeuvring level are also included in Michon's second level, the tactical level of driving. Taking Rasmussen's framework [22] into account, performance on this level is mainly skill based. The driver has to operate directly, on-site and does not have much time to act. But there are also conditions in which this task level may require rule-based or/and knowledge-based behaviour. For instance if the driver drives an unfamiliar car for the first time he/she acts more rule based than when he/she is already familiar with the vehicle and drives more or less automatically. In the case of a person who is a novice and drives a car for the very first time, he/she even has to act knowledge based.

8.1.1.2 Mastering traffic situations

There are many different road types including rural or urban roads, highways, intersections with or without traffic light regulations, different road widths, other road users (vehicles, cyclists, pedestrians). In order to drive safely, the driver has to adjust the driving performance to the different scenarios. This requires a close observation of the environment to detect and recognise particularly important aspects. The driver needs a certain level of awareness in order to be able to recognise potential hazards. He/she has to evaluate the situation and has to decide on how to (re)act. It is not sufficient to be able to steer a vehicle in order to complete the tasks on this second level of driver's behaviour. The driver also has to know the traffic rules and has to be trained in order to recognise, evaluate and respond appropriately to certain situations. Those tasks are represented on the 'mastering traffic situations' level. Preferably the driving behaviour on this level should be highly automated. For instance, the driver should reduce speed immediately and 'without thinking' but still considering the traffic behind him, in case a pedestrian suddenly intends to cross the road or in case another road user has priority and crosses his/her way.

In the driving tasks models of Moe [21] and of Michon [1] this level of driver behaviour is represented. Moe proposed three hierarchical steps that are included in this level: detecting, recognising and deciding. Michon's tactical level can be seen as equivalent. In Rasmussen's approach [22], this level of driving is usually reflected in rule-based actions. However, in some circumstances this may be seen as skill based, for example passing a familiar intersection may be accomplished in a rather automated way; or as knowledge based, for example if the driver experiences a situation rather rarely, like for instance to get a skidding vehicle under control.

8.1.1.3 Goals and context of driving

This third level of Hatakka's driver behaviour considers motives, goals and circumstances of driving. The question here is, why the person is driving (except for reaching places)? What function does driving have for the person? When is the person driving? What are the (usual) starting point and the (usual) final destination point of the driver? Which route does the driver choose, and why? Who joins the driver as passenger? In what condition is the driver? Is he/she tired? Is he/she driving under the influence of medicine, alcohol or drugs? What is the reason for this trip (e.g. private, professional, fun)?

Relevant issues of this third level are not considered in Moe's [21] driving process model and just partly reflected in Michon's [1] and Rasmussen's [22] driver's behaviour model. The decision processes are taken into account on the strategic level (Michon), for instance choosing the route can be understood as knowledge-based behaviour (Rasmussen). But attitudes towards any other matters are neglected. A reason for the neglect of certain aspects of this level might be that a person may have the skills that are required for this level, may have an opinion about all driving relevant issues, or may think strategically and be able to plan a route. It is still possible that this person is not able to drive a car safely. Thus, this level is 'somehow useful' for performing the driving task. But if the driving licence has to be obtained and necessary tests completed, no one will ask for the skills on this level. However, driving under the influence of alcohol or ignoring the speed limits may affect traffic safety negatively. It is therefore necessary to take this level of driver's behaviour into account when investigating traffic safety issues.

8.1.1.4 Goals for life and skills for living

This highest level of Hatakka's driver behaviour model [2] refers to the personal status, personal motives and attitudes. On this level the question is under what circumstances the driver lives. What lifestyle does he/she have? To which social, cultural or educational group does the driver belong? To establish those factors is relevant as earlier studies have shown. Personal variables like age group (e.g. [29]), social status [30] and gender (e.g. [12]) may have an influence on behaviour on all lower levels. This fourth level of behaviour is not considered in other hierarchical models. The reason might be the same like the neglect of some variables on the third level: everyone belongs to a social group, has certain age, gender and his/her own attitudes and goals in life. But not everyone is a driver.

8.1.2 How performance on the particular levels may influence drivers' performance on other levels

8.1.2.1 Vehicle manoeuvring

The performance on this level can be influenced by external factors (not driver-related), like components of the vehicle that are not working correctly, but it can also be affected by changes on one of the higher hierarchical levels of behaviour. Imagine a young male driver of 25 years. Young males are the drivers that are most often involved in fatal accidents [26]. Young men aged around 25 are usually strongly influenced by their peers [27]; the driver's friends may for instance insist that it is cool to drive without using the seat-belt, not to indicate direction changes, to speed or, to drink and drive. As he wants to be a part of this group, he might not even be convinced that they are right but still he will comply with what his peers suggest. In this way his performance on the basic level of driving can be influenced by certain characteristics of a higher level behaviour, in this case by 'giving in' into the social influence of the peer group. This social effect reflects the fourth and highest level '*goals for life and skills for living*' of the Hatakka's driver behaviour model. Considering the diamond interaction model of relevant areas [17], this driver behaviour level involves two 'diamond corners': *the individual* and *the vehicle.* Any change(s) in the individual (the driver) or/and in the vehicle may lead to changes in performance on this basic vehicle manoeuvring level.

8.1.2.2 Mastering traffic situations

Drivers' behaviour on this level may be affected by both internal and external variables. Returning to the example of the 25-year-old male driver; he is not only influenced by his peer group, but he also likes to take risks. He is single, has no kids, does not care that much about his future, does not think that he could get injured. His car and driving style may serve as a status symbol, addressed to the peers, to the grown-up world, or to the other sex. His attitudes towards life, towards risk taking and how he defines himself as a person may result in disregarding priority rules, speed limits or other road users' rights and needs. These aspects are part of the third level of the Hatakka's driving behaviour model named *goals and context of driving*. Behaviour on the *mastering traffic situations* level may be influenced by both goals and context of driving and by higher level issues related to goals for life and skills for living influencing behaviour in connection with manoeuvring. External variables that may affect the performance on the second level are infrastructure (e.g. road type, lights, road width), traffic rules (e.g. priority rules like 'right before left', speed limits) and other road users. These three factors contribute to determining the situation this driver may experience and which he consequently has to master.

Returning to the diamond model [17], this driver's behaviour level is influenced by the characteristics of the 'corners': *the individual*, *the infrastructure*, *the society* and *the interaction*.

8.1.2.3 Goals and context of driving

Any behavioural aspect that reflects this third level somehow affects the lower levels of driver behaviour, for example the choice of the route determines the situations the driver is going to experience and thus, he/she has to adapt to these 'chosen' situations. Driving under the influence of alcohol may affect *mastering traffic situations* (e.g. changes in situation awareness, field of view, reaction time [28]) and *vehicle manoeuvring* (e.g. having problems keeping the vehicle within the driving lane [28]). To accomplish the tasks on this third level of driver behaviour the driver needs to have decision and planning skills (to choose the destination, the route, the starting time). Further, the driver has to be clear about his/her motives, for example to answer the question 'Is it okay to drink and drive or not?'.

Within the interaction model of Chaloupka-Risser, Risser, and Zuzan [17], this third level of Hatakka's driver behaviour model is mainly represented in the 'corner' *the individual*. However, people develop opinions about most matters in the frame of both their physical and social environment. Thus, all 'corners' of the diamond model may influence the third level of driver's behaviour; the kind of vehicle at disposal (e.g. type, age, colour), the condition of infrastructure (e.g. road type, surface, width, design of priority issues), the social and societal preconditions (e.g. laws and rules and also how they are commented on by influential others), the presence or absence of other road users (e.g. other vehicles, vulnerable road users) and the characteristics of the driver (e.g. sensation seeking tendencies, health status).

8.1.2.4 Goals for life and skills for living

The variables on this fourth level, which also belong to the '*individual*' part of the diamond model, aren't sufficient conditions that could enable the performance of the driving task. Further, those factors are more or less 'invisible' on the road.

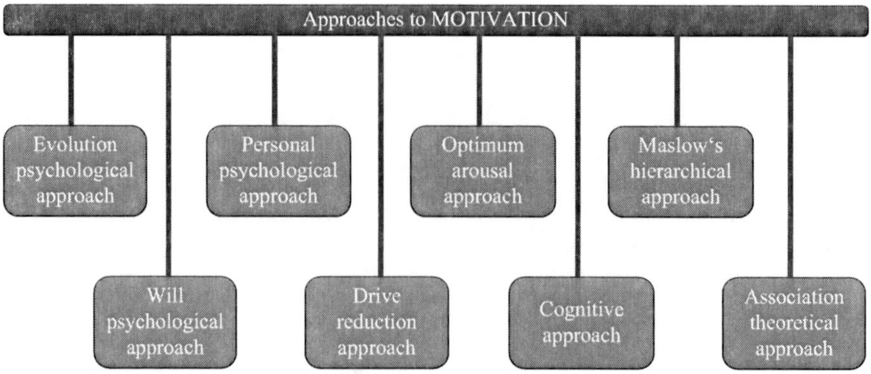

Figure 8.4 Approaches to motivation (adapted from Santrock [33] and Heckhausen & Heckhausen [34])

In research and traffic education, they are treated as influencing factors. However, they are rather neglected as variables that very much determine the visible behaviour in the public space, as a driver.

Thus, the third and the fourth levels mainly represent motivational factors and are highly relevant for the behavioural effects on the lower levels. They have an important influence on road safety, and it is therefore necessary to take these levels into account when driving behaviour is investigated.

8.2 Motivational factors

Motivation is related to internal processes that determine behaviour aiming to achieve a particular goal [32]. Motivation serves to manage available resources. In order to satisfy his/her needs and driven by a certain wish/goal, a person is motivated to act and make efforts. Thus, motivation drives people to act, behave, evaluate, judge, feel and to believe the way they do.

> 'Motivated behaviour is energized, directed, and sustained'. (Santrock [33, p. 425])

Attempts to explain human behaviour and internally related processes are diverse and so are the discussions and suggested explanations of the motivation concept (see Figure 8.4).

Kotler [35] defined a motive as a need that incites a person to perform a certain action. Motives are seen to be directly related to person's needs. A need is a:

> 'Trait that impels an individual to pursue a goal through an action that also gives purpose, meaning and direction for the behaviour of the individual.' (Ward & Lasen [36, p. 137])

Maslow's [37] work on motivation and needs was ground breaking. He proposed his famous hierarchical motivation model with five levels of human needs

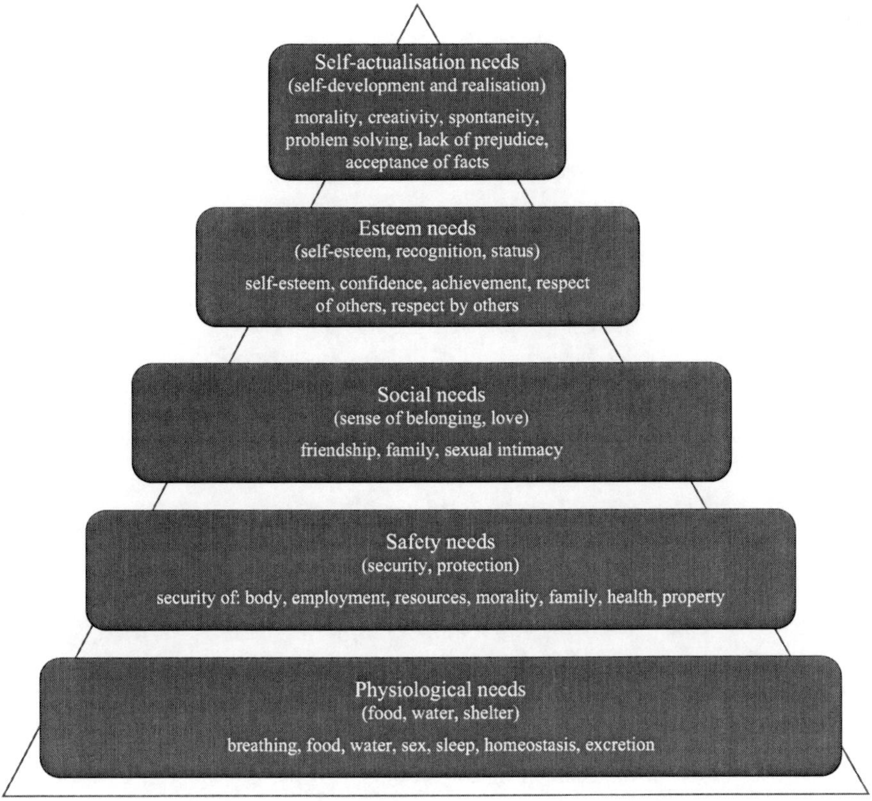

Figure 8.5 The five-level hierarchy model of Maslow [37]

([37], see Figure 8.5). The idea behind his hierarchical approach was that first the needs on lower levels have to be satisfied before needs on higher levels may be dealt with. Later, researchers criticised his postulation of the hierarchical approach and stated that human beings may wish and look for more than just one need at a time (e.g. [38]).

8.2.1 Physiological needs

Physiological needs include the access to water, food, a place to live, oxygen (air), the opportunity to sleep, excrete and to maintain homeostasis. Thereby, the principle of homeostasis reflects the body's automatic striving towards a constant level in the mentioned needs. The fulfilment of these needs serves and is required in order to survive (sex is not required for the individual but for the tribe to survive). Maslow [37] considered the needs on this level as most important, because they are vital. He claimed that the needs on this level have to be satisfied before a person can progress to look for the satisfaction of the higher level needs (therefore, the model is called hierarchical).

8.2.2 Safety needs

On this level the individuals' desire for safety and protection are taken into account. Safety issues as defined by Maslow are related to:

- health
- employment/financial security
- resources
- morality
- family
- property
- avoidance of accidents and injuries

The satisfaction of the needs included on this level is not sufficient to survive. However if the needs on this level are not met (the individual gets sick, unemployed, does not have enough resources, cannot rely on morality, has lost family members and/or property), it may have severe consequences. Needs on this level may differ in their characteristics between individuals. For example one person may have a different desire for owning property than another.

8.2.3 Social needs

The third level of needs considers a person's sense of belonging to 'somebody', to 'somewhere', to love and to be loved including having friends, romantic and intimate relationships and family. It is the desire to belong to a social group, share commonalities like religious beliefs, profession, hobbies or sports. If an individual feels alone and does not feel that he belongs to a social group and/or to other persons, this may also endanger the individual, making him or her unhappy or undermining health.

8.2.4 Esteem needs

These needs are a kind of increase in the third level of human needs. Individuals do not only strive for the feeling of being loved. They also want to be accepted, valued and respected, not only by others but also by themselves. They want to gain self-esteem and confidence. This level of needs covers also the desire of feeling strong, independent, competent and free. Esteem needs are closely related to social needs. A lack of esteem and/or self-esteem may induce inferiority complexes and other related problems that undermine well-being.

8.2.5 Self-actualisation needs

On this level, the needs for self-development and self-realisation are considered. These may strongly differ between persons. One individual may see it as self-realisation to start a family; another person may perceive their own career as his/her way to self-actualisation.

The hierarchical model could never be confirmed completely in empiric studies. Weinert [73] criticises Maslow's five defined levels as empirically not confirmed as classes of needs. Further, he questions Maslow's generalisation that the postulated needs exist for all human beings and the assumption that the needs depend on each other gradually. Weinert describes six important points of criticism that could be confirmed empirically:

- The time persons perceive a need as fulfilled differs individually.
- The hierarchical arrangement of needs does not meet the variety of how individual persons value needs.
- The levels of needs and the set categories may overlap and are not clearly separable.
- Striving to satisfy single needs depends on a person's real or subjective opportunities.
- It is possible to substitute needs that are not met.
- The perception of needs is determined by the individual's values, goals and norms.

But even if Maslow's theory has never been confirmed empirically, it is very because it is very transparent and a simple approach that summarises human needs. Illustrating this model in this context does not serve to explain the interplay between different motivations; rather to show the numerous varieties of existing human needs that may be fulfilled.

While attempting to understand the role of motivation in connection with driver's behaviour, we may question, 'Which role driving has in connection with these different types of needs'.

8.3 Motivation and driving

Mobility is an important basic need. Human beings fundamentally need to be mobile [39, 40]. In accordance with the definition by Mackensen *et al.* (cited in [41]), mobility is a person's movement from one defined starting point to a defined destination point. But mobility is more than this and so is driving. The vehicle [42] and also mobility may potentially serve as a status symbol [43]. To be mobile and to drive may contribute to fulfilling a person's need of freedom [29, 44], self-esteem and autonomy [45]. Thus, to establish the sense of being autonomous and free and to gain self-esteem can function as additional motives [7] linked to driving, beside the primary motive which is to reach a certain destination.

In order to answer questions about what direct and indirect roles the motivation and fulfilment of certain needs play in driving, and what function driving may have in connection with fulfilling needs, Maslow's [37] five types of needs discussed in the context of driving may be useful. Since some researchers criticised the hierarchical approach and retorted that human beings may wish and look for more than just one need at a time (e.g. [38]), the relation between certain needs and driving can be understood more as types of needs, rather than a hierarchy.

8.3.1 Physiological needs and driving

A certain level of physical fitness is a precondition for being able to drive a car safely. The driver has to be healthy and alert, free from fatigue and drowsiness. At the same time driving (as any other type of mobility) allows the satisfaction of physiological needs such as getting food and drink. The car can also be used as a tool to fulfil sexual needs (e.g. by impressing the other sex), which can be perceived as an additional motivation.

8.3.2 Safety needs and driving

Participating in traffic increases the risk of being involved in an accident. Drivers are compelled to evaluate how safe they feel while driving and what safety measures they want to implement in their 'micro-cosmos' in order to control the risk of being involved in an accident or getting seriously or even fatally injured. Safety measures are diverse from passive provisions like seat-belts, air-bags and adaptive coachwork to behaviour-related aspects like driving carefully, using ADAS consciously and respecting traffic rules. Special attention should therefore be placed on the perceived risk and drivers' safety needs. Car use also plays a role in connection with other issues in this group of needs; it is related to income and to the family situation. Not least, moral issues play a certain role in connection with car use.

8.3.3 Social needs and driving

The desire to have social relations and to belong to a social group is an important motive. Persons may be influenced not only by peer pressure but also by the opinion of closely related persons like family members or partners. For instance, drivers may behave differently when passengers are present from how they behave when they are alone in the car [70]. The attitudes of family members or close friends (social norms) may affect the driver's intention to behave in a certain way [46]. Thus, the social needs discussed here may well affect driver's behaviour.

8.3.4 Esteem needs and driving

As already mentioned, having a vehicle and being able to drive are closely related to the motives of feeling autonomous and free. It serves as one way to increase self-esteem [45]. Being the owner of a car and having a driver's licence is an instrument to reach satisfaction of needs on this level. However not only the vehicle, as property, and the opportunity to be able to drive induce the desired feelings, the way of driving does this also. To fulfil needs on this level, people may buy a certain make of a car or they may display a certain driving behaviour that emphasises, and maybe demonstrates, the sense of freedom, independence and confidence boosting their self-esteem. For example, Ellaway, Macintyre, Hiscock and Kearns [71] found in a questionnaire study that car users score significantly higher in self-esteem than users of public transport.

8.3.5 Self-actualisation needs and driving

Depending on the individual desire and the definition what self-actualisation means for the individual, driving may play an indirect role on this level from two possible perspectives: (1) driving may satisfy self-actualisation needs, for instance, if a person's desire is to become a bus or taxi driver or (2) driving may function as an instrument for self-development, for example by visiting a certain place (city, country) by car.

8.4 Motivational driver behaviour models

Some motivational driver behaviour models and the role of some known motives are now considered.

Homeostasis is the adaptive mechanism to achieve or obtain an equilibrium, or a steady state in response to the presence of stimuli. It is part of Maslow's [37] first level of human needs. Wilde [8, 47] proposed the risk homeostasis theory. Striving for homeostasis plays an important role in explaining driver behaviour. Wilde's basic assumption is that people have a stable, subjectively perceived level of risk that they accept. If the perceived risk changes due to any changes in the traffic system (see the diamond model, Figure 8.1), for instance if technical support systems are installed in the vehicle, the drivers' level of perceived risk may change [48]. It may either increase or decrease. In order to restore the usually accepted level of risk, the driver starts to behave differently. If the perceived risk increases, he/she drives more carefully (slower, pays more attention to potential hazards); if it decreases, the driver drives more riskily (faster, overtaking, paying less attention to the driving task).

Another motivational approach that also considers risk is represented in the risk threshold model (e.g. [7, 49, 50, 51]). In this approach, risk is taken into account from two perspectives; from the (1) subjective and (2) from the assumed objective point of view. The postulate is that drivers try to keep subjective and (assumed) objective risk in balance – they want to keep the risk under control. It is also assumed that drivers know a range of traffic situations that they perceive as safe. The upper limit of this safety margin serves as a threshold. In case the limit is exceeded, the situation is perceived as risky. Two issues determine the level of perceived risk: the perceived likelihood to experience a hazardous situation and how the driver assesses the potential consequences of such a situation. The most recent version of the risk-threshold model is the risk allostasis theory proposed by Fuller [72]. His theory proposes that drivers strive to maintain a level of risk that they subjectively perceive within a preferred range. The theory highlights the role of drivers' feelings and decision making. Representatives of risk-threshold models assume that the subjectively perceived risk determines the driver's behaviour. In general, drivers perceive a low risk in traffic ('Zero-risk theory'), and feel rather safe. If the perceived risk increases and exceeds the subjective threshold, the driver shows compensatory behaviour, like driving more carefully, slower, more attentively in order to bring perceived risk back below their subjective threshold.

Fuller [9, 51] proposed the risk/threat-avoidance model. He assumes that drivers have two main motives: to reach a destination and to avoid experiencing any hazardous situations. To participate in traffic increases the probability of being exposed to hazards. Thus, the motivation to reach a destination conflicts with the motivation to avoid hazards. In order to reach the destination the driver also has to interact with other vehicles, other road user types (e.g. vulnerable road users), and perhaps road construction sites or technical issues of the vehicle. The driver is confronted with potential hazards and obstacles the whole way from the starting point to the destination. Nevertheless, instead of avoiding these hazards, people move from start to the chosen destination. Driving and experiencing many different situations helps to learn what kind of situations can be assessed as 'hazardous', and how, with a necessary routine, these can be avoided.

Two motives described in this section that were also part of Maslow's [37] hierarchical approach of human needs, were: *perceived risk* and the *striving to homeostasis*. Other motivational issues that seem to play an important role in driving are social needs and esteem needs.

8.5 Motivational influence factors based on the theory of planned behaviour

A well-established theoretical approach that also reflects a hierarchical structure (like Keskinen's driver behaviour model [24] and Maslow's hierarchy of human needs [37]) and that potentially highlights the influence of motivational aspects on behaviour is the *Theory of Planned Behaviour* (TPB, [32]). The approach using the TBP refers to conscious decisions. Violations are 'conscious' types of errors and have been considered in the Generic Error Model System (GEMS) proposed by Reason [31]. Reason divided human failures into errors and violations (see Figure 8.6). He categorised three types of errors: (1) skill-related slips and lapses, (2) rule-related mistakes, and (3) knowledge-related mistakes. Whilst these kinds of errors more or less reflect cognitive processes, violations involve an important motivational issue.

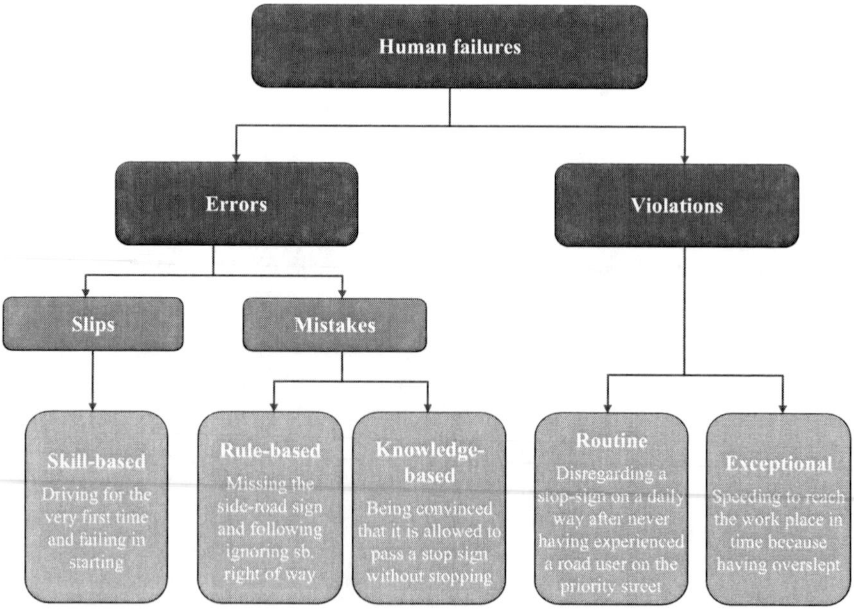

Figure 8.6 The categorisation of human failures (adapted from Reason [31])

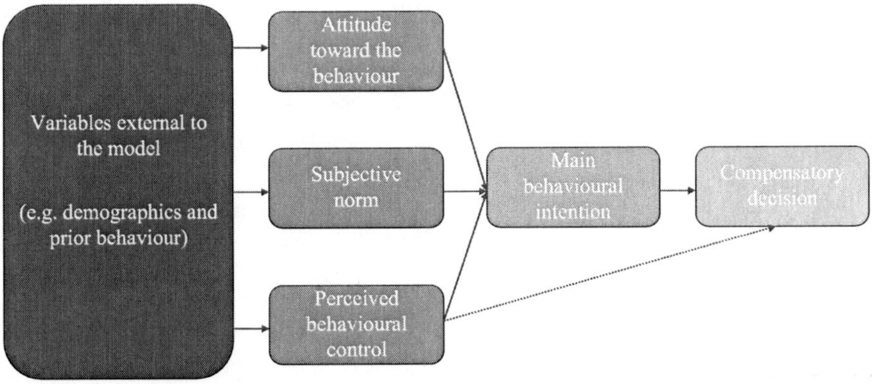

Figure 8.7 An extended version of the Theory of Planned Behaviour [52]

Here the individual knows how and what would be the correct behaviour but he/she does not comply with this knowledge. It is most important to understand a person's motivation lying behind such behaviour, and this point should be taken into account when traffic safety issues are investigated. The two main statements of the TPB that attempt to explain the predictors of this volitional behaviour are: (1) intentions are the best predictors of behaviour and (2) intentions are affected by perceived behavioural control, social norms and attitudes towards the behaviour. Different studies extend the theory of planned behaviour by adding factors like perceived risk, moral norms, descriptive norms, anticipated effects and experiences connected to past behaviour (e.g. [52, 53]). In their study, Zhou, Horrej and Ruifeng [52] added external variables to the variables originally included in the classical model of the TPB. Based on this, they created the structure model of compensatory intentions that also includes the consideration of risk compensation (see Figure 8.7).

In their study, Cestac, Paran and Delhomme [12] illustrated that drivers' intentions to show a certain behaviour are related to the perceived risk, perceived behavioural control and their attitudes. To have attitudes and express an opinion is important for self-confidence and self-esteem and also for the satisfaction of the esteem needs. Earlier studies have shown that social norms (e.g. perceived peer pressure) also affect driver's behaviour [54, 55].

Motivational factors play an important role in influencing the driving behaviour. In order to increase traffic safety, they should be considered when safety measures like driver assistance systems are developed and introduced.

8.6 The role of motivation in behavioural adaptation

Including ADAS to the driving task and driving context adds a new component to the traditional diamond interaction model (see Figure 8.8). This change may lead to certain adaptations in driving behaviour.

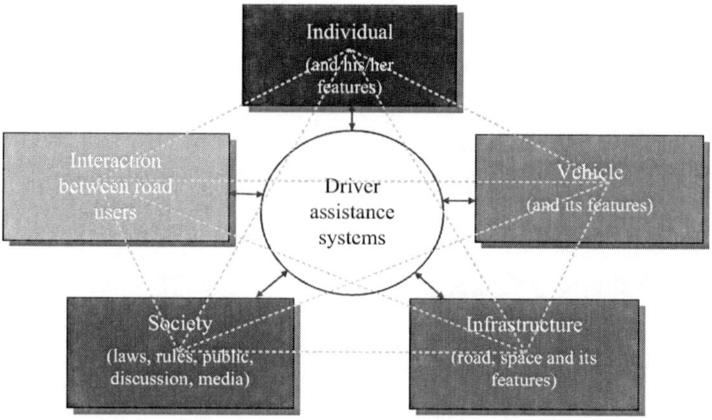

Figure 8.8 Addition of driver assistance systems as sixth component of the diamond interaction model (adapted from PROMETHEUS [56] and [17])

These adaptations can be related to traffic safety, positively, neutrally or even negatively. Behavioural adaptations were first defined by the OECD [4, p. 23] as:

> 'Those behaviours which may occur following the introduction of changes to the road-vehicle-user system and which were not intended by the initiators of the change. Behavioural adaptations occur as road users respond to changes in the road transport system in a way that their personal needs are achieved as a result. They create a continuum of effects ranging from positive increase in safety to a decrease in safety.'

Behavioural effects induced by the use of ADAS are of considerable research interest. A number of studies have investigated the influence of the use of ADAS on driver behaviour [57–59, 60]. In his thesis, Jenssen [5] concluded according to Nilsson, Stevens, Roskes and Heinrich [61] and his own research findings that changes in driver behaviour invoked by the ADAS use can be grouped into six categories. Table 8.1 gives an overview of these six categories, and links the levels of driver behaviour according to the hierarchical approach of Hatakka [25] (see Figure 8.3).

Jenssen [5] proposed five learning phases of behavioural adaptation, covering a time window of two years. Jenssen stated that after using the system for one year and longer, the shown behaviour is settled and stable. Below, the five phases of behavioural adaptation are introduced:

1. *The first encounter phase*. This first phase represents the first contact with the system. The driver learns how the system works and how to handle it while driving. In this phase drivers may be distracted by the actual handling of the system. A collateral effect that could arise is that drivers may not trust the system. This phase is characteristic for the initial one to six hours when the system is used, when the distance covered is less than 50 km. Both the length of this phase and the potential negative side effects that could occur depend heavily

Table 8.1 Categorisation of behavioural effects due to ADAS use (adapted from Nilsson et al. [61], Jenssen [5] and Hatakka [25])

Category	**Affected dimensions**	**Level of driver behaviour**
Perception	Auditive, visual, haptic, tactile	Mastering traffic situations
Cognition	Comprehension, interpretation, selection, decisions	Mastering traffic situations
Performance	Lateral- and longitudinal control, acceleration- & deceleration, system handling, driving errors	Vehicle manoeuvring
Driver state	Levels of attention, awareness, workload, stress, drowsiness	Mastering traffic situations
Attitudes	Acceptance, degree of reliance, trust and mistrust, rejection	Goals and context of driving
Adaptation to environmental conditions	Weather, infrastructure, visibility	Mastering traffic situations

on the human–machine- interface and interaction (HMI&I). If the system is self-explaining and intuitive in its use the duration of this period will be shorter than if the system confronts the driver with various problems and challenges.

2. *The learning phase*. The second phase is characterised by gaining control over the handling the system. The driver becomes more confident in the use and gets familiar with the limitations of the system. In this phase, too, depending on the HMI&I design, the driver may be distracted by the system use while driving. This learning phase directly follows the first encounter phase and lasts three to four weeks, corresponding to 10–40 hours of driving or a distance up to 1,000 km. The behaviour shown in this time window is still unstable.
3. *The trust phase*. During this period the driver starts to gain trust in the system and its functioning. Consequently, the driver allows the system to take control of the tasks for which it is intended. Gaining trust in the ADAS goes along with the danger of over-reliance on the system, which may result in a passive driving behaviour and low attention levels. Typically, this phase starts after one month of system use and ends with relatively stable driver behaviour around the sixth month of using the system.
4. *The adjustment phase*. In this phase the driver adapts the amount of trust acquired in the third phase. It can be assumed that it takes up to 12 months until the driver experiences all 'typical' or relevant situations. In a 12-month period the driver also experiences all seasons. A further assumption is that within one year the driver has the chance to drive on all kinds of roads more than once and to get to know how the system works under the various conditions and different driving situations that might arise. The driver may have encountered new situations that could potentially reveal system limitations which had not been experienced earlier. The trust that was gained in phase three is now blended with a certain amount of grudge against the system.

5. *The readjustment phase*. With further experience the driver learns how to handle system limitations. The 'grudge' turns into mistrust for distinct conditions. The driver learns when to trust the system and when it is necessary to stay alert in order to be able to intervene actively, if necessary. When trust develops, the risk arises that drivers lose skills, which is a problem in case of a system breakdown or system malfunction.

The six categories of Nilsson *et al.* [61] and the five learning phases of Jenssen [5] are based on studies that investigated the effect of ADAS on driver's behaviour. Depending on the kind of system, its human–machine interface design and the usability, the time-windows of the phases may be different from the suggested ones. The first two phases especially may vary due to the design and functionality issues. The focus of most studies on this subject was the two lower levels of driver behaviour proposed by Hatakka [2]. Mainly the effects of ADAS on cognitive and regulatory processes and on driving performance were investigated (e.g. [57–59, 60]). Motivational aspects were not dealt with so exhaustively. Two motivational variables that are considered in Nilsson's categories [61] and Jenssen's learning phases [5] are trust and acceptance. A closer look into Hatakka's hierarchical model of driver behaviour [2] will explain why it is important to take motivational factors into account.

8.6.1 Vehicle manoeuvring level and ADAS

When the driver starts to use ADAS, he/she has to adapt his/her skills and performance on the vehicle manoeuvring level. Functions, reactions of the vehicle, switches, human–machine interface and interaction may seem new with added ADAS. The driver has to acquire new skills that are necessary for operating this new technology and to perform the tasks on this level correctly; he has to adapt to this new situation in order to fulfil vehicle manoeuvring tasks correctly. Taking the learning phases of Jenssen [5] into account, changes on this level of driver behaviour will occur in the first two phases; the encounter and the learning phase. After driving for one week with the system the driver should have gained the skills to be able to fulfil this level of the driving task successfully.

8.6.2 Mastering traffic situations level and ADAS

Driver assistance systems 'interfere' in different traffic situations such as a car in front (e.g. Adaptive Cruise Control), intersection (e.g. Navigation System), snow and skid-risk (e.g. Electronic Stability Control), presence of pedestrians (e.g. Intelligent Video Surveillance) or speed limits (e.g. Intelligent Speed Adaptation). The idea behind many systems is to support the drivers in these contexts and to help them master certain traffic situations. Many studies investigated if ADAS might have negative effects on driver's behaviour on this level (e.g. [57–59, 60]). Drivers might react with a reduced level of attention (e.g. [62]) or might lose the skills to fulfil tasks on this level in case of system breakdown (e.g. [57]). So, on the one hand, an ADAS can tackle tasks that usually have to be completed by the driver. On the other hand, the driver must not lose those skills in order to be able to react in case of a system breakdown. Therefore, the use of ADAS also changes the preconditions for the driver to fulfil tasks on this level successfully. The driver has to adapt to this new situation and has to share parts of the task he was previously used to performing independently. He may not lose the needed skills but has to observe the system functionality all the time. How long the changes invoked by the use of ADAS on this level will take depends on the events that the driver experiences while driving, but Jenssen's [5] assessment (6 months) may well be assumed to be quite accurate in this respect.

8.6.3 Goals and context of driving and ADAS

As already mentioned, people form, or already have, an opinion about most matters. Even before a person starts using an ADAS in the vehicle, he/she may have an opinion about this particular ADAS. When the driver starts using ADAS, this opinion may change due to the experiences encountered such as different situations, experiencing system limitations, talking to friends and family about the systems and their experiences. According to Jenssen's learning phases [5], it can be assumed that within one year drivers will pass different stages of forming an opinion, of developing trust and of knowing when and how to rely on the particular system. For example, when drivers start to use a navigation system, they usually change their strategic procedure when planning a route. The system's task is to take over navigation and the driver just has to decide where to go and if he/she wants to take the shortest or quickest route. A certain amount of trust is required to follow the instructions given by the system. The level of trust may change: first, the driver might not have that much trust in the system because of negative rumours he/she could have heard (e.g. the old lady who arrived 1,400 km away from her destination [63]). After first experiences and after having recognised that the system works trust might increase. It could decrease again after a bad experience (e.g. being told to drive into a one-way-street in the wrong direction, or being led into a closed road because of construction works). After having learned how to deal with such situations, the trust might increase again. Based on different levels of trust, the driver may behave differently. He/she might develop strategies in his mind about how to get to the destination in case one of these 'horror stories' happens. The variables that are considered on the third level of driver behaviour are certainly influenced by the use of ADAS. Several studies investigated how trust in automation and overreliance may influence driver behaviour (e.g. [60]). Other issues that have been considered are acceptance of ADAS (e.g. [64]) and perceived risk, in connection with ADAS use (e.g. [48]). Generally, little research has been carried out in this area.

8.6.4 Goals for life and skills for living and ADAS

Several studies have shown the influence of these variables on the behaviour displayed on the road, and they are certainly relevant for the incorporation of ADAS into driving. Some ADAS aim to support particular groups of drivers, for instance the traffic sign detection system that supports older drivers in traffic sign recognition. The Alco-lock system is often related to the group of younger drivers.

Developers of ADAS try to work out the needs of certain groups in order to adapt the technique to their needs. It may be assumed that factors on this fourth level of driver behaviour are quite stable. Gender and social status do not change with the use of ADAS while driving. So, this level is different from the other three levels and can be seen more as a source of moderator variables that may influence driver behaviour and behavioural adaptation processes to ADAS on lower levels.

So, by respecting those levels of driver behaviour with the added ADAS and Jenssen's learning phases [5] the conclusion can be made that in long-term use motivational factors may play a more important role in driver behaviour and traffic safety issues than the cognitive and performance skills that are already developed within the first weeks of ADAS use.

8.7 A short view on impacts of various ADAS on driving

As mentioned before, by automating the driving task progressively the task itself changes: the driver is less involved in driving than before when electronic systems take over parts of the driving task. Since research often focuses on aspects of safety when ADAS is used, those behavioural adaptations that lead to reduced traffic safety are of special interest. Potential problems that may arise in response to driving with ADAS are, for instance, an insufficient level of situation awareness, reduced vigilance, transfer of responsibility to the system, potential loss of driving skills, and changes in acceptance of and attitudes towards ADAS [65]. Some studies (e.g. [9, 66, 67]) focused not only on cognitive or regulatory processes and driving performance but also on motivational aspects. Marell and Westin [66] investigated, in a long-term study, how the perceived risk of drivers changed as a response to nine months use of an information system that gives the driver information about speed limits. They found that drivers reported after that time that they perceived traffic rules more consciously and that they were more likely to respect them. Marell and Westin [66] also investigated how the system was accepted, if drivers indicated a high level of acceptance. The system used in this study was an assistance system that provides the driver information. So, how do drivers accept systems that not only provide information but also warn the driver of potential hazards or even intervene actively?

In a driving simulator study, Jamson, Lai and Carsten [67] investigated how drivers trust Forward Collision Warning (FCW) systems. These systems detect if the distance to the leading vehicle gets too small and warns the driver in this case. The variation of the systems investigated was determined by an adaptive system that adapted to the drivers' driving style and warned depending on that, and a non-flexible system that warned the driver depending on predefined critical distances. The results indicated that different driver types (aggressive vs. non-aggressive) showed different preferences. Non-aggressive drivers accepted both systems equally but aggressive drivers clearly preferred the adaptive system.

Adell and Varhelyi [10] conducted a long-term study that lasted from 2000 to 2002. They found that drivers had positive attitudes towards an Adaptive Accelerator

Pedal (AAP) after using it for two years. This system detects the current speed limits and adapts the speed automatically by changing the accelerator pedal function in such a way that no further increase in speed is possible. However, the driver can override the system by pressing the gas pedal five times more strongly than usual. Participants of the study had the feeling that the use of the system influenced performance and safety positively. The biggest perceived effect that the study indicated was that the subjects felt that the risk of being prosecuted because of speeding was much reduced.

A similar, but more active system, the Intelligent Speed Adaptation, was investigated in its long-term effect by Wallen Warner and Åberg [11]. They found that the amount of time when the drivers speeded was reduced initially, and that also the mean speed was partially lower. However, these effects disappeared over the time. Further, the results indicated that the participants attitudes, subjective norms, self-reported behaviour and perceived behavioural control influenced their behaviour when using the system.

The studies reported in this chapter show that ADAS may have many different effects on drivers. Many variables may have an interactive influence. Interestingly, many theories and models that discuss behavioural adaptation to ADAS are motivational theories (overconfidence or inappropriate estimation of risk as a result of automation processes in driving [6], risk compensation and risk homeostasis theories [7, 8], threat avoidance theory [9]). However, studies mostly focused on factors related to cognitive processes and to performance. There is a lack of research concerning motivational factors. Studies investigating motivational aspects usually just focused on drivers' level of trust, acceptance, attitudes towards the system and perceived risk. However, as has been shown in the extended 'diamond interaction model' (see Figure 8.8), the introduction of ADAS may not only have a direct influence on driver's behaviour but also on other factors that can affect driver's behaviour such as vehicle characteristics, infrastructure, driver's social environment and other road users, which affects communication among road users. Changes especially in the last two factors – social environment and communication among road users – will affect driver behaviour, mainly by influencing driver's motivations. Thus, there are more motivational aspects than just acceptance and trust that have to be taken into account. Maslow's pyramid of human needs [37] highlighted the relevance of social factors. These changes may definitely be expected to influence drivers' behaviour significantly.

8.8 Conclusion

When driver behaviour in response to ADAS use is investigated, the following human needs and possible changes should be considered:

- *Risk perception and risk homeostasis*: these are pointed out in different theories as factors that may moderate behavioural changes when measures for improved safety are to be introduced (e.g. [7, 8]).

- *The feeling of security and protection*: past studies have shown that drivers often pay more attention to other tasks than the driving itself (e.g. using the mobile phone) when they feel safe [68]. Thus, ADAS may have a negative impact on drivers' behaviour by making him/her feel safe and therefore tempting him/her to turn to concurrent activities.
- *Social needs*: the influence of the social environment on drivers' attitudes and behaviour is shown in investigations of driver behaviour. Attitudes and opinions concerning ADAS develop in society. The attitudes of social groups around the driver will most probably affect his/her attitudes and, further, his/her behaviour.
- *Esteem and self-actualisation needs*: driving and the kind of vehicle used, and its technical equipment may fulfil the function of gaining self-esteem, providing the feeling of freedom and self-confidence. ADAS may play an important role in this respect. ADAS may potentially support a human need for satisfaction of self-esteem and self-actualisation.

As has been pointed out in this chapter, motivational processes play an important role in driver behaviour. The motivational variables mentioned above that are potentially connected to ADAS use may have a significant impact on driver behaviour and therefore on traffic safety also. Decisions to drive safely or to violate traffic rules are related to these variables. Thus, when ADAS effects are investigated, in addition to cognitive and regulatory processes and performance measures, the influence of motives should be taken into account.

Acknowledgement

This research received funding from the European Community's Seventh Framework Program (FP7/2007-2013) under grant agreement no. 238833/ADAPTATION project (www.adaptation-itn.eu).

References

1. Michon J. A. 'A critical view of driver behaviour models. What do we know, what should we do?' in Evans L., Schwing R. (eds.). *Human Behaviour and Traffic Safety*. New York, NY: Plenum Press; 1985. pp. 485–525
2. Hatakka M. *Novice Drivers' Risk- and Self-Evaluations. Use of Questionnaires in Traffic Psychological Research Method Development, General Trends in Four Sample Materials, and Connections with Behaviour*. Annales Universitatis Turkuensis, ser.B, Humaniora. Turku: Painosalama; 1998
3. Keskinen E., Peräaho M., Laapotti S. *GDE-5PRO and GDE-5SOC: Goals for Driver Education in a Wider Context – Professional and Private Drivers in Their Environment* (unpublished manuscript). University of Turku, Finland; 2010
4. OECD. *Behavioural Adaptations to Changes in the Road Transport System*. Report Road Transport Research; 1990

5. Jenssen G. D. *Behavioural Adaptation to Advanced Driver Assistance Systems: Steps to Explore Safety Implications*. Doctoral thesis. Norwegian University of Science and Technology, Faculty of Engineering Science and Technology, Department of Civil and Transport Engineering; 2010
6. Bainbridge L. 'The ironies of automation' in Rasmussen J., Duncan K., Leplat J. (eds.). *New Technology and Human Error*. London: Wiley; 1987. pp. 271–284
7. Näätänen R., Summala H. 'A model for the role of motivational factors in drivers' decision-making'. *Accident Analysis & Prevention*, 1974;6(3-4): 243–261
8. Wilde G. J. S. 'Risk homeostasis theory and its promise for improved safety' in Trimpop R. M., Wilde G. J. S. (eds.). *Challenges to Accident Prevention: The Issue of Risk Compensation Behaviour*. Groningen: Styx Publications; 1994. pp. 9–24
9. Fuller R. A. (1984). 'Conceptualisation of driving behaviour as threat avoidance'. *Ergonomics*. 1984;27(11):1139–1155
10. Adell E., Varhelyi A. 'Driver comprehension and acceptance of the active accelerator pedal after long-term use'. *Transportation Research Part F: Traffic Psychology and Behaviour*. 2007;11(1):37–51
11. Wallen Warner H., Åberg L. 'The long-term effects of an ISA speed-warning device on drivers' speeding behaviour'. *Transportation Research Part F: Traffic Psychology and Behaviour*. 2008;11(2):96–107
12. Cestac J., Paran F., Delhomme P. 'Young drivers' sensation seeking, subjective norms, and perceived behavioural control and their roles in predicting speeding intention: How risk-taking motivations evolve with gender and driving experience'. *Safety Science*. 2011;49(3):424–432
13. Fishbein M., Ajzen I. *Belief, Attitude, Intention and Behaviour: An Introduction to Theory and Research*. Reading, MA: Addison-Wesley; 1975
14. Rouse W.B. 'Human-computer interaction in the control of dynamic systems'. *Computer Surveys*. 1981;13(1):71–99
15. Nilsson L. 'Automated driving does not work without the involvement of the driver' in Underwood G. (ed.). *Traffic and Transport Psychology: Theory and Application. Proceedings of the ICTTP 2004;* Nottingham, UK, Sep 2004. St Louis: Elsevier; 2005. pp. 273–301
16. Fastenmeier W., Gstalter H. 'Driving task analysis as a tool in traffic safety research and practice'. *Safety Science*. 2007;45(9):952–979
17. Chaloupka-Risser C., Risser R., Zuzan W.-D. *Verkehrspsychologie: Grundlagen und Anwendungen*. Vienna: facultas.wuv; 2011
18. Victoir A., Eertmans A., Van den Bergh O., Van den Broucke S. 'Learning to drive safely: social-cognitive responses are predictive of performance rated by novice drivers and their instructors'. *Transportation Research Part F*. 2005;8(1):59–74
19. Zhang Y., Kaber D. B. 'An empirical assessment of driver motivation and emotional states in perceived safety margins under varied driving conditions'. *Ergonomics*. 2013;56(2):256–267

20. Ranney T. A. 'Models of driving behavior: a review of their evolution'. *Accident Analysis & Prevention*. 1994;26(6):733–750
21. Moe D., Engen T., Jenssen G. D. *Utvikling av evalueringsmetodikk i Wise-Car:Del 1: Litteraturstudie og kunnskapsstatus (Evaluation Method in the WiseCar project) in Norwegian*. SINTEF Report A7758. Trondheim Norway; 2008
22. Rasmussen J. 'Reasons, causes and human error' in Rasmussen J., Duncan K., Leplat J. (eds.). *New Technologies and Human Error*. Chichester: Wiley; 1987. pp. 293–301
23. Dahlen E. R., White R. P. 'The big five factors, sensation seeking, and driving anger in the prediction of unsafe driving'. *Personality and Individual Differences*. 2006;41(5):903–915
24. Keskinen E. 'Why do young drivers have more accidents?' Presented at Junge Fahrer und Fahreinnen Conference; Cologne, Germany, 1996
25. Hatakka M. 'What makes a good driver? The hierarchical approach' in Bartl G. (ed.). *DAN-Report. Results of EU-Project: Description and Analysis of Post Licensing Measures for Novice Drivers*. Vienna: Austrian Road Safety Board (KfV); 2000. pp. 19–25
26. Horvath C., Lewis I., Watson B. 'The beliefs which motivate young male and female drivers to speed: a comparison of low and high intenders'. *Accident Analysis and Prevention*. 2012;45:334–341
27. Rienzi B. M., McMillin J. D., Dickson C. L., Crauthers D., McNeil K. F., Pesina M. D., et al. 'Gender differences regarding peer influence and attitude toward substance abuse'. *Journal of Drug Education*. 1996;26(4):339–347
28. Schnabel E. *Alcohol and Driving-Related Performance – A Comprehensive Meta-analysis Focusing the Significance of the Non-significant*. Doctoral thesis. Julius-Maximilians-University Würzburg, Faculty of Philosophy II; 2012
29. Balogun S. K., Shenge N. A., Oladipo S. E. 'Psychosocial factors influencing aggressive driving among commercial and private automobile drivers in Lagos metropolis'. *The Social Science Journal*. 2012;49(1):83–89
30. Taubman-Ben-Ari O., Katz-Ben-Ami L. 'The contribution of family climate for road safety and social environment to the reported driving behavior of young drivers'. *Accident Analysis and Prevention*. 2012;47:1–10
31. Reason J. *Human Error*. Cambridge: Cambridge University Press; 1990
32. Ajzen I., Fishbein M. 'The influence of attitudes on behaviour' in Albarracín D., Johnson B. T., Zanna M. P. (eds.). *The Handbook of Attitudes*. Mahwah, NJ: Lawrence Erlbaum; 2005. pp. 173–221
33. Santrock J. W. *Psychology*. 7th edition. Columbus, OH: McGraw-Hill College; 1997.
34. Heckhausen J., Heckhausen H. *Motivation und Handeln*. 3rd edition. Heidelberg: Springer Medizin Verlag; 2006
35. Kotler P. *Marketing Management*. 11th international edition. New Jersey: Prentice-Hall; 2003
36. Ward D., Lasen M. 'An overview of needs theories behind consumerism'. *Journal of Applied Economic Sciences*. 2009;4(7):137–155

37. Maslow A. H. *Motivation and Personality*. New York, NY: Harper; 1954
38. Neher A. 'Maslow's theory of motivation: a critique'. *Journal of Humanistic Psychology*. 1991;31(3):89–112
39. Geiger G. 'Der Mensch als animal migrans. Zur Psychologie der Mobilitat'. *Der Bürger im Staat*. **3**: Baden-Wurttemberg: Landeszentrale der politischen Bildung. [online]. 2002. Available from http://www.buergerimstaat.de/3_02/animal.htm [Accessed 26 July 2013]
40. Opaschowski H. W. *Umwelt. Freizeit. Mobilitat. Konflikte und Konzepte*. Opladen: Leske und Budrich; 1999
41. Bähr J. *Bevölkerungsgeographie*. Stuttgart: Verlag Eugen Ulmer; 2004
42. Steg L. 'Car use: lust and must. Instrumental, symbolic and affective motives for car use'. *Transportation Research Part A: Policy and Practice*. 2005; 39(2–3):147–162
43. Haslinger B. *Gendertypisches Mobilitatsverhalten in der Freizeit – automobile Fortbewegung als Handicap. Ein evolutionstheoretischer Ansatz*. Doctoral thesis. University of Vienna; 2006
44. Vogt W. 'Was ist und welchen Sinn hat Mobilität? Mobilität und gesellschaftliche Entwicklung'. *Der Bürger im Staat*. **3**: Baden-Württemberg: Landeszentrale der politischen Bildung. [online]. 2002. Available from http://www.buergerimstaat.de/3_02/sinn.htm [Accessed 26 July 2013]
45. Möser K. *Geschichte des Autos*. Frankfurt: Campus Verlag; 2002
46. Tseng C.-M., Chang H.-L., Woo T. H. 'Modeling motivation and habit in driving behavior under lifetime drivers' license revocation'. *Accident Analysis and Prevention*. 2013;51:260–267
47. Wilde G. J. S. 'The theory of risk homeostasis: implications for safety and health'. *Risk Analysis*. 1982;2(4):209–225
48. Rajaonah B., Tricot N., Anceaux F., Millot P. 'The role of intervening variables in driver–ACC cooperation'. *International Journal of Human–Computer Studies*. 2007;66(3):185–197
49. Näätänen R., Summala H. *Road User Behaviour and Traffic Accidents*. New York, NY: North Holland Publishing Company; 1976
50. Summala H. 'Modeling driver behavior: a pessimistic prediction?' in Evans L., Schwing R. C. (eds.). *Human Behavior and Traffic Safety*. New York, NY: Plenum Press; 1985. pp. 43–65
51. Summala H. 'Risk control is not risk adjustment: the zero-risk theory of driver behavior and its implications'. *Ergonomics*. 1988;31(4):491–506
52. Zhou R., Horrej W. J., Ruifeng Y. 'The effect of conformity on pedestrians road-crossing intentions in China: an application of the theory of planned behavior'. *Accident Analysis and Prevention*. 2009;41(3):491–497
53. Holland C., Hill R. 'The effects of age, gender and driver status on pedestrians' intentions to cross the road in risky situations'. *Accident Analysis and Prevention*. 2007;39(2):224–237
54. Lee C., Abdel-Aty M. 'Presence of passengers: does it increase or reduce driver's crash potential?' *Accident Analysis and Prevention*. 2008;40(5): 1703–1712

55. Simons-Morton B., Lerner N., Singer J. 'The observed effects of teenage passengers on the risky driving behavior of teenage drivers'. *Accident Analysis and Prevention*. 2005;37(6):973–982
56. PROMETHEUS. *Introducing Information and Assistance Systems in the Driver-Vehicle-Environment Interaction.* Presentation material, PROMETHEUS HMI working group; 1989
57. Brouwer R. F. T., Hoedemaeker D. M. *On-Road Study on Short-Term Effects of ADAS on Driving Behaviour*. AIDE Deliverable D1.2.3 ('Driver support and information systems: Experiments on learning, appropriation and effects of adaptiveness'); 2006
58. Stanton N. A., Young M., McCaulder B. 'Drive-by-wire: the case of driver workload and reclaiming control with adaptive cruise control'. *Safety Science.* 1997;27(2/3):149–159
59. Buld S., Krüger H.-P. 'Die Auswirkung von Teilautomation auf das Fahrverhalten' in Grandt M. (ed.). *Entscheidungsunterstützung für die Fahrzeug- und Prozessführung.* Bonn: DGLR; 2003. pp. 241–253
60. Popken A. *Drivers' Reliance on Lane Keeping Assistance Systems as a Function of the Level of Assistance*. Doctoral thesis. Chemnitz University of Technology, Faculty of Behavioural and Social Sciences, Department of Cognitive and Engineering Psychology; 2009
61. Nilsson L., Stevens A., Roskes A., Heinrich J. 'Behavioural risk analysis' in Bekiaris E., Papakonstantinou G. (eds.). *D3/8.1 of ADVISORS project*. Thessaloniki: Aristotle University of Thessaloniki, Laboratory of Transport Engineering; 2001
62. Hoedemaeker M., Kopf M. *Visual Sampling Behaviour When Driving with Adaptive Cruise Control*. Presented at the 9th International Conference on Vision in Vehicles, Australia; 2001
63. Die Welt. 'Navi leitet Belgierin 1400 Kilometer in die Irre'. [online]. 2013. Available from http://www.welt.de/vermischtes/kurioses/article112781187/Navi-leitet-Belgierin-1400-Kilometer-in-die-Irre.html [Accessed 20 January 2013]
64. Hoedemaeker M., Brookhuis K. A. 'Behavioural adaptation to driving with an adaptive cruise control (ACC)'. *Transportation Research Part F*. 1998;1:95–106
65. Dragutinovic N. M., Brookhui K. A., Hagenzieker Marchau V. A. W. J. 'Behavioural adaptation in response to ADAS'. Presented at the 3rd International Conference on Traffic & Transport Psychology (ICTTP), Nottingham, UK; 2004
66. Marell A., Westin A. 'Intelligent transportation system and traffic safety – drivers perception and acceptance of electronic speed checkers'. *Transportation Research Part C: Emerging Technologies*. 1999;7(2-3):131–147
67. Jamson A. H., Lai F. C. H., Carsten O. M. J. 'Potential benefits of an adaptive forward collision warning system'. *Transportation Research Part C: Emerging Technologies*. 2008;16(4):471–484

68. Carroll J., Howard S., Peck J., Murphy J. 'A field study of perceptions and use of mobile telephones by 16–22 year olds'. *Journal of Information Technology Theory and Practice*. 2002;4(2):49–61
69. Skippon S. M., Reed N., Luke T., Robbins R., Chattington M., Harriso A. H. 'Relationships between driving style, Self-Reported Driving Behaviour and Personality' in Dorn L. (ed.). *Driver Behaviour and Training Vol. IV*. Farnham: Ashgate; 2010. pp. 89–104
70. Skippon S. M., Diels C., Reed N. 'Driving style as a fitness indicator' in Dorn L. (ed.). *Driver Behaviour and Training Vol. V*. Farnham: Ashgate; 2012. pp. 359–374
71. Ellaway A., Macintyre S., Hiscock R., Kearns A. 'In the driving seat: psychosocial benefits from private motor vehicle transport compared to public transport'. *Transportation Research Part F: Traffic Psychology and Behaviour*. 2003;6(3):217–231
72. Fuller R. 'Driver control theory' in Porter B. E. (ed.). *Handbook of Traffic Psychology*. Amsterdam: Elsevier; 2011. pp. 13–26
73. Weinert A. B. *Organisations- und Personalpsychologie* (5. Auflage). Weinheim: Beltz (AO); 2004

Chapter 9

Drivers' adaptation to mobile phone use: Interaction strategies, consequences on driving behaviour and potential impact on road safety

Véronique Huth[1] and Corinne Brusque[1,2]

Abstract

The increasing use of the mobile phones while driving raises a safety concern due to its distractive potential and its consequent effects on crash risk. The way phone use affects driving depends on the usage behaviour of the driver. First, drivers can actively regulate their exposure to phone interactions. Second, they can make choices on the strategic level of the driving task so as to ensure the compatibility of the phone use with driving. Strategies can aim at lowering the demands in one of the concurrent tasks, for example, by using assistance systems or a hands-free device. Third, the phone use while driving can impair the driving performance because of the distractive nature of the dual-task situation. With the aim to mitigate this impairment and to uphold an acceptable driving performance, drivers can deliberately adapt their behaviour on the tactical level of the driving task.

These driver adaptation strategies to mobile phone use are discussed along with their actual implementation and effectiveness. An extensive literature review has been complemented with findings from naturalistic driving studies and in-depth interviews carried out within the ADAPTATION project. A discussion on the potential safety impact of phone use and drivers' adaptation to it concludes this chapter.

9.1 Introduction

The use of mobile phones has increased exponentially over the last decade. Mobile phone subscriptions accounted for 15% of the global population in 2001 and reached 85% in 2011 [1]. Concerns on how in-car phone use and related driver distraction are evolving have been expressed in a report recently published by the World Health Organization [2]. The report cites studies from several countries

[1] IFSTTAR-LESCOT, Bron, France
[2] University of Lyon, Lyon, France

revealing that 60–70% of drivers use their mobile phones while driving [e.g. 3, 4]. Further studies have found that usage rates are especially high among younger drivers [e.g. 5, 6].

With mobile technologies becoming more and more advanced and varied, nowadays the phone offers a wide range of interaction opportunities. As a consequence, using the mobile phone in the car is no longer restricted to making calls [7]. Hence several types of interactions need to be distinguished when analysing in-car phone use. Drivers can accept an incoming call and engage in a conversation or they can initiate a call themselves, which implies dialling a number or retrieving it from the directory before conducting the conversation. They can read or write text messages and use several other mobile applications. Calls can be made using a hands-free device, whereas all other interactions usually imply hand-held use of the phone.

As a consequence, mobile phone interactions involve different types of distraction [e.g. 8]:

- *Cognitive distraction*: Take the mental attention away from the driving task.
- *Visual distraction*: Take the eyes off the roadway to obtain information on the phone.
- *Manual distraction*: Take a hand off the steering wheel to manipulate the phone.

Given that driving can be a highly demanding task, the distractive effects of engaging in a secondary task related to the mobile phone may impact on the driving performance and accident risk. A great number of studies have dealt with this issue and recent reviews of the literature on the distractive effects of phoning while driving underline its importance for driving safety [2, 9, 10]. The factor by which crash risk increases has been estimated at 4 in a review study carried out by Brace et al. [11], confirming earlier results of Redelmeier and Tibshirani [12]. Visual–manual interactions with the phone represent a special safety concern, since they promote vehicle trajectory deviations. A review of naturalistic driving studies by Martin [13] highlights that text messaging most drastically increases the odds for drivers to be involved in safety-critical events, followed by dialling a number and reaching for a mobile phone or headset.

It needs to be considered, however, that the use of the mobile phone is based on drivers' choices and that drivers can influence both the interaction task and the driving task to a certain extent. Summala's hierarchical 'sieve' model postulates that exposure to risk is regulated by higher level decisions, which depend on motivational and cognitive factors [14]. Drivers can adapt to phone use at different levels in an attempt to control its potential risks [15]. First, they can take decisions in order to shape their exposure to phone use, for example, by completely or selectively rejecting phone use while driving or by regulating temporal aspects of the interaction. This type of driver adaptation to mobile phone use is addressed in Section 9.2 of this chapter.

Moreover, drivers can focus their decisions on the strategic or tactical level of the driving task, as defined in the hierarchic model by Michon [16]. At a strategic level, they can strive for compatibility of phone use and the driving task by ensuring low demands in one of the tasks. Choosing specific driving situations to interact with the phone or using assistance systems during the interactions are possible strategies that are discussed in the Section 9.3 of this chapter, along with the effects of using a hands-free kit so as to lower the demands of the phone task.

Finally, drivers can counteract the effects of mobile phone-induced distraction at a tactical level. Impairments in the driving performance that result from the interaction with the phone are presented in the Section 9.4 of this chapter, and drivers' behavioural adaptation that is aimed at mitigating these negative effects by increasing safety margins is discussed.

This chapter integrates findings from the literature into a comprehensive picture of drivers' behavioural adaptation to mobile phone use. Previous findings have been enriched by results obtained within the ADAPTATION project. Naturalistic studies analysed within ADAPTATION have provided data on the driving habits related to phone use of 16 drivers including the frequency and length of calls, the movement of the car during calls, or the road type on which calls take place. Furthermore, 10 drivers have participated in a more detailed naturalistic study, which allowed observation of more specific aspects of the phone use, in particular the differences between incoming and outgoing calls. Lastly, in-depth interviews have collected 12 drivers' statements on their habits and strategies regarding phone interactions while driving. Specific comments are included into this chapter in order to support and further illustrate the findings presented.

9.2 Regulating the exposure to phone use while driving

On the assumption that using a mobile phone while driving constitutes the drivers' choice, the drivers' willingness to engage in this secondary task plays a decisive role. Lerner and Boyd [17] refer to this aspect in their *deciding-to-be-distracted* approach. This perspective considers that the risk generated by a secondary task is not only determined by the distractive potential of the activity itself, but also by the frequency and duration of its execution and by the criteria that time the drivers' engagement in the activity [8]. The strategies drivers can apply to regulate their exposure to phone use while driving are presented in Section 9.2.1 and some relevant influence factors that are likely to modify these regulation strategies are discussed in Section 9.2.2.

9.2.1 Exposure regulation strategies

The first and fundamental option of exposure regulation consists of the general decision on whether to use the phone at the wheel. Some drivers prefer not to use the phone at all when they are driving [18, 19]. For example, older drivers may take this decision if they judge phone use as too demanding [20, 21]. An interview study by Pöysti et al. [15] found that younger age, a higher annual mileage, being male, higher perceived driving skills and a lower safety orientation predicted in-car phone use. Further studies confirmed that the share of drivers who use phone while driving decreases with age [6, 22] and that men report to use the phone more often than women [5].

Those drivers who decide to use their phone while driving can still limit and regulate their exposure as phoning habits from everyday life are not transferred one-to-one into the car [23]. Although incoming calls or text messages are triggered externally and have considerable potential to influence driver decisions to interact with the phone due to social pressure (see 9.2.2), the drivers can decide to reject the task. This decision may be based on the expected characteristics of the secondary task (caller identity, content of conversation) or the judgement of the current driving task (complexity, expected time until arrival). Research shows that drivers' intention to answer an incoming call can be predicted by factors related to the driving context, such as the vehicle speed and headway distance [24] and the connection to a hands-free device [25]. The willingness to use the phone also depends on the type of interaction. For instance, drivers have reported higher willingness to answer a call than to place a call, and sending a text message was the kind of interaction the drivers were least willing to carry out [26].

In the detailed ADAPTATION naturalistic driving study [28], outgoing calls represented two-thirds of the calls. This suggests that phone use while driving is not mainly externally triggered, but is based on proactive decisions by the drivers. The ADAPTATION in-depth interviews [28] revealed that other types of interactions with the phone also occur during driving, although these might be less actively sought by some drivers. A comment hints towards a more restricted engagement in visual–manual interactions with the phone while driving:

- *I usually read text messages while driving, but would rather not write them.*

Schömig et al. [27] found that highly demanding driving contexts led the drivers to reject secondary tasks or to delay beginning them. More precisely, the deliberate decision to engage in the secondary task was based on an anticipation of the development of situational demands and the judgement of their compatibility with the execution of a concurrent task. The study showed that these strategies are effective in maintaining safe levels of driving performance. Yet, all drivers might not apply them equally. Brusque and Alauzet [23] showed that some drivers never make or answer calls when they are driving whereas others regularly use phone while driving, irrespective of the driving context. A high level of phone use while driving could be explained by work-related reasons, high mileage and extensive mobile phone use in everyday life [6]. The majority of drivers, however, do not adopt any of these extreme behaviours and well take the driving context into account before engaging in a phone call. Usage strategies related to the driving situation are further discussed in Section 9.3.1 of this chapter.

The length of the interaction constitutes a factor of exposure regulation. Zhou et al. [25] examined drivers' willingness to limit the conversation time and to inform the caller that they are driving. They have found that drivers tend to engage in these compensatory actions particularly when using the phone hand-held. Further evidence for the drivers' length regulation of phone calls is provided by the results from the ADAPTATION naturalistic driving studies. Short call durations (up to one minute) were more frequent among outgoing calls than among incoming calls. Assuming that drivers are more likely to influence the duration of those calls they have initiated themselves, this result shows the drivers' intentions to limit call durations. Whilst this finding could be influenced by the motives of the calls [28], indicators of length regulation to reduce risk in driving have been found in the naturalistic driving data with respect to the call mode. The frequency and duration of hand-held calls was significantly reduced compared to hands-free calls. Section 9.3.3 of this chapter includes additional comments on this finding.

Another way of adapting the exposure of phone use to driving is to interrupt the performance of the secondary task and to pace it according to the conditions. Drivers may try to divide the task into several chunks and pay attention to the traffic situations at regular and appropriate intervals [29]. In a driving simulator study, it could be shown that drivers who accepted performance of a secondary task, in a highly demanding situation, delayed beginning until they could anticipate how the situation would evolve [27]. Furthermore, it has been found that drivers take the opportunity to interrupt secondary tasks in order to execute controls of the primary task [30]. Concerning mobile phone use in particular, the drivers can stop the conversation or inform the interlocutor about the driving conditions so as to facilitate a correspondingly adapted conversation. However, a situation-aware management of the secondary task may pose some difficulties in the case of phone use. First and foremost, the driver may not be able to delay the beginning of the task easily, for example, if the phone is ringing. A possible technical solution to this issue is to redirect the incoming calls to a mailbox, which can be accessed by the drivers at the time of their choice [31].

Phone interactions might not necessarily be 'chunkable'. Contrary to a passenger in the car, the phone interlocutor does not experience the driving situation and will not automatically adapt the conversation to the conditions. Informing the caller about a critical driving situation might take too much time, especially when the driver has to react to unexpected changes in the driving situation. A comment from the ADAPTATION in-depth interviews makes reference to the difficulty of giving explanations to the caller:

- *When I have to interrupt the call quickly, I say 'There are the police'.*

It seems to be easier to chunk visual–manual interactions with the phone. However, their interruption can lead to interaction errors, which extend the interaction beyond the drivers' intentions and chosen moments. This concern has been raised in the ADAPTATION in-depth interviews.

Lerner [32] made reference to these possible limitations of exposure regulation to interactions with the mobile phone. He states that temporal aspects of the secondary task, chunking, error potential and recovery, as well as the ability to self-monitor distraction are potentially relevant issues that should be investigated more thoroughly.

9.2.2 Factors influencing exposure regulation

In addition to the characteristics of the activity, cognitive and motivational factors have been found to alter the drivers' decision-making process regarding exposure regulation. These influence factors can affect judgements on the secondary task or the primary driving task [27]. On the one hand, *motivations* for the mobile phone use and *social pressure* can overrule the intention to apply risk-minimising strategies, and on the other hand, *anticipation failures* and *biased risk assessment* can distort the anticipatory control the strategies are based on. Research on the willingness to engage in distracting activities indicates that task-related determinants are more important than driving-related factors [32].

It has been suggested that the motivation to engage in a secondary task can outweigh arguments against its execution, such as unfavourable situational demands [27]. The caller identity, for example, can convince the driver to attend a call even in complex driving environments. Data gathered in the detailed ADAPTATION naturalistic driving study show that phone use often takes place in urban areas, contradicting the strategies indicated by the same participants in the in-depth interviews. Instead, the observed behaviour corresponds to the call motives reported by the participants [28]. The motivation to use the phone may be externally triggered, for example, by its ringing, but the usage behaviour is often driven by internal motives concerning the management of time and resources, the enjoyment of technology use, socialising and general lifestyle attitudes [32, 33].

Schömig et al. [27] explained the absence of strategic adaptation of driver performance to secondary tasks that Horrey and Lesch had observed in their experiment [34] by a possible influence of subjective pressure to perform the concurrent tasks, despite high workload. Similarly, social pressure to stay in contact with colleagues, family and friends might incite drivers to use the phone even though they are busy with the driving task. Correspondingly, social pressure has been identified as a predictor of the intention to make a call in scenarios that include time pressure with a concern for reassuring one's social environment [35]. Furthermore, over recent years, mobile phone use in everyday life has assumed a shape of heavy use from an early age. Having evolved into a daily habit, it also takes place when driving, and distracted driving has become a normative behaviour [36]. Accordingly, a study on social norms regarding phone use while driving revealed that drivers who were confronted with crash scenarios recognised the greater responsibility of a distracted driver for an accident, but they did not suggest greater punishments (fines and jail time) for this driver than for an undistracted driver [36]. Using a hands-free kit and thus acting in conformity with the law might also favour social acceptability. Drivers regulate their phone use in accordance with perceived social norms, which may overcome the perceived barriers originating in the driving context, the accident risk and the probability of getting fined [33].

In the event that drivers apply the regulation strategies, the anticipation process they are based on is decisive. For the strategies to work, the situation has to evolve in the way the drivers had expected. Since the drivers have limited attentional resources they can allocate to the intermittent controls of the situation, they will only monitor the aspects of a situation they have previously defined as relevant and will not be able to detect any completely unexpected changes in the situation [27].

Strategy-related decision-making is grounded on combined risk assessment of the driving task and the phone task. Drivers' subjective estimates of objective risk lead to a feeling of risk once the threshold that provokes an emotional response is passed [37]. This implies that the application of strategies depends on a cognitive process and on drivers' individual thresholds. Research has shown that usage rates of the mobile phone while driving are related to subjective risk estimates. Non-users are more convinced about the risk of phoning while driving than users [38] and the more drivers use their phone while driving, the less they feel it is dangerous [5, 39]. Generally speaking, risk estimates of phone use while driving are rather low [32, 40]. This may be explained by the fact that phoning while driving is a risk that drivers are familiar with, and may be related to an optimism bias that has been detected – drivers perceive their own risk when using the phone as lower and more controllable than the risk of others when doing the same [33, 38]. Accordingly, the general acknowledgement of phoning while driving as unsafe is often not reflected in the drivers' behaviour [36]. Thus, the underestimation of the personal risk may prevent the driver from executing a situation-adaptive interaction with the mobile phone [27]. Later in this chapter, this type of effect is discussed with regard to the use of hands-free devices. Drivers' risk assessment of cognitive distraction is especially poor compared to visual–manual distraction, and consequently strategies applied to hands-free phoning differ from those applied to hand-held phone use (see 9.3.3). The risk of getting a police fine, however, does not seem to affect drivers' intention to interact with the phone, for example, by sending text messages [35]. Likewise, the awareness of the safety risk implied in text messaging does not prevent the great majority of young drivers from reading, nor replying to, or initiating texts [36, 41]. In these cases, the motivations of the phone use, as well as the social pressure involved, seem to outweigh the disadvantages given by the perceived risk.

9.3 Seeking after compatibility of driving and mobile phone use

In addition to the issue of exposure rates, it is crucial to observe under what conditions the exposure takes place. The driver has to take decisions on the strategic level of the combined task of driving and using the phone. Wilde's risk homeostasis theory [42] postulates that drivers accept a certain level of risk, which they seek to maintain. Similarly, Fuller [37] states that drivers try to keep task difficulty (i.e. the interaction between the task demands and the driver capabilities) within certain boundaries and that they may take decisions on driving-related behaviour based on the feeling of risk that informs them about the current task difficulty. Performing a secondary task, such as using the mobile phone, increases the driver demands. It entails attention-sharing, which represents higher levels of mental workload for the driver [43, 44]. In order to maintain an acceptable task difficulty, drivers have to find an appropriate match of driving situations and secondary task characteristics, for example, by choosing specific moments and locations for their phone use and by adapting it to the traffic situation [45]. A safety problem arises if the drivers misjudge the compatibility of the secondary task with the driving task. In a more general context, Lerner et al. [46, p. 500] state: 'The actual risk associated with some device will be a joint function of how use of that device interferes with driving and the circumstances under which the drivers are willing to use it.'

Sections 9.3.1 and 9.3.2 deal with the choices the drivers take while seeking after compatibility of driving and phone use, both regarding the demands of the driving task and Section 9.3.3 with the demands of the phone use.

9.3.1 Driving situation

Previous studies have concluded that the engagement in secondary tasks depends on the characteristics of the driving situation [47], with a decreasing willingness to perform distracting in-vehicle activities at increasing situational demands [48]. In a study by White et al. [33] participants reported that they were likely to restrict their phone use depending on the demands of the road situation. Drivers might take advantage of a situation that has low demands in order to use the mobile phone, for example, by choosing a section with lower traffic density to make a call [27], or they might create such a situation. For instance, some drivers interact with the phone only while waiting at red traffic lights, and others stop the car before they start a phone call [18, 19]. In addition, it was found that the task type influenced the willingness to engage less in high-demand than in low-demand situations, and that drivers were only willing to engage in highly demanding tasks such as sending a text message when the driving tasks had low demands [26]. In line with Fuller's task capability interface model, if drivers are already close to their desired task demand capacity, they will not add demands that would exceed this level of resource allocation.

It cannot be assumed, however, that drivers put these behavioural choices systematically into practice. Naturalistic driving data from the Netherlands reveal that the majority of interactions with the mobile phone occur in a moving car [49]. Similarly, results of the ADAPTATION naturalistic driving study suggest that hands-free calls typically take place when the car is moving, and that less than 30% of hand-held calls are made when the car is stopped [28]. In the same study, hands-free calls that were held in a moving car were not necessarily initiated while the car was moving. Instead, a significant share of calls was started when the car was stopped or moving slowly, suggesting that the drivers take the chance to dial in those situations. This strategy might be based on the judgement that visual–manual interactions with the phone are more demanding than carrying on a conversation. Correspondingly, the ADAPTATION in-depth interviews reveal that strategies referring to stopping the car are above all purposefully applied to visual–manual interactions with the phone (in particular text messages), but not to calls. Comments include:

- *I write text messages only when the car is stopped (parked or red traffic lights).*
- *I revise my agenda when the car is parked.*
- *I try to take the opportunity at red traffic lights to dial when I want to make a call.*
- *Writing texts or using the Internet while driving leads to a loss of vigilance, but when the car is stopped there is no inconvenience (including short stops or traffic jams).*

The ADAPTATION in-depth interviews also provide evidence for drivers' strategies to choose the moment of interaction with the mobile phone depending on the complexity of the driving situation. Example comments are:

- *I make outgoing calls in situations that do not require so much concentration.*
- *I take the chance to make calls on the highway. I avoid using the phone in the city.*
- *I avoid making calls in heavy traffic.*
- *I avoid interacting with the phone on unfamiliar roads.*

In-car technologies can support the driver in the application of such strategies. For example, a message system interface that gives driver control over the timing of speech-based e-mail presentation has been tested in a simulator study [50]. While the system had a positive effect on driving performance under most circumstances, it proved to be detrimental in highly demanding situations. The authors assume that the cognitive costs associated with the decision-making process imposed by the system use might be too high. This flaw could be eradicated by technical solutions based on a real-time diagnosis of the driving situation that have been designed to assist the driver in deciding when to carry out secondary tasks by considering situational demands. These so-called workload managers assess the workload of the driver and regulate the information input that is potentially distractive [51]. For instance, the system would divert an incoming call automatically to the mailbox if the driver is performing a manoeuvre at an intersection at the time the phone rings. This means that external requests are only presented to the driver when the system assumes that the driver is able to react to them without compromising driving safety [52]. For the drivers' self-paced interactions with the phone, for example, outgoing calls, a workload manager can anticipate the levels of workload based on the upcoming infrastructure (such as a sharp curve) and give advice to the driver when not to perform a distractive task. Simulator tests on this system revealed that drivers positively rate the support in combining the secondary task with driving, but that they often overrule the system messages [53]. The authors conclude that the attention-leading component of the workload manager that indicates high load situations is decisive, while extensive messages should be avoided.

However, it has to be considered that drivers may differ in their appreciation of situation complexity and in their individual thresholds regarding the criticality assessment of driving situations. Some drivers do not seem to be concerned about incompatibility of phoning and driving, as the following comments from the ADAPTATION in-depth interviews illustrate:

- *I make a call as soon as I need it.*
- *Generally, I use the phone regardless of the traffic. For interactions that require more visual attention (texts, emails), I avoid situations that demand high concentration.*
- *I read text messages anytime.*
- *There is no problem with short calls. You can make an effort and concentrate on the call and the road. On the highway, you can also make longer calls.*

Furthermore, there can be discrepancies between the intentions the drivers express and their actual behaviour. For example, driving on the highway was repeatedly mentioned as a preferred situation for phone use in the in-depth interviews, but the naturalistic driving data shows the opposite result. The participants used their phone most frequently on urban roads, even more often than their exposure rate on road types would predict [28]. Especially intensive phone users have shown less adaptation of their usage behaviour to the driving context in naturalistic driving observations [54].

9.3.2 Assisted driving

In addition to the choice of specific traffic situations, lowering the demands of driving by using an assistance system could be a strategy that helps making phone use compatible with driving. Being assisted frees attentional resources [55], which may be used to carry out secondary tasks. In line with this, it has been found that the activation of an Adaptive Cruise Control can promote non-driving related activities, such as calling or surfing on the Internet [56].

In a focus group study on the use of in-vehicle technologies, drivers have reported that they take the opportunity to use the mobile phone when they travel assisted by Adaptive Cruise Control [54]. Similarly, the ADAPTATION in-depth interviews suggest that some drivers have strategies related to the use of speed regulation assistance in order to compensate for attentional resources that are allocated for making a call. Participants commented as follows:

- *It happens that I put Cruise Control on for a call.*
- *I activate Cruise Control or the Speed Limiter when I prepare an outgoing call.*
- *Generally, I try to stop the car or call back later. But if I am on the highway and the traffic is flowing, I keep talking on the phone with Cruise Control on.*
- *I take advantage of Cruise Control on the highway in order to compensate the decrease of attention to the road.*

Contrary to these subjective results, naturalistic driving data does not confirm an integrated use of speed regulation systems and the mobile phone. The analyses were centred on free flowing traffic conditions on the highway, in accordance with the results on the almost exclusive choice of this setting for the use of Cruise Control [54]. In the French naturalistic driving data set from the INTERACTION project, the rate of mobile phone use (calls) was twice as high when the speed regulation system was not active compared to the periods when Cruise Control was engaged [54]. The analysis of naturalistic driving data within ADAPTATION has detected that about 50% of drivers reduce their speed when using their mobile phone. However, no systematic strategies could be identified for integrating assisted driving with phone use [28].

More extensive analyses on the time spent with Cruise Control and the time spent on the phone again detected an overlap of Cruise Control use and mobile phone use, but still no hints towards a systematically integrated use of assistance and the phone have been found. Against expectations, drivers were not more likely to compensate for the demand of the distractive task by activating an assistance system when using the phone hand-held than hands-free. Possibly, the support by Cruise Control is not sufficient for the driver to reallocate attentional resources, or there may be other factors that influence the simultaneous use of Cruise Control and the mobile phone.

9.3.3 Use of hands-free devices

Hands-free devices have been developed to minimize manual distraction of the driver by holding the mobile phone. Several studies have shown that phoning hands-free is judged as carrying much less risk than manipulating the keyboard of the phone in order to write a text message or to dial a number, answering a call or conversing hand-held [23, 38, 57]. Often the hands-free use of the phone while driving is even perceived as non-hazardous [3]. Comments collected during the ADAPTATION in-depth interviews correspond with these previous findings. For instance, drivers stated:

- *Hands-free phoning is much less dangerous than hand-held phoning.*
- *Holding the phone in the hand is very dangerous, I only do it very rarely for important calls and I keep them very brief.*
- *I only use the phone at the wheel when my hands-free kit is on.*
- *If the hands-free kit is not connected, I do not answer the call; I stop to connect the kit and call back.*
- *Attention suffers from using the phone, also when having a call conversation. But it is much better with the hands-free kit. It is not too dangerous because I make simple phone calls in low complexity situations (e.g. highway) and I use the hands-free kit.*

Previous research revealed that hands-free kit users deem the risks of phoning while driving higher than hand-held phone users [35]. It can be assumed that those drivers install a hands-free kit in order to feel safer. On the other hand, using a hands-free kit is associated with high usage rates of the phone while driving. Perhaps intensive phone users tend to install hands-free kits, or the possibility of phoning with a perceived lower risk increases phone usage. Results from the ADAPTATION naturalistic driving studies [28] suggest that the perception of hands-free calls as less dangerous than hand-held calls leads to differences in the application of strategies. First, drivers reduced exposure to hand-held calls compared with hands-free calls. Hand-held calls were less frequent than hands-free calls and they tended to be short (<1 min), whereas hands-free calls were mostly of medium length (1–5 min). Second, drivers had different criteria when choosing the driving situation regarding the movement of the car. Hand-held calls were typically made when the car was stopped or moving slowly, whilst hands-free calls took place more often when the car was moving.

Drivers' behaviour might not only be guided by the feeling of risk, but also by the avoidance of police fines when using a hands-free kit [5]. The fact that hands-free phoning is allowed in many countries might even encourage drivers to consider that it does not increase crash risk [33]. Behavioural reactions to restrictions on hand-held phone use seem to differ among user types. After the hand-held ban in Finland, for example, occasional users mainly ceased to use the phone while driving, whereas frequent users chose to install a hands-free kit, and drivers with an intermediate usage frequency kept making hand-held use of the phone [58]. Drivers with hands-free kits still report being influenced by the presence of the police, probably due to their occasional hand-held use of the phone or the use of other phone functions [33]. Hand-held phone users justify their usage as optimising their travel time and keeping them informed (c.f. 9.2.2). For them, the advantages of using the phone outweigh the risk of distraction and fines. However, occasional hand-held phone users are sensitive to the disapproval of other drivers and the police [35].

Observational and self-report studies have shown that the number of hands-free phone users is two to seven times higher than the number of hand-held phone users [3, 4]. However, the number of hand-held users is not negligible. A survey in Australia even revealed that about a quarter of drivers were using a hand-held phone while driving [59]. This behaviour seems to be more frequent among younger drivers (<45 years) than older drivers [60]. In the ADAPTATION naturalistic study 32% of all calls were hand-held, despite the availability of a hands-free device in the car. This number decreased to 12% in the detailed naturalistic study, which included filling phone diaries and used a slightly different participant sample [28].

There is a growing body of research evidencing that hands-free phone use does not impair driving performance less than its hand-held use [3, 11, 57, 61, 62]. Hands-free calls were found to reduce visual monitoring of the dashboard and the traffic situation, and to impair vehicle control [63]. Whilst an increased number of off-road excursions when using the phone, hand-held, compared to hands-free were observed in a simulator study [43] and stronger lateral position deviations were found during dialling hand-held [64], the effects of the conversation itself on lateral vehicle control did not significantly differ between the two call modes [64, 65]. No significant differences between call modes have been found regarding mental workload [65, 66] nor reaction times [67–70]. Against common beliefs and drivers' perceptions, cognitive distraction is of such importance that the improvement by the hands-free kit is marginal and the impact on drivers' behaviour ends up being comparable to hand-held phone use [3, 11, 12, 45, 57, 71]. Interviews of drivers confirm that visual–manual distractions are judged as hazardous, whereas cognitive distraction by a conversation is widely underestimated [23, 57]. In conclusion, it seems that the benefit of eliminating visual–manual distraction when using hands-free devices is overruled by the fact that drivers consider hands-free calling as safe and, by consequence, do not apply appropriate strategies to compensate for the distraction given by the (underestimated) cognitive demand.

9.4 Adapting driving behaviour in response to distractive effects of mobile phone use

Despite drivers' efforts to regulate the exposure to phone use and to make it compatible with driving, the combined demands of driving and phoning can exceed the available attentional capacities and hence entail impairments of driving performance and safety [30]. Results on such effects are presented in the first part of this section (9.4.1). Self-regulatory behaviour can occur on the tactical level of the driving task. According to Fuller's task capability interface model [37], drivers would increase safety margins when they recognise that their desired task difficulty is exceeded due to the incompatibility of the two concurrent tasks. Adaptations of the driving behaviour, which aim at counteracting the detrimental effects of phone use on the driving performance, are outlined in the second part of this section (9.4.2).

9.4.1 Impairment of driving performance due to distraction

Impairment of the driving performance has been detected for in-vehicle entertainment systems [72]. More specifically, the use of the mobile phone is associated with the experience of hazardous situations [15, 23]. An extensive body of research exists on the distractive effects of phoning while driving, analysing the different types of interactions with the mobile phone. These include visual–manual interactions such as dialling [73, 74], answering a call [75], holding the phone [43, 64] and texting [76]. Likewise, the effects of holding a conversation hands-free have been studied [21, 43, 67, 77–79]. As discussed earlier in this chapter, evidence suggests that during phone calls cognitive distraction plays a decisive role in the deterioration of driving performance [3, 66]. Specific aspects of driving performance deterioration are now presented. They concern the vehicle control, driving related information processing and reaction times.

Whilst studies on phone conversations only have shown a slight impairment of the lane-keeping performance [80], no effects on lane-keeping [68, 81] or even decreased lateral position deviations (see 9.4.2), it has been repeatedly demonstrated that dialling a number and composing a text message have a negative effect on the lateral control of the vehicle [64, 65, 82]. Similarly, mobile phone-induced inattention leads to failures in keeping a safe distance when following a vehicle and to smaller gap acceptance when crossing or merging into traffic [2, 22, 83]. Also text messaging is associated with particular increases in the variability of following distances [82]. Moreover, the speed control is affected when using the phone, resulting in difficulties in maintaining a constant speed [68] and in disturbances of the traffic flow by considerably reduced speed [22].

Situation awareness, as defined by Endsley [84], is also impaired by the use of the mobile phone. Drivers fail to perceive and process traffic signs and hazards that are present on the road [22, 59]. This may be due to an altered visual information uptake when using the phone. Studies have concluded that drivers made more in-vehicle glances when writing text messages than when driving without distraction and have detected corresponding impairment of driving performance [76, 82]. Confirming this result, Drews et al. [85] found that text messaging was more detrimental than holding a phone conversation while driving. A reduction of the perceptual field of view has been observed during mobile phone interactions, mainly originating from fewer mirror and peripheral checks [83, 86]. This reduction seems to be provoked by cognitive distraction [9] and does not necessarily relate to visual–manual phone interactions. Further effects of the increased mental workload are inattentional blindness [79] and change blindness [87], where the information is within the field of view, but its processing is defective, although the element might be relevant to the driving task. The described reductions in situation awareness provoked by phone use represent impairments regarding the identification of hazards and the reactions to them, consequently affecting crash avoidance capabilities [88]. Correspondingly, it has been stated that distraction-related crashes occur when the driver's attentional resources are absorbed by the phone interaction, which prevents the driver from sufficiently processing an unexpected traffic hazard [89]. Reaction times to sudden and unexpected events have repeatedly proven to be delayed when using the phone while driving [65, 90, 91], a finding that has been endorsed by meta-analyses of relevant studies [61, 80]. Holding a conversation, with a passenger or on the phone, is not comparable to listening to music on the radio in terms of cognitive demand, given that the latter does not increase reaction times [69].

The participants of the ADAPTATION in-depth interviews made reference to the distraction provoked by the phone use while driving and its effects on the driving performance. The description of the effects on driving remained vague, though:

- *It is difficult to concentrate on the road and on the phone conversation. In the end, you do worse in both; neither the driving performance nor the quality of the conversation is good.*
- *You are less vigilant and reactive because the conversation catches your attention.*
- *You forget to use the indicator; you do not pay attention to what happens outside.*
- *You do not concentrate enough on driving. It is not only the eyes – your mind is in the conversation and less on the road, although you don't realise it.*
- *It is very dangerous for other road users, for example, motorcyclists. You don't pay attention to them and do not have good reactions.*

It has been suggested that the impairment of the driving performance and consequent impact on driving safety may vary with the complexity of the secondary task [2]. However, some findings point towards distractive effects that are independent of the difficulty level of the phone conversation [21, 78, 92]. Possibly, the threshold for the impact on driving performance is relatively low, or the fact of initiating a call and engaging in a conversation is more decisive than its content.

9.4.2 Compensation strategies to mitigate effects of distraction

Various studies on compensation strategies on the tactical level of the driving tasks show that drivers increase their safety margins by adapting the vehicle control. Increased safety headways to lead vehicles have been observed in several studies on the use of mobile phones [79, 93], both for conversations [90] and text messaging [82]. Drivers also reduce their speed when using the phone [72, 93, 94]. In their study on the use of smartphones while driving, Basacik et al. [7] however showed that the compensatory speed reduction observed did not allow the drivers to perform the driving task as good as when driving without distraction.

A further compensation strategy that has been detected is the simplification of trajectory control – drivers avoid lane changes when they are phoning [e.g. 95], especially in dense traffic conditions [76]. Contrasting with the deteriorated lateral vehicle control when manually interacting with the phone (9.4.1), decreases in lateral position deviation have repeatedly been observed related to the phone conversation [64,65], indicating that drivers might increase their alertness and modify their safety margins in high workload situations.

In order to compensate for the impairments in situation awareness, drivers can add periodical short control glances during the performance of a visual secondary task [27]. Furthermore, the adaptation of speed can be regarded as an indirect compensation for longer reaction times to hazards. Increased reaction times when using the phone while driving also lead to more intensive braking and shorter stopping distances [59, 83, 96]. This effect seems to be a necessary consequence so as to avoid crashes, whereas the previously described compensation strategies are more likely to be consciously intended by the drivers.

The ADAPTATION in-depth interviews revealed some compensatory strategies on the tactical level of the driving task. Comments included:

- *I slow down.*
- *I drive on the right lane.*
- *I perform fewer manoeuvres, and if I have to perform a manoeuvre I ask the person on the phone to hang on for a while.*
- *I follow the vehicles without overtaking.*

Nevertheless, these strategies are not always applied by the drivers. For instance, Alm and Nilsson [77] did not detect any compensation through increased following distances, and Cooper et al. [97] found that drivers only reduced their speed while phoning if the traffic was sufficiently dense. Thus, the implementation of compensation strategies may depend on the risk the drivers perceive from the combination of the driving task and the mobile phone task. This might explain why some studies found a compensatory speed reduction for hands-free phoning [43, 92, 96] and others only for hand-held phoning [65, 66, 68]. Correspondingly, a meta-analysis on speed reduction [61] concludes that it is stronger for hand-held than for hands-free phone use. As discussed earlier in this chapter (Section 9.3.3), the perception of hand-held phone use as a safety threat and the underestimation of the cognitive distraction by hands-free phone use might incite the drivers to show less compensatory behaviour when using the hands-free kit [44, 61]. Moreover, the drivers need to be aware of the impairments in driving performance that are provoked by phone use. Lesch and Handcock [40] concluded that drivers indeed tend to overestimate their performance and that they could benefit from educational campaigns on the effects of phone use while driving.

In the ADAPTATION in-depth interviews, comments on the adaptation of the driving behaviour were often accompanied by the acknowledgement of the distractive effects of phone use, or by a condition for their implementation. Comments qualifying the adaptation of the driving behaviour are:

- *I adapt my driving because I know that my attention will be less sustained.*
- *It depends on the level of the communication.*
- *It depends on the situation.*
- *I do not change my driving behaviour for calls, but for text messages or mobile apps.*
- *I slow down on unfamiliar roads only.*

9.5 Conclusions on potential safety impact

The use of the mobile phone while driving constitutes a dual-task situation that entails increases in crash risk. Three types of driver adaptation strategies to this situation have been observed which aim at regulating this risk: regulation of exposure to phone use, regulation of the combined demands of the phone and driving task at strategic level and behavioural adaptation on tactical level during phone use.

The application of strategies regulating the exposure to phone use was found to depend on user profiles and the type of interaction with the mobile phone. Overall, phone use that includes visual–manual distraction is more regulated than the use of a hands-free device for calling. Importantly, several obstacles to exposure regulation have been identified. They include characteristics of the interaction itself, in particular issues related to the 'delayability' and 'chunkability' of the phone interaction. Moreover, the strong motivations for phone use, which are likely to outweigh safety related arguments, are combined with a biased risk perception on phone use while driving. It represents a risk that is familiar and is perceived as controllable. Drivers underestimate the risk stemming from the cognitive distraction to a particularly large extent. The social acceptance of phone use and the social pressure to be contactable are further factors that distort risk-minimising exposure regulation.

Although drivers report their intention to make strategic choices for phone use depending on the driving context, this behaviour is not always observed. Again, the obstacles described above may help explaining this discrepancy. Reducing the demands of the driving task by using assistance systems is a strategy that could not yet be validated with the currently available data. However, the use of a hands-free device does not have the expected positive effects on driving safety. Given that the risks of calling with a hands-free device are misperceived, fewer compensatory strategies are implemented.

Negative effects of the phone use on driving performance have been found regarding the vehicle control, driving related information processing and reaction times. Compensatory strategies for these aspects have been identified at the tactical level of the driving task, but their systematic implementation and their effectiveness is still questionable at the current state of knowledge. Under which conditions such mitigating behavioural adaptation is carried out needs to be investigated. Furthermore, it is unclear whether the drivers consciously implement this compensation strategy in order to increase their safety margins, or whether it is an effect of the high dual-task demands and the consequent difficulty to perform the driving task correctly.

Interacting with the phone while driving has become a wide-spread habit, despite legal restrictions on its hand-held use, and there is a growing need to address its effects on driving safety. Considering the range of behavioural adaptation strategies presented in this chapter, efforts should be made to induce drivers to use the phone in a risk-minimising way. In doing so, obstacles to implementation of self-regulation strategies should be used as a starting point for design of driver education actions. They can also assist in design of phone interactions as well as related support technologies that help the driver to appropriately manage both phone use and the driving task. Moreover, research on behavioural changes in response to mobile phone use while driving should consider the factors of the Joint Conceptual Theoretical Framework (JCTF), presented in Chapter 2 of this book, including the underlying processes of behavioural adaptation, as well as the influence of external factors and driver characteristics. With mobile technologies developing rapidly into more and more sophisticated devices, the cognitive, visual and manual demands are expected to increase considerably in the near future and positively influencing the safety impact of mobile phone use while driving will gain greater importance.

Acknowledgement

The research leading to these results has received funding from the European Community's Seventh Framework Programme (FP7/2007-2013) under grant agreement no. 238833/ADAPTATION project (www.adaptation-itn.eu).

References

1. International Telecommunication Union. *Mobile-cellular telephony* [online]. Available from http://www.itu.int/ITU-D/ict/statistics/ [Accessed 28 Jan 2013]
2. World Health Organization. *Mobile phone use: A growing problem of driver distraction.* Geneva, Switzerland: WHO; 2011
3. Dragutinovic N., Twisk D. *Use of mobile phones while driving – Effects on road safety; A literature review.* R-2005-12, SWOV, Leidschendam, 2005
4. Narine S., Walter L.K., Charman S.C. *Mobile phone and seat belt usage rates in London 2009.* TRL Project Report 418, 2009

5. Sullman M., Baas P. 'Mobile phone use amongst New Zealand drivers'. *Transportation Research Part F: Traffic Psychology and Behaviour.* 2004;7(2):95–105
6. Brusque C., Alauzet A. 'Analysis of the individual factors affecting mobile phone use while driving in France: Socio-demographic characteristics, car and phone use in professional and private contexts'. *Accident Analysis and Prevention.* 2008;40:35–44
7. Basacik D., Reed N., Robbins R. *Smartphone use while driving – A simulator study.* TRL Published Project Report, PPR 592, 2011
8. United Stated Department of Transportation. *Visual-Manual NHTSA Driver Distraction Guidelines for In-Vehicle Electronic Devices.* Federal Register. 2012;77(37):11200–50
9. Bruyas M.P. 'Impact of mobile phone use on driving performance: Review of experimental literature' in Regan M.A., Lee J.D., Victor T.W. (eds.). *Driver Distraction and Inattention. Advances in Research and Countermeasures, Volume 1.* UK: Ashgate; 2013. pp. 293–309
10. Kircher K., Patten C., Ahlström C. *Mobile telephones and other communication devices and their impact on traffic safety.* VTI Report 729A, 2011
11. Brace C.L., Young K.L., Regan M.A. *Analysis of the literature: the use of mobile phones while driving.* Monash University Accident Research Centre, 2007 & Vagverket, Swedish Road Administration, 2007
12. Redelmeier D.A., Tibshirani R.J. 'Association between cellular telephone calls and motor vehicle collisions'. *The New England Journal of Medicine.* 1997;336(7):453–58
13. Martin J.L. 'A review of epidemiological data on the risks of using a telephone while driving' in Regan M.A., Lee J.D., Victor T.W. (eds.). *Driver Distraction and Inattention. Advances in Research and Countermeasures, Volume 1.* UK: Ashgate; 2013. pp. 171–193
14. Summala H. 'Hierarchical model of behavioral adaptation and traffic accidents' in Rothengatter T., Carbonell Vaya E. (eds.). *Traffic and Transport Psychology.* Amsterdam: Pergamon Press; 1997. pp. 41–52
15. Pöysti L., Rajalin S., Summala H. 'Factors influencing the use of cellular (mobile) phone during driving and hazards while using it'. *Accident Analysis and Prevention.* 2005;37:47–51
16. Michon J.A. 'A critical view of driver behavior models: what do we know, what should we do' in Evans L., Schwing R.C. (eds.). *Human Behaviour and Traffic Safety.* New York: Plenum Press; 1985. pp. 485–524
17. Lerner N., Boyd S. *On-road study of willingness to engage in distracting tasks.* DOT HS-809-863, National Highway Traffic Safety Administration, Washington DC, 2005
18. Thulin H., Gustafsson S. *Mobile phone use while driving: Conclusions from four investigations.* VTI Report 490A, Swedish National Road and Transport Research Institute, 2004

19. Esbjörnsson M., Juhlin O. 'Combining mobile phone conversations and driving – Studying a mundane activity in its naturalistic setting'. Presented at ITS World Congress; Madrid, Spain, 2003
20. Brouwer W.H., Waterink W., Van Wolffelaar P.C., Rothengatter T. 'Divided attention in experienced young and older drivers: lane tracking and visual analysis in a dynamic driving simulator'. *Human Factors.* 1991;33:573–82
21. McKnight J.A., McKnight S.A. 'The effect of cellular phone use upon driver attention'. *Accident Analysis & Prevention.* 1993;25:259–65
22. Lamble D., Rajalin S., Summala H. 'Mobile phone use while driving: Public opinions on restrictions'. *Transportation.* 2002;29(3):223–36
23. Brusque C., Alauzet A. 'L'utilisation du téléphone mobile au volant en France: entre déni du risque et autorégulation du comportement'. *Recherche Transports Sécurité.* 2006;91: 75–97
24. Kian M.P., Mahabadi H.A., Yazdi S.M., Hajizade E., Nahvi A. 'Impact of headway distance and car speed on drivers' decision to answer an incoming call'. *Traffic Injury Prevention.* In press, DOI 10.1080/15389588.2012.749464, 2012
25. Zhou R., Rau P.P., Zhang W., Zhuang D. 'Mobile phone use while driving: Predicting drivers' answering intentions and compensatory decisions'. *Safety Science.* 2012;50:138–49
26. Hancox G., Richardson J., Morris A. 'Factors affecting willingness to engage in mobile phone functions while driving'. *Proceedings of the European Conference on Human Centred Design for ITS*; Valencia, Spain, June, 2012; Lyon: HUMANIST 2012. pp. 39–46
27. Schömig N., Metz B., Krüger H.-P. 'Anticipatory and control processes in the interaction with secondary tasks while driving'. *Transportation Research Part F: Traffic Psychology and Behaviour.* 2011;14(6):525–38
28. Huth V., Lancelle V., Gabel C., Bonnard A., Brusque C. 'Exploring mobile phone usage and its context with naturalistic driving observations' in de Waard D., Brookhuis K., Dehais F., Weikert C., Röttger S., Manzey D. et al (eds.). *Proceedings HFES Europe Chapter Conference Toulouse* [online]. 2012. Available from http://hfes-europe.org [Accessed 20 July 2013]
29. Salvucci D.D., Macuga K.L. 'Predicting the effects of cellular-phone dialling on driver performance'. *Cognitive Systems Research.* 2002;3:95–102
30. Rauch N., Gradenegger B., Krüger H.-P. 'User strategies for the interaction with in-vehicle devices while driving'. *IET Intelligent Transport Systems.* 2008;2(4):266–75
31. Bruyas M.P., Brusque C., Debailleux S., Duraz M., Aillerie I. 'Does making a conversation asynchronous reduce the negative impact of phone call on driving?'. *Transportation Research Part F: Traffic Psychology and Behaviour.* 2009;12(1):12–20
32. Lerner N. 'Deciding to be distracted'. *Proceedings of the Third International Driving Symposium on Human Factors in Driver Assessment, Training and Vehicle Design*; Rockport, Maine, June 2005. pp. 499–506

33. White K., Hyde M., Walsh S., Watson B. 'Mobile phone while driving: An investigation of the benefits influencing drivers' hands-free and hand-held mobile phone use'. *Transportation Research Part F: Traffic Psychology and Behaviour*. 2010;13(1):9–20
34. Horrey W.J., Lesch M.F. 'Driver-initiated distractions: Examining strategic adaptation vor in-vehicle task initiation'. *Accident Analysis and Prevention*. 2009;4:115–22
35. Brusque C. 'Drivers' perceptions of risk linked to mobile phone use while driving and implications for the design of driver awareness campaigns' in Regan M.A., Lee J.D., Victor T.W. (eds.). *Driver Distraction and Inattention. Advances in Research and Countermeasures, Volume 1*. UK: Ashgate; 2013. pp. 195–208
36. Atchley P., Hadlock C., Lane S. 'Stuck in the 70s: The role of social norms in distracted driving'. *Accident Analysis & Prevention*. 2012;48:279–84
37. Fuller R. 'Towards a general theory of driver behaviour'. *Accident Analysis and Prevention*. 2005;37:461–72
38. Wogalter M.S., Mayhorn C.B. 'Perceptions of driver distraction by cellular phone users and nonusers'. *Human Factors*. 2005;47(2):455–67
39. Hallett C., Lambert A., Regan M.A. 'Cell phone conversing while driving in New Zealand: Prevalence, risk perception and legislation'. *Accident Analysis and Prevention*. 2011;43(3):862–69
40. Lesch M.F., Hancock P.A. 'Driving performance during concurrent cell-phone use: are drivers aware of their performance decrements?'. *Accident Analysis and Prevention*. 2004;36:471–80
41. Atchley P., Atwood S., Boulton A. 'The choice to text and drive in younger drivers: Behaviour may shape attitude'. *Accident Analysis and Prevention*. 2011;43(1):134–42
42. Wilde G.J.S. 'The theory of risk-homeostasis: Implications for safety and health'. *Risk Analysis*.1982;2:209–55
43. Haigney D.E., Taylor R.G., Westerman S.J. 'Concurrent mobile (cellular) phone use and driving performance: Task demand characteristics and compensatory processes'. *Transportation Research Part F: Traffic Psychology and Behaviour*. 2000;3(3):113–21
44. Matthews R., Legg S., Charlton S. 'The effect of cell phone type on drivers' subjective workload during concurrent driving and conversing'. *Accident Analysis and Prevention*. 2003;35(4):451–57
45. McCartt A.T., Hellinga L.A., Braitman K.A. 'Cell phones and driving: review of research'. *Traffic Injury Prevention*. 2006;7:89–106
46. Lerner N., Singer J., Huey R. *Driver strategies for engaging in distracting tasks using in-vehicle technologies*. NHTSA Report No. DOT HS 810 919, Washington DC, 2008
47. Huemer A.K., Vollrath M. 'Driver secondary tasks in Germany: using interviews to estimate prevalence'. *Accident Analysis and Prevention*. 2011;43:1703–12
48. Horrey W.J., Lesch M.F. 'Factors related to drivers' self-reported willingness to engage in distracting in-vehicle activities'. *Proceedings Human Factors*

and Ergonomics Society Annual Meeting. New York, 2008. CA:Human Factors and Ergonomics Society; 2008

49. INTERACTION project. *Methodology, results and recommendations*. Deliverable 7; ed. N. Reed, 2012
50. Jamson A.H., Westerman, S.J., Hockey G.R.J., Carsten O.M.J. 'Speech-based e-mail and driver behaviour: Effect of an in-vehicle message system interface'. *Human Factors: The Journal of the Human Factors and Ergonomics Society*. 2004;46:625–39
51. Green P. 'Driver distraction, telematics design, and workload managers: Safety issues and solutions'. *Proceedings of the 2004 International Congress on Transportation Electronics*. Detroit, USA. 2004; pp. 165–80
52. Tattegrain-Veste H., Bruyas M.P., Bellet T., Pachiaudi G., Forzy J.F., Martini D. et al. 'Evaluation of a voice interface management system'. *Transportation Research Record*. 2001;1759:46–51
53. Mühlbacher D., Totzke I., Schömig N., Rothe S. 'Workload management while driving: Do subjects' judgements reflect driving safety?'. *Proceedings of the European Conference on Human Centred Design for ITS*. Berlin, Germany. 2010. Lyon: HUMANIST; 2010. pp. 289–99
54. Lancelle V., Hugot M., Brusque C., Bonnard A. *Description and results of the French naturalistic driving observation*. Report of the FP7 project INTERACTION, European Commission, Belgium, 2012
55. Stanton N.A., Young, M., McCaulder B. 'Drive-By-Wire: The case of driver workload and reclaiming control with Adaptive Cruise Control'. *Safety Science*. 1997;27:149–59
56. Bianchi Piccinini G., Simoes A., Rodrigues C.M. 'Effects on driving task and road safety impact induced by the usage of Adaptive Cruise Control (ACC): A focus groups study'. *The International Journal of Human Factors and Ergonomics*. 2012;1(3):234–53
57. McEvoy S.P., Stevenson M.R., McCartt A.T., Woodward M., Haworth C., Palamara P., Cercarelli R. 'Role of mobile phones in motor vehicle crashes resulting in hospital attendance: A case-crossover study'. *British Medical Journal*. 2005;331:428–30
58. Rajalin S., Summala H., Pöysti L., Anteroinen P., Porter B. 'In-car cell phone use and hazards following hands free legislation'. *Traffic Injury Prevention*. 2005;6(3):225–29
59. Regan M. 'Driver distraction: Reflections on the past, present and future' in Faulks I.J. et al. (eds.). *Distracted driving*. Sydney: Australasian College of Road Safety; 2007. pp. 29–73
60. Johnson M.B., Voas R.B., Lacey J.H., McKnight AS. 'Living dangerously: Driver distraction at high speed'. *Traffic Injury Prevention*. 2004;5:1–7
61. Caird J.K., Willness C.R., Stelle P., Scialfa C. 'A meta-analysis of the effects of cell phones on driver performance'. *Accident Analysis and Prevention*. 2008;40:1282–93
62. Ishigami Y., Klein R.M. 'Is a hands-free phone safer than a handheld phone?'. *Journal of Safety Research*. 2009;40:157–64

63. Transport Canada. *The impact of cognitive distraction on driver visual behaviour and vehicle control.* Transport Canada Publication TP# 13889E, Ottawa, 2002
64. Brookhuis K.A., de Vries G., de Waard D. 'The effects of mobile telephoning on driving performance'. *Accident Analysis and Prevention.* 1991;23(4): 309–16
65. Törnros J.E.B., Bolling A.K. 'Mobile phone use-effects of handheld and handsfree phones on driving performance'. *Accident Analysis and Prevention.* 2005;37:902–9
66. Patten C.J.D., Kircher A., Öostlund J., Nilsson L. 'Using mobile telephones; cognitive workload and attention resource allocation'. *Accident Analysis and Prevention.* 2004;36:341–50
67. Strayer D.L., Johnston W.A. 'Driven to distraction: dual-task studies of simulated driving and conversing on a cellular telephone'. *Psychological Science.* 2001;12(6):95–102
68. Burns P.C., Burton S., Smith R.K., Burch D. *How dangerous is driving with a mobile phone? Benchmarking the impairment to alcohol.* TRL Report TRL547, TRL, Berkshire, UK, 2002
69. Consiglio W., Driscoll P., Witte M., Berg W. 'Effect of cellular telephone conversations and other potential interference on reaction time in a braking response'. *Accident Analysis and Prevention.* 2003;35:495–500
70. Hendrick J.L., Switzer J.R. 'Hands-free versus hand-held cell phone conversation on a braking response by young drivers'. *Perceptual & Motor Skills.* 2007;105:514–22
71. Breen J. *Car telephone use and road safety. Final report. An overview prepared for the European Commission.* Skipton, UK, Jeanne Breen Consulting. [online]. 2009. Available from http://ec.europa.eu/transport/road_safety/pdf/car_telephone_use_and_road_safety.pdf [Accessed 23 July 2013]
72. Horberry T., Anderson J., Regan M.A., Triggs T.J., Brown J. 'Driver distraction: The effects of concurrent in-vehicle tasks, road environment complexity and age on driving performance'. *Accident Analysis and Prevention.* 2006;38:185–91
73. Kantowitz B., Hanowski R. 'Simulator evaluation of heavy-vehicle workload: II. Complex secondary tasks'. *Proceedings of the Human Factors and Ergonomics Society 40th Annual Meeting*; Philadelphia, PA, 1996. pp. 877– 81
74. Reed M.P., Green P.A. 'Comparison of driving performance on-road and in a low-cost simulator using a concurrent telephone dialing task'. *Ergonomics.* 1999;42(8):1015–37
75. Waugh J.D., Glumm M.M., Kilduff P.W., Tauson R.A., Smyth C.C., Pillalamarri R.S. 'Cognitive workload while driving and talking on a cellular phone or to a passenger'. *Proceedings from the IEA 2000/HFES 2000 Congress*; San Diego CA, July–August 2000. New Jersey: Prentice Hall; 2000
76. Cooper J., Yager C., Chrysler S.T. *An investigation of the effects of reading and writing text-based messages while driving.* Report SWUTC/11/47666000024-1, Southwest Region University Transportation Center, 2011

77. Alm H., Nilsson L. 'The effects of a mobile telephone task on driver behaviour in a car following situation'. *Accident Analysis and Prevention*, 1995; 27(5):707–15
78. Briem V., Hedman L. 'Behavioural effects of mobile telephone use during simulated driving'. *Ergonomics*. 1995;38(12):2536–62
79. Strayer D.L., Drews F.A., Johnston W.A. 'Cell phone-induced failures of visual attention during simulated driving'. *Journal of Experimental Psychology: Applied*. 2003;9(1):23–32
80. Horrey W.J., Wickens C.D. 'Examining the impact of cell phone conversations on driving using meta-analytic techniques'. *Human Factors*. 2006; 48(1):196–205
81. Sayer J., Devonshire J., Flannagan C. *Effects of secondary tasks on naturalistic driving performance.* UMTRI-2005-29, University of Michigan, 2005
82. Harbluk J.L., Noy Y.I., Trbovich P.L., Eizenman M. 'An on-road assessment of cognitive distraction: Impacts on drivers' visual behavior and braking performance'. *Accident Analysis and Prevention*. 2007;39(2):372–79
83. Endsley M.R. 'Design and evaluation for situational awareness enhancement'. *Proceedings of the Human Factors Society 32nd Annual Meeting*. HFES, Santa Monica, 1988. CA:SAGE;1988. pp. 97–101
84. Hosking S. G., Young K. L., Regan M.A. 'The effects of text messaging on young drivers'. *Human Factors.* 2009;51(4):582–92
85. Drews F.A., Yazdani H., Godfrey C.N., Cooper J.M., Strayer D.L. 'Text messaging during simulated driving'. *Human Factors*. 2009;51(5):762–70
86. Recarte M.A., Nunes L.M. 'Effects of verbal and spatial-imagery tasks on eye fixation while driving'. *Journal of Experimental Psychology: Applied.* 2000; 6(1):31–43
87. McCarley J.S., Vais M.J., Pringle H., Kramer A.F., Irwin D.E., Strayer D.L. 'Conversation disrupts change detection in complex traffic scenes'. *Human Factors*. 2004;46:424–36
88. Gugerty L., Rakauskas M., Brooks J. 'Effects of remote and in-person verbal interactions on verbalization rates and attention to dynamic spatial scenes'. *Accident Analysis Prevention.* 2004;36(6):1029–43
89. Ranney T.A., Mazzae E., Garrott R., Goodman M. *NHTSA driver distraction research: past, present and future.* Driver Distraction Internet Forum (http://www_nrd.nhtsa.dot.gov/driver_distractio), Washington, DC: National Highway Traffic Safety Administration, 2000
90. Ishida T., Matsuura T. 'The effect of cellular phone use on driving performance'. *IATSS Research.* 2001;25:6–14
91. Lee J., Caven B., Haake S., Brown T. 'Speech-based interaction with in-vehicle computers: The effect of speech-based email on drivers' attention to the roadway'. *Human Factors*. 2001;43(4):631–40
92. Rakauskas M.E., Gugerty L.J., Ward N.J. 'Effects of naturalistic cell phone conversations on driving performance'. *Journal of Safety Research*. 2004; 35:453–64

93. Young K., Regan M. 'Driver distraction: A review of the literature' in Faulks I.J et al. (eds.). *Distracted Driving.* Sydney: Australasian College of Road Safety; 2007. pp. 379–405
94. Jamson A.H., Merat N. 'Surrogate in-vehicle information systems and driver behavior: Effects of visual and cognitive load in simulated rural driving'. *Transportation Research Part F.* 2005;8:79–96
95. Beede K.E., Kass S.J. 'Engrossed in conversation: The impact of cell phones on simulated driving performance'. *Accident Analysis & Prevention.* 2006;38:415–21
96. Hancock P.A., Lesch M., Simmons L. 'The distraction effects of phone use during a crucial driving maneuver'. *Accident Analysis & Prevention.* 2003;35:501–14
97. Cooper J.M., Vladisavljevic I., Medeiros-Ward N., Martin P.T., Strayer D.L. 'An investigation of driver distraction near the tipping point of traffic flow stability'. *Human Factors.* 2009;51:261–68

Chapter 10

Evaluating the potential for workload based driving assistance systems from a psychological, technological and physiological perspective

Wilfried Hajek

Abstract

Advanced Driving Assistance Systems (ADAS) support the driver during highly automated and continuous driving tasks (e.g. adaptive cruise control) and provide warnings in potentially dangerous situations (e.g. forward collision warning). As ADAS do not work flawlessly, and because of legal issues, the driver needs to supervise these systems in order to intervene when necessary. This means that the driver has to continually monitor information given by the systems, even though the human brain is not optimised for prolonged and monotonous control tasks, which can be interrupted by highly critical situations at any point in time. As has been shown by several validations of the Yerkes–Dodson law, the driver is vulnerable to 'out-of-the-loop' problems in low workload conditions and may not realise quickly enough that a situation is critical. Conversely, in high workload conditions the driver loses sight of the control task or needs more time to initiate an appropriate reaction in a dangerous situation.

This chapter describes how a new theory has been developed, and a compensation strategy for the use of adaptive cruise control in high workload conditions has been designed and evaluated in a simulator and in on-road conditions. The technological possibility of detecting high workload conditions using physiological data has been established; the workload when the drivers' reaction time is influenced by a secondary task has been evaluated. In addition, as the usage of a system is strongly dependent on behaviour and impressions, the chapter describes how acceptance, as well as awareness of the system, has been examined.

10.1 Introduction

In a world full of technical development, especially in the automotive, space and air flight sectors, humans are becoming reliant on the support of technological systems.

BMW Group Research and Technology, Munich, Germany

These developments open a new world of possibilities but on the other hand raise new issues that have to be addressed. Technology is developing 'intelligent' systems designed to replicate human ways of thinking and behaviour. Especially for unpredictable situations, it may not be possible to program rules within a system and this is where the human brain is usually far superior. In contrast with such situations of human superiority, people show poor performance when undertaking continuous and monotonous tasks, which can lead to the so-called 'out-of-the-loop' problem [1, 2].

In the automotive sector the circumstances above described have led to the development of systems that support the driver in monotonous tasks but that require the driver to retake control in critical situations, such as when an accident might occur. One of the main challenges lies in defining the points in time when a driver should be taking over and in the definition of the right security and time threshold that give the driver enough time to regain control and undertake appropriate actions. Currently these thresholds are calculated mainly with respect to environmental conditions (such as the distance to the leading car, speed of the vehicle, negative acceleration of the leading car, distance to the lane markings etc.). Furthermore, some systems like the Adaptive (or Active) Cruise Control (ACC) provide drivers with the possibility of manually changing certain safety-related parameters.

Nevertheless, there are stressful situations, e.g. when the driver is getting disturbed by other passengers or when thinking about complex emotional or logical problems, which could lead to high workload and where the driver would need a higher safety margin to react in a critical situation than would normally be the case. In such situations it would not normally be possible for the driver to change the safety margins him/herself and such an action would increase the already high workload. So, an action by the car itself to increase safety margins (through the parameters) or warn the driver of a risk, based on the actual driver workload, could lead to the prevention of accidents in critical situations. With such a system the driver would be given a longer time gap to retake control and more time for initiating appropriate action.

This chapter summarises knowledge gained during three years of research into driver state detection in the automotive sector with physiology, the development of driver state algorithms and the design of compensatory measures for high workload situations. The approach draws on the MIT AgeLab 'AwareCar' [3], which is described further below. However, the approach is adapted with a new 'flower model' that is used as a theoretical foundation for a compensation strategy focusing on the overload situation. The flower model is inspired by nature and raises safety margins if the external conditions (driver workload) are getting more adverse.

A major issue, to date, has been lack of research to develop unobtrusive physiological measurement of driver workload. Therefore one of the aims of the work reported in this chapter has been to implement a physiological measurement device in a driving simulator as well as in a real car. Further work reported here is the validation of other studies concerning offline detection of workload and development and implementation of a real time detection algorithm in a car. This chapter describes how the flower model approach was validated for an ACC.

10.2 Theoretical foundation

This section describes how the 'flower model', developed by the author and used as theoretical foundation for the compensation strategy, builds upon previous work.

10.2.1 The Yerkes–Dodson law

The connection between arousal and performance has been researched for decades [4, 5] and even before the Yerkes–Dodson law was introduced in the early years of the 20th century [6] it became one of the most famous laws in psychology. Yerkes–Dodson postulates an inverted U-shaped relationship between arousal and performance: when arousal is too high, humans are in an overloaded state that decreases their performance; whereas, if arousal is too low performance again decreases. Therefore, with a medium level of arousal optimal performance should be possible.

10.2.2 The adapted Yerkes–Dodson law

Coughlin, Reimer & Mehler from the MIT Age Lab postulated a so-called AwareCar [3], which continually detects the driver's state and, according to fixed implemented guidelines, would display his/her state and, by constantly refreshing (i.e. alerting or calming) the driver, would keep him/her in the optimal performance range and therefore ensure the best reaction time in a critical situation. The authors adapted the original Yerkes–Dodson Law for the driving task but instead of arousal uses workload/stress [7]. Workload/stress that is too high leads to too high activation which is labelled as overload and leads to performance degradation. Workload/stress that is too low leads to fatigue which is classified as under load in the original Yerkes–Dodson work.

They also postulated compensational measures: if the driver is in an overload state, he has to be calmed down to lower the activity level; if the driver is fatigued because of too low activity, he/she should be activated to return to the optimal range of performance.

A problem with this concept is that useful compensatory measures are strongly dependent on the reason for the workload; it is also questionable whether such compensatory measures exist and if they can be implemented without leading to more problems. For example, if as compensatory measures for fatigue a window is opened to provide fresh air, this could be disturbing on a rainy day. Similarly, if information displayed on a screen is reduced to avoid distraction, such changes could lead to more workload and confusion if the driver does not understand the reason for changes to the usual setting. To address some of these problems, a new model has been developed by the author; this does not aim to bring the driver back to their optimal state, but instead adapts the car to the actual workload level. This model is described in the next section.

10.2.3 The flower model

As already described, the Yerkes–Dodson law postulates a decrease in performance under high workload conditions and the MIT AgeLab provides a theoretical model

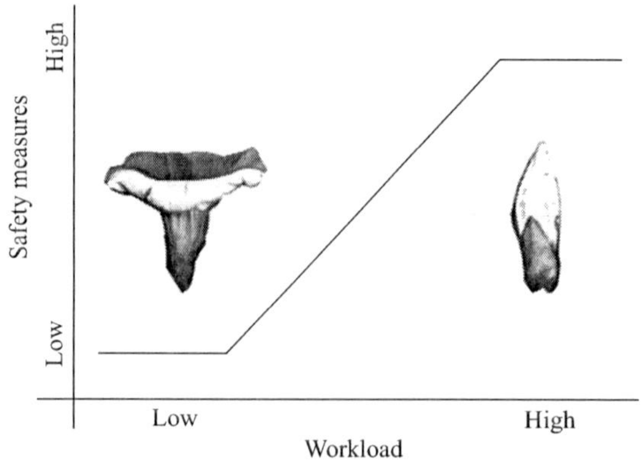

Figure 10.1 Flower model for safety support from ADAS

for measures to bring the driver back to optimal performance. However, in the author's opinion, this approach requires that the level of (cognitive) workload be known before an optimal compensation strategy usable for all high workload periods can be designed. This is not currently possible and so alternate approaches need to be explored. In particular, if a critical situation arises in which the driver has to take control, there must be enough time for an appropriate reaction. The idea for the flower model that addresses this situation has been inspired by nature: the hedge bindweed (*Calystegia sepium*) opens its petals when the sun is shining but closes them under unfavourable circumstances, such as when it is raining [8]. According to this behaviour, the relevant parameters of an ADAS providing safety margins and assistance should be low under low workload and high under high workload (Figure 10.1).

As a first validation of the flower model it has been evaluated with ACC under high and low workload condition.

10.3 Measuring workload

Whilst a precise definition of workload is elusive, in a driving context it is experienced in relation to the demands of the driving task and other tasks in which drivers engage. Previous research has demonstrated that cognitive workload can have a measurable effect on physiological arousal [9–12]. However, at the beginning of this study, an online workload algorithm was not available and so research began on validating the effects of workload on physiological arousal. In later stages physiological data was collected and used as a foundation for new algorithm development.

10.3.1 Workload induction by secondary task: The n-back task

The n-back task is an orally presented secondary task using a sequence of stimuli. The MIT AgeLab developed a refined version of the original n-back task of Kirchner [13] with usage as a secondary task for driving situations in mind. There are three difficulty levels of workload (cf. [14]): Zero back, one back and two back. At the zero-back task level (low workload) participants have to repeat the number they actually heard; at the one-back task level (medium workload) participants have to keep the number heard in their short-term memory for later presentation and at the same time repeat the number they heard before the actually presented number; in the two-back task level (high workload) participants have to retain two numbers in their short-term memory; therefore after the presentation of the first two numbers, participants have to verbally repeat the number they heard two numbers before and remember the actually presented numbers for later presentation.

Zeitlin [15] postulated that an ideal secondary task should interfere as little as possible with the primary task, be easy to use and be easy to learn. The n-back task meets all these requirements and was used in a verbal presentation and auditory response manner as the cognitive workload induction method in the experiments presented below.

10.3.2 Measuring arousal over physiology

Measuring physiological data has a long history in research, with some studies showing promising results for establishing a high quality connection between data and workload [16–23]. Nevertheless, no fully validated workload detection algorithm has been discovered during literature reviews and several reasons for this can be postulated: First, a lot of the research has involved only a small number of participants; second, most physiological data are very sensitive to movements, time and environmental issues; and, third, the secondary tasks used may not have been as sensitive and stable as the adapted n-back task.

Ongoing research by the MIT AgeLab concerning the n-back task and further developments in the simulation sector now provide the possibility of a more robust approach. The studies discussed here involved a substantial number of participants (60–100), medically certified physiological measurement devices, low movement artefacts (through clear instructions concerning no major movements apart from driving), low environmental distraction (from very detailed simulator environment programming) and detailed protocols with sub-second analysis of data (because of scripted time synchronising of simulator and physiological data streams). The following physiological data were used for recording arousal and algorithm development.

10.3.2.1 Heart rate

Heart rate (HR) is the frequency of heart contractions. Each contraction forces blood through the circulatory system of the human body and is produced by electrical impulses that can be detected by an electrocardiogram (ECG) measuring electrical impulses from the heart. HR is measured in beats in a given time interval (usually minutes) and modulated by sympathetic and parasympathetic activity of the autonomous nervous system, which is dependent on physical and/or mental effort [9, 18]. Higher effort leads to an increase of HR compared to the resting situation [9, 12].

10.3.2.2 Heart rate variability

Heartbeats have variable time durations with changing oscillation patterns, leading to patterns and frequency content that is called heart rate variability (HRV). HRV, like heart rate, can be retrieved from an ECG. With increasing mental effort, HRV decreases compared with the resting situation [9, 12].

10.3.2.3 Electro-dermal activity

Because of autonomous nervous system activity, there are electric changes in the skin that are called Electro-dermal Activity (EDA). Two different types of EDA are generally distinguished: one before exposure to a stimulus and another after that exposure. Therefore, Tonic EDA (also electro-dermal level or skin conduction level) means the average or baseline EDA level in resting situations, whereas Phasic EDA includes the Electro-dermal Response (EDR) and refers to the EDA level after exposing the participant to a stimulus. EDR is the response to a stimulus that occurs with a long latency time of 1.3–2.5 s and is usually measured at the hand or on the foot [9].

10.3.2.4 Respiration

Respiration is needed to supply the body with oxygen and to expel carbon dioxide. In general there are two different assessments of respiration. The first measures the depth and frequency of breathing whereas the second concerns parameters connected with gas exchange during the breathing process [24]. Respiration rate, which refers to the frequency of breathing in a certain interval (usually a minute), is the most common measure [9].

10.4 Experiments and results

During the planning stages of the experiments reported here considerable time was devoted to ensuring high quality, repeatable and valid data based on experience from the problems reported in prior experiments (10.3.2). In the following sections a short overview of the main experiments and outcomes is provided:

- Experiment 1 – Relationship between Brake Reaction time and Workload level
  For the development of a Workload-Adapting ACC (WACC), the foundations, especially the physiological detection of workload levels with simple statistical offline methods, have to be developed. A simulator experiment was conducted with a cue task and three levels of workload. Physiological data was recorded. Furthermore, the relationship between workload levels and brake reaction time was studied.
- Experiment 2 – Wizard of Oz simulation of the WACC in a simulator
  Based on the first experiment, a WACC seems feasible if it adjusts headway from 1 s to 2 s in high workload conditions; that is, where participants are showing a brake reaction time increase. Therefore, a simulator experiment was scripted in such a way that the ACC automatically changes the time-headway parameter under high workload condition. This would be the same behaviour as a WACC would show. The system was used in both critical and non-critical driving situations. Acceptance, system awareness, physiological and simulator data were recorded.

- Experiment 3 – Wizard of Oz simulation of the WACC on the road
  According to the supportive results of the first two simulator experiments, a Wizard of Oz simulation of the WACC system was implemented in a BMW 5er series. Acceptance and system awareness under realistic conditions were researched and physiology data was recorded for the development of an online detection algorithm.
- Experiment 4 – On road study with WACC
  As a final step, a working WACC was implemented in a car that had a real time online detection algorithm of the actual driver workload. System mode was developed from binary (baseline and high workload level) to gradual (baseline–low–medium–high workload level) and acceptance for both system modes was established. Furthermore, workload detection rates were examined in greater detail and confounding variables identified as explanations for incorrect workload classification.
- Experiment 5 – Emotion and physiology
  As an outlook for future work, the possibility of detecting emotional states through physiological data was investigated in a small concluding study. Following literature research it was decided to use short films to induce emotional reactions in participants whilst recording physiological data.

10.4.1 Experiment 1 – Relationship between brake reaction time and workload level

Before developing an online workload detection algorithm, it has to be proven that offline detection of different workload levels is possible. This experiment aimed to explore how physiological changes indicate different levels of cognitive workload and to determine whether multiple levels of workload could be distinguished beyond a baseline/high workload separation. In addition, this experiment aimed to indicate whether there is an influence (and how strong this is) of different cognitive workload levels on brake reaction time. As a workload-adapting ACC could change the following distance in high workload situations to provide the participant with more reaction time, this is a critical issue for establishing the usefulness of a WACC.

10.4.1.1 Experimental setting

The study design was modified from a previous workload study [25]. A straight track with white cars was implemented in a simulator setting. Only white cars were used to eliminate a potential confounding factor of contrast between cars and brake lights. The driver received the instruction to stay in the right lane, to turn on ACC and to follow a leading car always with the same speed of 70 km/h. To ensure easier maintenance of this speed, the gas pedal was programmed such that between certain pedal angles, the speed always stayed at 70 km/h. The participants had to drive and complete the three n-back task levels and a cue task (only repeating one number presented in the same way as the n-back task) several times (they had previously received training until they had achieved over 80% of correct answers on the n-back task). According to a strict timed protocol, a 'foreign' car ahead was switched from the middle lane to the right lane and adjusted until it reached a headway of 1.5 s to the participants' car. It then performed one of three actions:

1. Decelerating slowly to 55 km/h and then speeding up again
2. Accelerating, or
3. Decelerating till it reached a speed of 25 km/h

Only the last action required an intervention of the participant to prevent an accident. The other two actions ensured that the requirement for braking could not be predicted. All n-back task levels were completed by the participant during critical brake events as well as without these events. Workload levels were counterbalanced over the actions of the foreign cars, as well as in their order over participants to eliminate effects of same sequence. The responses from the participants were recorded for further validity checks and the numbers of correct answers were logged by the experimenter.

Physiological data were measured continuously with the simulator data and analysed as within participant measurements.

10.4.1.2 Results

During the study, 88 participants were tested. Fifteen had to be excluded: 10 because of technical problems and 5 because they were doing the wrong n-back task for the required sequence. A within ANOVA on physiological data for the 73 participants showed highly significant increases in beat to beat heart rate, highly significant decreases in heart rate variability and significant increases in skin conduction over all workload levels compared with the reference period. Respiration was excluded from further analysis because the verbal answering behaviour for the n-back task would be too high a confounding factor for establishing the effect of workload in this measurement.

Concerning brake reaction time, two different groups within the 73 participants were studied. One group ($n = 36$) was without warning and one group ($n = 37$) received an auditory warning as it would have been given by a forward collision warning system. Concerning the condition without warning sound, an ANOVA for the 36 participants showed no significant effect over all workload levels. A detailed in-depth contrast analysis showed a significant effect of brake reaction time increase between cue-task and the two-back task level.

10.4.1.3 Conclusion

This offline analysis showed that physiological data, particularly heart rate and heart rate variability have great potential for implementation in an online algorithm for detecting different workloads. Furthermore, detailed analysis showed potential not only for the detection of a binary but also for a gradual workload detection algorithm, if required.

Other analysis shows that brake reaction data only shows differences between high and low workload periods so it seems that from a driver's performance point of view, a binary WACC seems more feasible.

*10.4.2 Experiment 2 – Wizard of Oz simulation of a WACC in the simulator**

According to the previous experiment, brake reaction time increases in high workload situations and because the WACC increases the distance to the leading car, this should compensate. However, with risk homeostasis theory in mind this assumption has to be validated in a critical situation under high workload conditions and compared with normal ACC. The main research issue in this experiment was an evaluation of the acceptance of a WACC compared with ACC. In particular, the influence of prior system information on acceptance was studied.

10.4.2.1 Experimental setting

A car following procedure was designed in a static simulator. Participants received an overview of the experiment and instructions appropriate for driving with a normal ACC but they did not get any prior information concerning the WACC. At the beginning and before every experimental part, participants got the instruction that the system has its limitations. That is, it could not detect standing objects. In the introductory part of the experiment they also experienced such a situation at low speed (60 km/h) with a truck blocking the right lane. Furthermore participants got the instructions that they were not allowed to change lane during the whole experiment, and therefore always had to stay in the right lane.

A simulator protocol was scripted such that the WACC system would adapt the following distance according to the task workload. Participants also received instructions and a training session on the n-back tasks. In the experiment participants only had to do the two-back task, representing the high workload situation.

Participants experienced four non-critical situations separately with ACC and WACC, where the system was able to handle the slow braking manoeuvre of the leading car. Participants were instructed accordingly that there is no reason for intervention in these cases. In part 2 a traffic jam in the middle lane was blocking this lane and the leading car changed from the right lane to the middle lane to give way to a car blocking the right lane. Participants then had to brake to prevent an accident. The critical situation always appeared in the high workload condition. Therefore participants with ACC had a smaller distance (1 s headway) to the standing object than participants with the WACC (2 s headway).

The order of experiencing the WACC and ACC systems was counterbalanced to eliminate effects of sequence in within participant analysis. Afterwards, a between participants setting ensured that half of the participants drove with ACC and half of the participants had WACC in a critical situation. Therefore, participants experienced the critical situation just once.

After each experimental session with one of each system, acceptance was rated. At the end of both experimental sessions, system awareness was requested. Afterwards participants got information concerning the system modes and its functionality and a second acceptance rating was performed.

*For detailed information consult the relevant paper [26]

10.4.2.2 Results

No significant difference was found in acceptance between the ACC and WACC systems. This is (most probably) explained by the result that only 7 out of 47 participants noticed any change in following distance (and no prior information was provided on this). After participants received an explanation of system differences the WACC showed highly significant increases in the subjective variables (such as reasonable, helpful, comfortable). No significant difference between ACC and WACC system were found in subjective stress level, which was as expected since the two-back task was the main stress factor and had to be completed in both conditions. Another question was asked concerning the anticipated purchase price and (although these measures do not really correspond with real spending behaviour) it was highly significant that more money would be paid for the WACC system.

Simulator data was logged and analysed concerning safety aspects of the WACC. Brake reaction time was measured from the point of the lane change of the leading car (and thereby providing an unobstructed view of the blocking vehicle) to the point of first brake pedal pressure. No significant results were found concerning brake reaction time although this might be expected from risk homeostasis theory. This means that participants react in the WACC and ACC conditions at the same point in time. However highly significant smoother deceleration was found in the WACC condition compared with the ACC condition.

10.4.2.3 Discussion

In this experiment the main focus was on establishing acceptance and safety of the WACC system and, as a first step, these issues were tested in a simulator setting. Although the results cannot be directly transferred to reality, they give a first hint of what a WACC could provide.

The subjective acceptance, in general, and for most of the subscales, suggest a clear preference for the WACC when people are provided with prior information. Apart from that, the finding suggests that most drivers would not detect any changes in the mode of an ACC system in high workload conditions. The reason could be that the high workload masks awareness of changes such as the distance to a leading car, which is not a critical issue, as long as there is no need for a sudden braking manoeuvre.

Concerning safety, there was no increase in brake reaction time in a critical situation with WACC compared with ACC. However, highly significant smoother deceleration was found in the WACC condition compared with the ACC condition, which implies a possible lower risk for both rear end and in-front collisions in real life scenarios involving such critical braking manoeuvres.

As already noted, these results have to be validated in real driving situations, especially the questions concerning system mode awareness. As this experiment was performed in a static simulator, only the sound provided information concerning increase and decrease of speed, and distance perception also cannot be directly compared with reality.

10.4.3 Experiment 3 – Wizard of Oz simulation of the WACC on the road†

The promising results of the simulator experiments suggest that participants accept the WACC at least as well as ACC and that they do not detect a difference in performance between the two systems. This experiment aimed to study these issues in on-road conditions by simulating performance of a WACC (with an operator-selected headway). Concerning the workload detection algorithm, prior simulator experiments have shown that detection of different workload levels is possible. As heart rate showed stable results in both of the previous two experiments, detection algorithms based on heart rate were used here.

10.4.3.1 Experimental setting

The setting was an equipped BMW 5 series car on a 19 km track of the A9 to the north-east side highway out of Munich. After a short introductory session with an explanation of the n-back task, an introduction to the car and completing demographic, control variables (trust in automation, sensation seeking) and a general ACC acceptance questionnaire, the participants were connected with electrodes for measuring physiological data. This was followed by a short training period in the city with the car and the ACC. The experimental setting started after one minute driving time on the highway.

Participants were instructed to follow another vehicle with a maximum speed of 120 km/h and to stay in the right lane of the highway. Furthermore, they were asked to execute the two-back task whenever it occurred and if they felt safe doing so. After 18.8 km participants were instructed to leave the highway for the purpose of completing a questionnaire.

Participants drove with the system twice: the first part of the experiment was without prior information concerning the WACC and the second, return part, included information. Appreciation of system mode and acceptance were evaluated using a short interview for later categorisation. For establishing acceptance a project-specific 10-point acceptance scale was used as well as the Van der Laan acceptance scale [28] for a more detailed insight.

After driving back on the same route and with the same procedure but with information concerning the system mode, the acceptance questionnaires were repeated. A short interview concerning perceived advantages and disadvantages of the system was also undertaken.

To help promote better comparison of physiological data, such that only the two-back task was a significant factor in workload, the participants were asked not to undertake additional cognitive thinking themselves.

10.4.3.2 Results

Thirty-eight participants were tested from which eight had to be excluded: one because of technical reason, two because they did fulfil requirement concerning knowledge of ACC and five who were not able to complete more than 80% of the two back tasks during the whole experiment.

†For detailed information consult the diploma thesis of Hanna Bellem [27]

Concerning system mode awareness, 17 (56.67%) of the participants did not detect any change in distance selected by the system, whereas 6 (20%) did realise how the system was working and another 7 (23.33%) detected something concerning the WACC, but which did not correspond with the actual behaviour of the system.

Concerning acceptance in general, the scores for the ACC as well as the WACC were in the positive area above 7 out of 10. There was also a significant increase in acceptance scores for participants who did not detect the change in headway (and therefore the way the system was working) after they drove the WACC system, but there were no significant results with and without information in acceptance for this group. For the group who did detect the different systems' behaviour there were no significant results. No significant results were found using the Van der Laan score and for the demographic variables.

It was found possible to classify over 60% of the low workload periods and over 80% of the high workload periods correctly using (commercially confidential) detection algorithms based on heart rate.

10.4.3.3 Discussion

The experiment showed that in on-road conditions over 50% drivers did not realise the change of headway implemented by the WACC. This was thought to be promoted by how slowly the WACC system changes its distance to a leading car. Interestingly, the significant increases in acceptance for the participants who did not detect change in headway suggest that undetectable automation is preferred by participants.

Concerning the detection algorithm that classifies high and low workload, the results show promise. One of the remaining important issues for the good, but not perfect, detection rates was the question of whether there was other induced workload detected by the algorithm but not logged by the experimental protocol.

To get deeper insights into both the issues of undetectable automation and other induced workload a further experiment was conducted. Regarding a non-detectable WACC system, the headway algorithm was modified to continually adapt to workload. Regarding the issue of other workload sources, it was planned that all events that could have an effect on workload apart from the two-back task would be logged.

10.4.4 Experiment 4 – On road study with WACC[‡]

Results from the simulator studies showed good acceptance of the WACC system under 'perfect' conditions (when the system was directly coupled with the beginning of the two-back task). However, a WACC based on heart rate would not only react to the two-back task but also to all workload. This experiment first aimed to study acceptance of such a system. Further, and as an expansion of the first research question and based on the experimental results of system mode awareness, a gradual WACC was implemented, which would lead to a continual increase and decrease in the distance because of stepwise changes in workload (low, medium, high workload). In this experiment, the acceptance of this gradual mode compared with the binary mode was investigated.

Also, as already mentioned in discussion of experiment 3, an increase in workload (identified as an increase in heart rate) could result not only from the two-back

[‡]For detailed information consult the diploma thesis of Andreas Trzuskowsky [29]

task but also from other changes in the environment or behaviour of the participants. Although this aspect cannot be resolved completely, some elaboration was attempted within this experiment.

10.4.4.1 Experimental setting

The same route was taken as in experiment 3, but with the length expanded to 32.3 km. Participants were informed about the behaviour of the system and received an introduction to the car and to the different types of n-back tasks. Participants were connected to the physiological measurement device and the data were streamed to a notebook with the working workload classification algorithm. The algorithm communicated the desired headway to the car's ACC system, which implemented a change in speed until the appropriate headway was achieved.

In the city segment, before reaching the highway, the participants had time to get a feeling for the ACC system. Altogether there were three workload periods (instead of two as in experiment 2). In one direction of the route participants experienced the binary WACC system (changing distance between high and low workload level) and in the other direction, they experienced the gradual system (changing continually between each workload level). A demographic questionnaire was completed before the experiment and questionnaires were administered at the end of each direction to establish acceptance of the system and preferences for system mode and system parameters.

The experimenter, on the back seat, also logged certain pre-defined events with keystrokes (one at their beginning and one at their end – to be called 'extratrigger') for later correlation with the occurrence of potential additional workload supplementing the levels induced by the n-back task.

10.4.4.2 Results

Ten participants took part in this experiment. With this small number of participants results have to be interpreted very carefully only as indicative of the direction in which experiments with a larger sample could go.

In general, participants realised that both the binary and gradual WACC systems changed the headway in accordance with the workload situation, as they had been told. Furthermore the results of the questionnaires suggested that participants felt that the system was mainly working correctly and also that it was too sensitive at the moment. Acceptance was fair at around 7 out of 10 points, which was in line with the results of the previous experiments.

Concerning system mode preference, the findings showed no significant results and therefore no explicit preference for one system. However, an in-depth analysis of further variables showed significant decreases for transparency, credibility, reliability, trustiness and comfortable distance to the leading car for the gradual distance mode and therefore a clear preference for the binary system mode.

Concerning the detection algorithm, the results of experiment 3 could be validated for the online setting and the overall detection rate was increased to 72%. Furthermore, if the extratrigger events are included as possible occurrence of additional workload (and the previous workload classification is thereby corrected) the detection rates increase to 86%.

10.4.4.3 Discussion

The overall finding suggests that participants accept the WACC system fairly well and that it is also perceived as working correctly in most cases, even if the algorithm is not working absolutely perfectly. Nevertheless, one has to keep in mind that an algorithm with our definition of correctness (percentage of the complete workload induced time) will never reach 100% because of time lags and because of other influences on heart rate that occur for a variety of reasons. Furthermore, this research suggests that binary adaptation of headway from low to high workload conditions is better accepted than a continuous change of headway.

In general this experiment shows that the heart rate WACC can be implemented in a car and therefore is technically possible. Furthermore, it is well accepted by the participant even if the system is not working correctly in every case. Also, participants prefer a system that is adapting only for high workload conditions rather than continually over time.

There remain issues that have not been resolved through this research. A major question is how the different workload conditions detected by the algorithm also correlate with a decrease in reaction time. This work only investigated the n-back task that is one kind of workload induction method. In addition there are illnesses such as heart problems and diabetes that have an influence on physiological measurements. Also, there are other influences that could lead to increases in heart rate apart from workload; one of these sources could be emotions. It has not been evaluated how these factors influence the workload detection algorithm. Finally, there is the question of long-term adaptation to the system, which has yet to be addressed.

Looking forward and to gather deeper insight on emotional influences on heart rate, a short final experiment is reported.

10.4.5 Experiment 5 – Emotion and physiology

The possibility exists that physiological parameters are not only influenced by workload but also by emotional stimuli. Hence a future research issue is whether workload can be distinguished from different emotions and whether it is possible to execute actions according to the classified workload or emotion, or a mixture of both. As this would be the start of a significant research endeavour, a more limited initial investigation was initiated to see if basic emotions can be distinguished, based on physiological data.

10.4.5.1 Experimental setting

This initial work used short (2-minute) films designed to elicit emotions matching one of the four quadrants according to the circumplex model of affect theory [30–32]. This model suggests that emotions are distributed in a two-dimensional space, containing arousal and valence (positive or negative) dimensions where Arousal represents the vertical axis and Valence represents the horizontal axis.

After a short introduction where participants received the instruction to not look away from the screen for the 2-minute periods, they filled out a demographic questionnaire and received information concerning the procedure of the experiment. Afterwards, participants were seated and connected with physiological sensors. Film sequences followed a fixed procedure for ensuring that one emotion with strong arousal or valence would not influence one with low arousal or valence. At the end, positive emotions were induced, engaging the participant in a good mood before leaving the experiment.

Participants were seated in front of a large television screen connected to a notebook with a scripted procedure running. Light was dimmed to allow focus on the film experience and the experimenter was seated behind a wall so as to not disturb the participants.

Physiological data from respiration, ECG and skin conductance sensors were recorded for analyses after the experiment and subjective measurement derived by the self-assessment-manikin test without dominance scale [33] ensured identification of the induced emotions as categorised by the circumplex model.

10.4.5.2 Results

A review of numerous studies concerning emotional effects on physiological data [34] suggests increased heart rate for negative emotions such as anger, anxiety, contamination related disgust, fear and crying sadness as well as for positive emotions like happiness and joy. Decreases in heart rate seem to be an indicator for negative emotions with passivity aspects and mutilation related disgust, imminent fear, non-crying sadness and suspense. Increased respiratory rates are correlated with emotional activation derived from feelings such as anger, anxiety, disgust, fear, sadness, happiness and amusement. Increases as well as decreases in respiration rates can be induced by sadness and contentment. Decreased EDA only is part of non-crying sadness, acute sadness, contentment and relief whereas other emotions lead to increases in EDA.

The problem in deriving specific emotions from physiological data is very complex and requires a combination of changes in different measurements. At the time of writing no immediate conclusions could be reached, suggesting a need for further experimentation and analysis.

10.5 Conclusion

The research presented in this chapter has considered the implementation of adaptive driving assistance system using physiological data. This has been approached from a theoretical perspective as well as through an in-depth analysis of a prototype workload-adapting ACC (WACC). The results presented suggest that the flower model described in the chapter is a useful supplement to the adapted Yerkes–Dodson law and has application in design of the WACC system.

The experiments on brake reaction time in the simulator appear to suggest that participants are not necessarily adapting their behaviour when driving with a WACC, thus experiencing an additional margin of safety. However it is appreciated that simulator results need validation in real driving situations. Furthermore, it was possible to develop a reasonably effective algorithm that was able to establish an appropriate level of workload in the specific scenarios studied. It was also possible to establish that participants, in general, preferred a binary headway-adjusting WACC to a continuously variable system.

It became obvious during the real-driving work that changes in physiological data are not only dependent on the imposed workload task but also can be influenced by many other events on the street. A question then arises as to whether all workload can be said to increase risk.

Concerning sensors and data quality, this research was undertaken with medically certified equipment. Clearly, technological work is required to develop, for example, contactless measurement devices that are able to collect data without interfering with the normal procedure of driving and getting in and out of the car.

In general, this research can be seen as a first step in the direction of physiological based driving assistance systems. It has shown promising results but it has also identified a range of new issues that have to be studied and evaluated in future research before such systems can be introduced in the driving context.

Acknowledgement

This research received funding from the European Community's Seventh Framework Program (FP7/2007-2013) under grant agreement no. 238833/ADAPTATION project (www.adaptation-itn.eu).

References

1. Kaber D. B., Endsley M. R. 'Out-of-the-loop performance problems and the use of intermediate levels of automation for improved control system functioning and safety'. *Process Safety Progress*. 1997;16(3):126–131
2. Endsley M. R., Kiris E. O. 'The out-of-the-loop performance problem and level of control in automation'. *Human Factors*. 1995;37(2):381–394
3. Coughlin J. F., Reimer B., Mehler B. *Driver Wellness, Safety & the Development of an AwareCar* [online]. 2009. Available from http://agelab.mit.edu/system/files/file/Driver_Wellness.pdf [Accessed 1 March 2013]
4. Duffy E. 'The psychological significance of the concept of "arousal" or "activation"'. *Psychological Review*. 1957;64(5):265–275
5. Freeman G. L. 'The relationship between performance level and bodily activity level'. *Journal of Experimental Psychology*. 1940;26(6):602–608
6. Yerkes R. M., Dodson J. D. 'The relation of strength of stimulus to rapidity of habit-formation'. *Journal of Comparative Neurology and Psychology*. 1908;18:459–482
7. Coughlin J. F., Reimer B., Mehler B. 'Monitoring, managing and motivating driver safety and well-being'. *Pervasive Computing*, IEEE. 2011;10(3):14–21
8. Mao Y. Y., Huang S. Q. 'Pollen resistance to water in 80 angiosperm species: flower structures protect rain-susceptible pollen'. *New Phytologist*. 2009;183(3):892–899

9. De Waard D. *The Measurement of Drivers' Mental Workload.* PhD thesis, University of Groningen. Haren, The Netherlands: University of Groningen, Traffic Research Centre; 1996
10. Veltman J. A., Gaillard A. W. K. 'Physiological workload reactions to increasing levels of task difficulty'. *Ergonomics*. 1998;41(5):656–669
11. Mayser C., Piechulla W., Weiss K. E., König W. 'Driver workload monitoring'. In Strasser H., Kluth K., Rausch H., Bubb H. (eds.). *Quality of Work and Products in Enterprises of the Future. Proceedings of the 50th-Anniversary Conference of the GfA and the XVII Annual ISOES Conference*, Munich, Germany, May 7th–9th. Stuttgart: Ergonomia; 2003. pp. 41–44
12. Brookhuis K. A., De Waard D. 'Measuring physiology in simulators'. In Fisher D. L., Rizzo M., Caird J. K., Lee J. D. (eds.). *Handbook of Driving Simulation for Engineering, Medicine and Psychology*. Boca Raton, FL: CRC Press; 2011. pp. 17-1–17-10
13. Kirchner W. K. 'Age differences in short-term retention of rapidly changing information'. *Journal of Experimental Psychology*. 1958;55(4):352–358
14. Mehler B., Reimer B., Dusek J. A. *MIT AgeLab delayed digit recall task (nback)* [online]. 2011. Available from http://agelab.mit.edu/system/files/Mehler_et_al_n-back-white-paper_2011_B.pdf [Accessed 1 March 2011]
15. Zeitlin L. R. 'Subsidiary task measures of driver mental workload: A long-term field study'. *Transportation Research Record*. 1993;1403:23–27
16. Wang F., Sagawa K., Inooka H. 'Time domain heart rate variability index for assessment of dynamic stress'. *Computers in Cardiology*. 1998;25:97–100
17. Hof T., Driel C. V., Arem B. V., Hoedemaeker M., Brookhuis K. *Mental workload while driving with a congestion assistant* [online]. n.d. Available from http://altmetis.eldoc.ub.rug.nl/FILES/root/2008/mental_hof/IEA_Congestion_Assistant.pdf [Accessed 1 March 2011]
18. Mulder L. J. M., De Waard D., Brookhuis K. A. 'Estimating mental effort using heart rate and heart rate variability'. In Stanton N., Hedge A., Brookhuis K. A., Salas E., Hendrick H. (eds.). *Handbook of Human Factors and Ergonomics Methods*. Boca Raton, FL: CRC Press; 2005. pp. 20-1–20-8
19. Katsis C. D., Ganiatsas G., Fotiadis D. I. 'An integrated telemedicine platform for the assessment of affective physiological states'. *Diagnostic pathology*. 2006;1:1–16
20. Liu B., Lee Y. 'In-vehicle workload assessment: effects of traffic situations and cellular telephone use'. *Journal of Safety Research*. 2006;37:99–105
21. Mehler B., Reimer B., Coughlin J. F., Dusek J. A. 'The impact of incremental increases in cognitive workload on physiological arousal and performance in young adult drivers'. *Transportation Research Record: Journal of the Transportation Research Board*. 2009;2138:6–12
22. Mehler B., Reimer B., Coughlin J. F. 'Physiological reactivity to graded levels of cognitive workload across three age groups: an on-road evaluation'. *Proceedings of the 54th Annual Meeting of the Human Factors and Ergonomics Society*, San Francisco, CA, September–October; 2010. pp. 2062–2066

23. Mehler B., Reimer B., Coughlin J. F. 'Sensitivity of physiological measures for detecting systematic variations in cognitive demand from a working memory task: an on-road study across three age groups'. *Human Factors.* 2012;54(3):396–412
24. Brookhuis K. A., De Waard D. 'Monitoring drivers' mental workload in driving simulators using physiological measures'. *Accident Analysis & Prevention.* 2010;42(3):898–903
25. Engström J., Aust M. L., Viström M. 'Effects of working memory load and repeated scenario exposure in emergency braking performance'. *Human Factors.* 2010;52(5):551–559
26. Hajek W., Gaponova I., Fleischer K. H., Krems J. 'Workload-adaptive cruise control – A new generation of advanced driver assistance systems'. *Transportation Research Part F: Traffic Psychology and Behaviour.* 2013;20:108–120
27. Hanna B. An on-road study of system awareness, acceptance, and trust of a simulated Workload Adaptive Cruise Control, Technische Universität Chemnitz
28. Van der Laan J. D., Heino A., De Waard D. 'A simple procedure for the assessment of acceptance of advanced transport telematics'. *Transportation Research Part C: Emerging Technologies.* 1997;5(1):1–10
29. Trzuskowsky A. Adaption eines Fahrerassistenzsystems basierend auf einer Online Workload-Detektion, RWTH Aachen University, Diploma Thesis. Germany, December 2012
30. Posner J., Russell J. A., Peterson B. S. 'The circumplex model of affect: an integrative approach to affective neuroscience, cognitive development, and psychopathology'. *Development and Psychopathology.* 2005;17(3):715–734
31. Russell J. A. 'A circumplex model of affect'. *Journal of Personality and Social Psychology.* 1980;39(6):1161–1178
32. Russell J. A., Weiss A., Mendelsohn G. A. 'Affect grid: a single-item scale of pleasure and arousal'. *Journal of Personality and Social Psychology.* 1989;57(3):493–502
33. Lang P. J., Ghman A., Vaitl D. 'The international affective picture system' [photographic slides]. Gainesville, FL: The Center for Research in Psychophysiology, University of Florida; 1988
34. Kreibig S. D. 'Autonomic nervous system activity in emotion: a review'. *Biological Psychology.* 2010;84(3):394–421

Chapter 11

Distraction and inattention prevention by combining Behaviour-Based Safety with Advanced Driver Assistance Systems

Claudia A. Wege[1] and Trent W. Victor[1]

Abstract

Development of Behaviour-Based Safety (BBS) and Advanced Driver Assistance Systems (ADAS) has been carried out largely independently of each other. Although both approaches have the same goal of improving traffic safety, they operate at different timescales to achieve accident prevention. This chapter examines how both ADAS and BBS approaches can be combined into a more holistic framework and applied to preventing distraction and inattention.

The combination of ADAS (immediate feedback) with BBS (long-term feedback) is illustrated by using the analogy of 'team play'. Both ADAS and BBS are players of the same team on the 'accident prevention playing field' and united, they become a better team. With the BBS-ADAS team, traffic safety has the opportunity to advance into an entirely new league.

11.1 Time for kickoff

In this chapter, two accident prevention approaches are summarised: the Advanced Driver Assistance Systems (ADAS) approach and the Behaviour-Based Safety (BSS) approach. The scope of the chapter is (a) to summarise the literature on ADAS and BBS, (b) to identify ADAS and BBS principles, (c) to examine their impact on traffic safety, (d) to identify barriers to implement ADAS and BBS techniques and (e) to discuss how to apply these approaches when providing inattention prevention feedback.

Development of ADAS and BBS has been carried out largely independently of each other. To our knowledge, an effort to combine the benefits of both has not yet been developed; there is an apparent scarcity of research on how to prevent inattention whilst it is occurring, *and* in the long-term through aggregated behavioural feedback.

[1] Volvo Group Trucks Technology (GTT), Sweden

There is a need to further enhance the current ADAS and BBS approaches, both independently and in an integrated fashion. A novel integrated approach, based on modern technological innovations, combines ADAS and BBS, into *one* Behaviour-Based-Safety-Advanced-Driver-Assistance-System (BBS-ADAS) approach. As an exemplification of the combined BBS-ADAS feedback approach, this chapter focuses on attention feedback through the combination of immediate attention feedback with long-term attention feedback. BBS driver behaviour management is effective, but can be more effective when drivers are also immediately made aware of their behaviour and when the immediate ADAS feedback is delivered in a coordinated way along with the long-term BBS feedback. The independent and combined approaches are presented in Figure 11.1 together with their respective driver feedback properties.

Research on human–machine interaction and automated ADAS has suggested the theory of 'team play' [2]. In this chapter, the theory of 'team play' is adapted to the field of accident prevention and the novel BBS-ADAS approach. The analogy of team play is visualised in Figure 11.2. As shown in the figure, the playing field is accident prevention with a focus on attention enhancement. Team ADAS and BBS have the same goal but different playing strategies. Both teams are already successful independently of each other, but integrating the best players of each team into one united team could provide the opportunity to increase success, a key reason being that players from both teams complement each other's shortcomings.

11.2 The playing field: Accident prevention

The estimated number of road traffic fatalities worldwide is about 1.24 million [3] and 39,300 in the EU each year [4]. Road traffic injuries were the eighth leading cause of death worldwide in 2012 [5]. Trends by the World Health Organization suggest that by 2030 road traffic injuries will become the fifth leading cause of death worldwide, if nothing is changed to prevent traffic accidents. Most of the accidents, however, are preventable. Accident causation research has greatly progressed from the time when Bortkiewicz [6] stated that accidents occurred at random and were thus inexplicable. Scholars now believe that accidents are primarily (90%) due to human failure [7–9]. It is therefore only logical to put the 'human factor' at the forefront of accident research and to examine underlying risky behaviours that lead to accidents. Once the risky behaviours have been examined, it becomes necessary to advance initiatives about *how to change* them. This chapter focuses on one risky behaviour, that of visual attention away from the forward roadway; [10–12]. When drivers' visual attention is not allocated to the source of a conflict in the forward roadway, this can lead to incidents, near-crashes and crashes. Distracted driving is one of the major causes of accidents with 10 people being killed, 960 people being injured and 2,040 properties being damaged each day in the United States alone [13]. Damage of $43 billion per year has been estimated due to driver distraction related crashes in the United States [14].

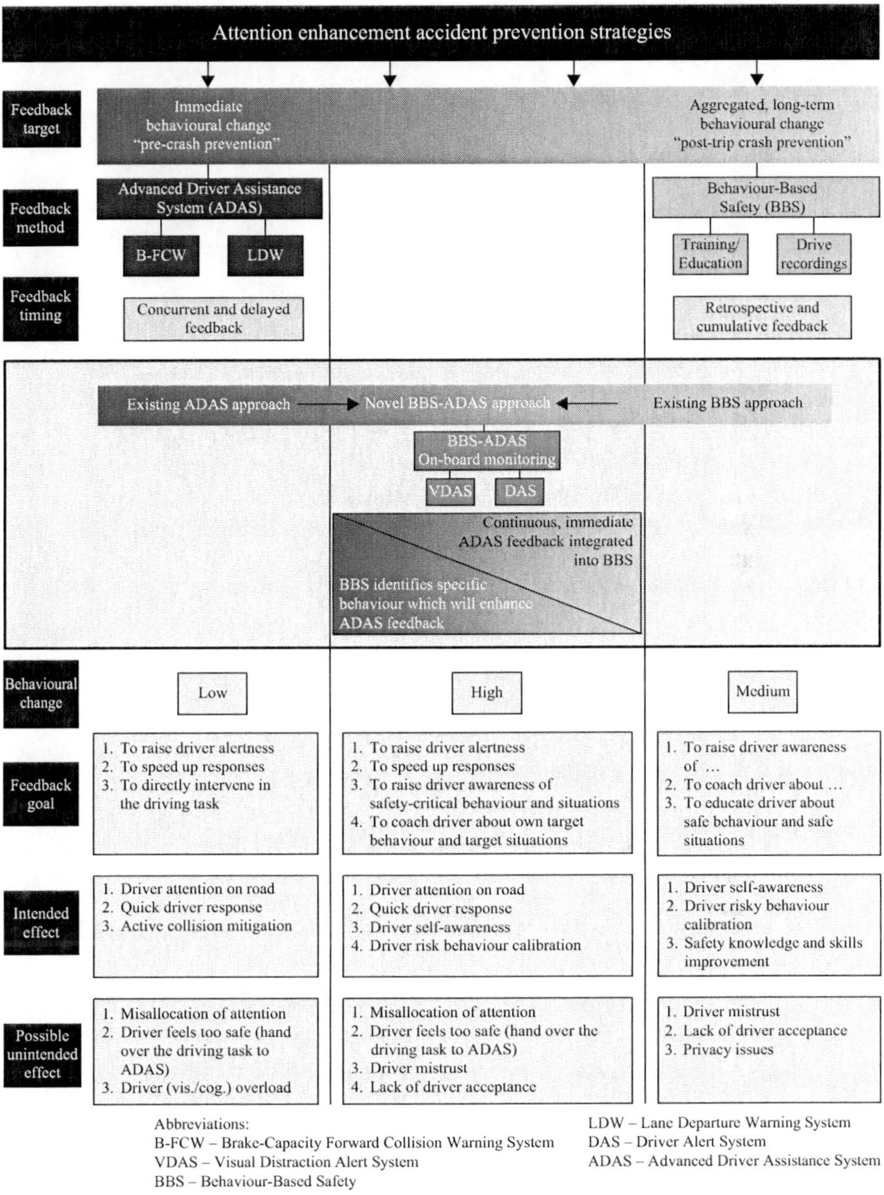

Figure 11.1 Attention enhancement accident prevention strategies: ADAS, BBS and BBS-ADAS

A naturalistic driving study found that 78% of crashes and 65% of near-crashes included distraction as a contributing factor [15]. Recently, algorithms which are claimed to be able to predict crash risk from risky glance patterns in naturalistic driving have been developed [16–18]. Research on distraction and inattention detection is very active [e.g. 19, 20]; however, in order to make studies comparable,

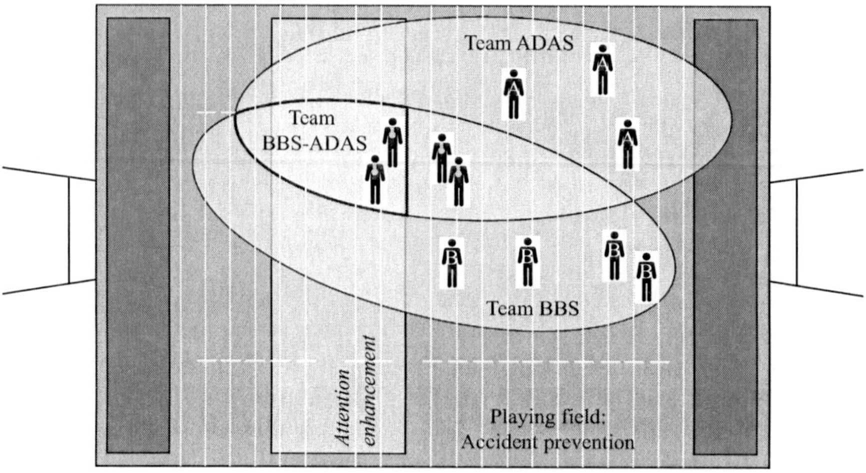

Figure 11.2 *Uniting a winning team: BBS and ADAS applied to attention enhancement on the accident prevention playing field*

a common terminology for driver distraction and inattention is important. A number of key experts have met since 2009 as the United States and European Union Bilateral Intelligent Transportation Systems Technical Force with the objective of establishing a conceptual framework and taxonomy for understanding and categorising driver inattention [21]. The key assumption is that there are two general forms of inattention: insufficient attention (e.g. sleep related attentional impairment and insufficient attentional effort) and misdirected attention. Misdirected attention can be categorised into more specific processes, such as *driver distraction* and *incomplete selection of safety-critical activities.*

Driver distraction can be defined to occur when the driver diverts attention away from activities critical for safe driving to one or more activities that are not critical for safe driving [22]. This can occur due to various contributing factors. Typically, these factors can be clustered into visual distraction (e.g. looking away from the roadway), auditory distraction (e.g. responding to a ringing cell phone), biomechanical distraction (e.g. manually adjusting the radio volume) and cognitive distraction (e.g. being lost in thought) [23] or some combination of these. Visual distraction occurs when, instead of focusing visual attention on the road, distracted drivers look at other targets for a certain period of time. The most common driver initiated contributing factor to visual distraction is interaction with mobile devices brought into the vehicle, which usually involves drivers taking their eyes off the road [8]. Several studies show that drivers are willing to interact with mobile devices while driving (e.g. using a cell phone) [24] and are often unaware of the extent to which distracted driving is dangerous [25]. A common glance pattern associated with distracted driving is, so called, 'visual time sharing' when visual attention is allocated back and forth between the forward roadway and a secondary task such as dialling a phone number. Elvik [26] identified statistical regularities of risk factors that are referred to as 'laws of accident causation' that are associated with road accident occurrence. One of the attempts to explain visual overload when driving and texting simultaneously is the 'law of complexity'. It states that the more units of information per unit of time to which a road user must attend, the higher the probability of an error. This 'law' implies the accident rate will increase as more elements of information a driver must process also increase (e.g. visual time-sharing when text messaging).

Incomplete selection of safety-critical activities refers to 'situations where the driver allocates sufficient resources to one or more activities critical for safe driving, or believed by the driver to be critical for safe driving, while the resources allocated to other activities critical for safe driving do not match the demands of these activities' [21, p. 35]. As an example, consider the situation where a driver is following another vehicle and approaches too close resulting in a forward collision warning (FCW) that alerts the driver to the impending collision. If the FCW is designed correctly it will direct and enhance attention to the appropriate information needed for forward collision avoidance; for example, it will cause the driver to direct attention towards a lead vehicle. However, if this warning is inappropriately designed, it may cause the driver to look towards the vehicle interior displays seeking explanatory information. This type of inattention may become more important in the future as safety systems are increasingly designed to alert drivers to upcoming potential dangers [e.g. 27, 28]. Thorough evaluations of ADAS Human–Machine Interface (HMI) design with suitable methods [e.g. 29], along with the formulation of ADAS design guidelines and principles based on human capabilities are needed.

11.3 Team player line-up

11.3.1 Team Advanced Driver Assistance Systems (ADAS)

As defined in Chapter 2, ADAS have emerged in modern vehicles to support the driver in the driving task. The key point for the present chapter is that ADAS focus on providing *immediate* feedback to influence *current* driving behaviours. Such ADAS are designed to change behaviour immediately and part of the behaviour change includes direction of attention towards relevant parts of roadway traffic. Thus, for the purpose of this chapter, ADAS are limited to pre-crash *driver warning systems* that alert the driver by providing immediate risk-relevant feedback targeting a near-crash or crash event. A near-crash is defined as an event whereby a vehicle comes 'dangerously close' to another vehicle, object, person or animal [31]. For example, a brake-capacity forward collision warning (B-FCW) system can provide a warning in order to requiring some kind of decision making and/or action to prevent a crash [29].

In general, according to drivers' needs and requirements, ADAS enhance safety, comfort, entertainment and awareness with the overall objective of avoiding drivers' errors and accidents. It is speculated that the implementation of ADAS may lead to a fatality decrease of 40% [32]. Today, a number of ADAS are on the market, varying in function, design and manufacturer [33, 34]. A large body of literature on ADAS functionality was categorised [35], including longitudinal support systems, lateral support systems, driver awareness support systems, visibility support systems and off-board information systems. Categories of ADAS systems can be further distinguished by the *type of intervention* they provide to the driver [36]:

(a) *Driver information systems* (which aim to increase the driver's situation awareness; examples being night vision systems, navigation systems and e-horizon systems)
(b) *Driver warning systems* (which warn in case of lapses, such as unintended lane changes, or provide information concerning blind spots and forward collisions)
(c) *Intervening systems* (which provide active support to the driver; for example, an adaptive cruise control system or a queue assistance system or emergency braking system)
(d) *Integrated active safety systems* (which encompass all systems including some of the above that work towards vehicle safety in a cooperative manner)

The timing of ADAS feedback is either 'concurrent feedback' within milliseconds (e.g. FCW system) or 'delayed feedback' within seconds (e.g. Driver Alert System) [37].

The feedback goals of ADAS are to:

1. raise driver alertness to a safety-critical situation
2. speed up responses in safety-critical situations
3. sometimes (not always) directly intervene in the driving task

The intended ADAS feedback effects are to:

1. achieve driver attention to a relevant part of the driving scene
2. facilitate quick driver responses
3. avoid or mitigate accidents

Primarily, ADAS do not intend to provoke long-term driver behavioural change but rather focus on changing current behaviours.

There are potential shortcomings in techniques used to develop and evaluate ADAS; they are usually evaluated over a short time period and/or in a simulated driving environment, not allowing drivers to use the systems as they would if they were permanently installed in their vehicles. The overall expectation is that drivers will learn to interact with ADAS and will adapt to new systems in the ways that were intended by the developers; however, ADAS can have effects that are not intended by the designers. Unintended ADAS effects can include the driver feeling too safe (overreliance) or too stressed about interacting with the ADAS (cognitive overload). Although engineering research for technical improvement of ADAS is ongoing, a recent study [29] has demonstrated the challenges in designing HMI for ADAS. The results of this study indicate that drivers might experience two problems: a 'need for comprehension' when a B-FCW warning is discontinued, and a 'confirmation from the system' that a collision warning is over. Overall, the safety impact of ADAS depends to a great extent on their interaction with the driver. For example, in order to provide efficient support for the driver to avoid crashing into an obstacle ahead, it is crucial that the system warning intuitively generates the appropriate response, e.g. eyes towards the obstacle and initiation of an avoidance manoeuvre [38]. For a natural, effective and intuitive user interaction with ADAS, human factor specialists need to study drivers' natural reactions to these systems operating in real world naturalistic conditions to ensure that they do not adversely affect attention [29]. Different ADAS HMI designs may cause different behavioural effects over time, including unexpected behavioural adaptations. An overview of the above mentioned properties of ADAS feedback is provided in Figure 11.1.

There are a variety of major telematics guidelines, practices and standards for ADAS design. UMTRI [39] provides a general overview of in-vehicle telematics principles literature and legislation. However, guidelines for visual presentations in ADAS HMI design are not well developed. Some user interface requirements exist for ADAS for voluntary use by manufacturers [30, 40–44]. Additionally, technical reports on ergonomic aspects of in-vehicle presentation for transport information and control systems [45] exist. However, specific ADAS HMI ISO standards are still lacking although work on guidelines are ongoing [e.g. 46].

In general, HMI principles build upon principles from psychology and cognitive science. Likewise, the principles of psychology (the cause and effect of human behaviour) and cognitive science (the working mechanisms of the human mind) are strongly connected to BBS principles (see the following section). However, human factors and BBS principles are not yet strongly connected to ADAS design. In this chapter, we argue that there is a need to integrate BBS principles into the ADAS design process. Natural, intuitive and long-term effective ADAS–user interaction is achievable if known human factors and BBS principles are applied. The development of the existing ADAS approach towards a more integrated BBS-ADAS approach is visualised in Figure 11.1 by the arrow pointing from the existing ADAS approach to the novel BBS-ADAS approach.

11.3.2 Team Behaviour-Based Safety (BBS)

In this section, an overview on the main principles of behaviour-based safety (BBS) programmes is given and barriers for their application in the automotive domain are identified. BBS program is used as a synonym for behavioural management technique. The general BBS principle, adapted from Skinner's [47] operant conditioning learning theory, is that behaviours followed by desirable consequences are more likely to be repeated in the future and those followed by undesirable consequences are less likely to be repeated in the future [1]. BBS strategies are based on the original four-step BBS process developed by Geller [1] called DO-IT. DO-IT is an acronym for **D**efine, **O**bserve, **I**ntervene, and **T**est trageted at-risk and/or safe behaviours. It was originally conceptualized to target immediate behaviour. Safety managers in industrial work settings can usually systematically observe safe versus at-risk behaviour of employees and provide feedback later. Geller [1] was among the first who introduced the BBS concept into industrial work safety and occupational behaviour research. Safety program techniques following the BBS principle have proven to be relatively easy to implement, cost-effective and highly efficient for reducing occupational injuries and fatalities in a variety of industrial domains. Examples include introduction in a pizza delivery company [48], a paper mill [49], the mining industry [50], a gas pipeline company [51], a food manufacturing plant [52], an oil and natural gas company and a glass manufacturing company [53] as well as in the rail industry such as Amtrak's ticket handling system [54]. Meta-analyses of health and safety BBS studies have found a significant reduction in injuries ranging from 96.6% [55], to 59.6% [56], to 69% [53]. Krause, Seymour and Sloat [57] conducted an analysis comparing the injury reduction due to BBS strategy implementation at 43 different sites over time. In Figure 11.3, results are shown with a continuously increased reduction of work related injuries over a five-year period.

Research on BBS programmes applied to traffic safety is limited. And, no BBS-study has specifically focused on attention enhancement yet. Only a few studies have examined the effects of BBS principles applied in the automotive domain. Thus, the literature review on BBS programmes best practices in this chapter is largely based on reviews by Hickman *et al.* [58], Hickman and Hanowski [31], and Victor *et al.* [59]. It is hypothesised that adapting BBS principles techniques and processes to the automotive industry, and ultimately to ADAS design, can yield lasting changes in improving road safety, in general, and driver attention support management, in particular. BBS programmes, which aim to coach or

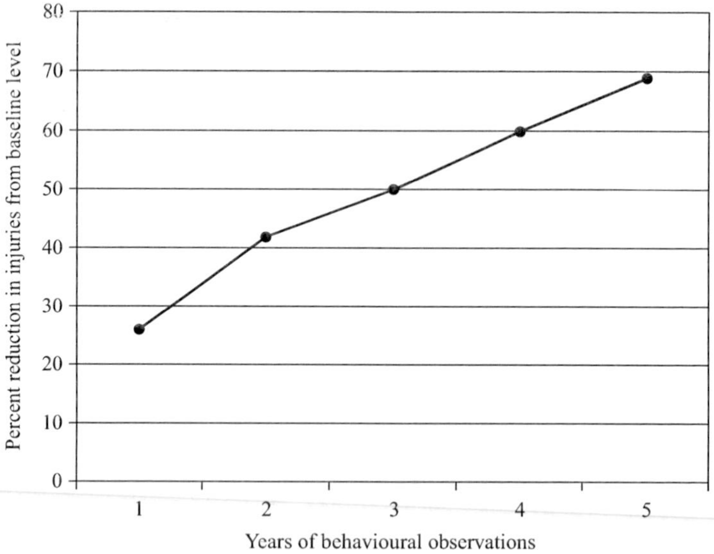

Figure 11.3 The average percent injury reduction across 73 sites using a BBS program (reported by Krause, Seymour and Sloat [57, table 3])

educate drivers, have the opportunity to operate as attention enhancing accident prevention strategies. However, until recently, the primary problem with implementation was the difficulty in obtaining good quality, continuous measures of behavioural data. Studies evaluating the benefit of BBS in the motor transport domain are rare because implementation of BBS programmes in fleets is a relatively new field [50, 60, 61]. The study reported by Hickman *et al.* [58] as well as a knowledge-review paper [62] provide a comprehensive overview of existing BBS approaches in commercial vehicle operations. Hickman and Hanowski [63] investigated the safety benefits of on-board safety monitoring devices over a 17-week period. Results showed a significant reduction of recorded safety-related events by 52.2%. In another report on the same dataset, Hickman and Hanowski [31] showed that the same device affects the prevalence of distracting activities while driving, for example through a significant effect for cell phone usage compared with baseline. Toledo and Lotan [64] investigated the effects of an on-board driver monitoring device over a five-month period. In this study driver feedback was presented on a personal webpage giving drivers access to information about previous trips. It was found that drivers only reduced their at-risk behaviour for the first month after performance feedback was provided. After the first month, at-risk behaviour remained stable. Initially, feedback improved safety, but this effect diminished over time as drivers accessed their webpages less frequently. This pattern of behaviour was not further explored by the authors. An interpretation of the results is that drivers may have developed an awareness of safety-critical events which caused the initial decrease in at-risk behaviours.

In order to closer compare BBS with the previously characterised ADAS approach, the same properties that were used to describe ADAS are used here to describe BBS (see Figure 11.1).

In general, the BBS approach is defined as 'post-trip accident prevention'. The primary aim of BBS is to coach a driver by providing risk-relevant feedback targeting a safety-critical behaviour event, near-crash event or crash event.

The feedback goals of BBS are to:

1. raise driver awareness of safety-critical behaviour and situations
2. coach drivers about their own target behaviour and target situations
3. educate drivers about safe behaviour and safe situations

The intended BBS feedback effects are:

1. driver self-awareness [25]
2. driver calibration [65]
3. improvement in driver safety knowledge and skills [1]

The BBS approach intends to provoke a long-term driver behavioural change. The unintended effects of BBS feedback can include driver mistrust, scepticism, privacy issues and disciplining of drivers without a driver policy in place. If the BBS program is improperly managed, it may give rise to negative attitudes within the workforce. The feedback about risky behaviour is usually delivered to the driver after a trip; the timing of the feedback of BBS is either 'retrospective feedback' within minutes or hours after the trip or 'cumulative feedback' within days, weeks or months [37] after a behavioural event, a near-crash event or a crash event occurred. The general target users of BBS are safety managers of commercial fleets (truck, bus, taxi, and distribution or delivery fleets). The proposed move of the traditional BBS perspective towards a more BBS-ADAS approach is visualised by the arrow pointing from an existing BBS approach to the novel BBS-ADAS approach in Figure 11.1.

Hickman and Hanowski [63] have identified four main BBS techniques that can be applied to the automotive domain:

1. training and education
2. behaviour-based incentives and goal-setting
3. behavioural observation
4. feedback and coaching

These BBS techniques are interpreted for application to attention enhancement techniques and are explained in more detail below.

11.3.2.1 Training and education

Driver attention training includes driver education for novice drivers [e.g. 66]. Safety-related training is generally applied using driving simulators [67, 68] or computer-based training to improve attention selection in driving [e.g. 69]. Distraction, as an issue, has been largely neglected in the design of driver education and training programmes. This fact is surprising because, as stated in [19, p. 560], research shows that 'the driving public has little understanding of what activities are distracting, of the relative risk associated with different sources of distraction, of the impact of distraction and the need to self-regulate in response to distraction'. In general, attention training and education is a necessary, but not sufficient, safety management technique. Drivers need the knowledge on how to drive safely and to identify unsafe behaviours; however, the existence of knowledge does not necessarily result in safe driving. If a driver has no knowledge on how to drive safely, he/she has a knowledge gap. However, if he/she has the knowledge but still engages in unsafe behaviour, then it becomes a motivation gap (i.e. 'I know how to drive safe, but I choose not to do it.'); behavioural feedback and coaching address this latter issue.

11.3.2.2 Behaviour-based incentives and goal-setting

The importance of goal setting and incentive strategies is discussed in literature from many different research domains. Barton, Tardif, Wilde, and Bergeron [70] provide a general framework for the implementation of incentive schemes in the automotive domain. Bandura [71] examined why feedback motivates behavioural change. According to Bandura, individuals who are dissatisfied with their performance will be motivated to increase their effort in the future if given proper feedback. In order for feedback to have an effect on individual performance, the individual must have a performance goal. A performance goal provides the individual with a standard and expected level of performance. Furthermore, goals allow individuals to compare their performance with their standard performance level.

11.3.2.3 Behavioural observation

Behavioural observation can be carried out in different ways. It can range from observation by a safety manager, usually referred to as 'human watchdog observation', to peer observation and covert observation. When observations are performed, a behavioural checklist can help to systematically record and track behaviours. Without behavioural observation, no performance feedback can be provided. Behavioural observation in an in-vehicle setting can be carried out using vehicle sensors and cameras. One example of this observation type is on-board distraction monitoring devices (see [72] for a review). Technological constraints have prevented observation of high quality behavioural data derived from driving behaviours thus far. Most on-board safety monitoring devices are kinematic-trigger-based (e.g. acceleration triggers, excessive speed triggers, unplanned lane departures, frequent hard braking, close following distances, failure to yield at intersections) and are not based on driver visual behaviour (e.g. off-road glances, or eye-closures). Furthermore, current technology can generally not distinguish between safe and at-risk behaviour during a safety-related event as it occurs. Hickman and Hanowski [31] analysed data from DriveCam, a vendor of on-board safety monitoring systems for professional fleets aimed at reducing risky driving behaviours using in-vehicle video technology. With the DriveCam system, it is possible to 'flag' those triggered driving events, to save 12 s of video (e.g. 8 s prior to the event trigger and 4 s after) and to automatically send them to trained data analysts. Analysis of distraction-related behaviours is thus performed by human review and classification of behaviours from the video material and associated telematics data. This procedure also allows drivers and fleet safety managers to review data and video at a later date; thus, the feedback is delayed. Driver safety performance scores and feedback are maintained in a database so that a manager can rank drivers to determine who most needs coaching. A key difference between the DriveCam system and other types of driver recorders is that a long-term coaching service is provided, instead of only providing in-vehicle feedback or capture of video recordings.

11.3.2.4 Feedback/coaching on behavioural events

The role of feedback in changing behaviour has been studied in various domains. In general, BBS approaches are based on human learning theories and their underlying cognitive, motivational and energetic processes. One particular theory is that of operant conditioning by Skinner [47]. Skinner influenced the 'behavioural approach', which means that once behaviour can be operationally defined, and reliably tracked, it can be influenced. In Chapter 2, several learning theories are reviewed that explain factors underlying behavioural change. With drive-recording systems that also provide a coaching service (e.g. DriveCam or SmartDrive), events can be reviewed and feedback on safe and at-risk driving behaviours provided.

In a coaching session, video events examples of risky behaviours (e.g. near-crashes or close following behaviours) and feedback on aggregated driving performance data (e.g. last three months) can be provided. The primary benefit of coaching is that users are collaboratively engaged in the behaviour improvement process.

11.4 Advancing into a new league: Uniting ADAS and BBS team players

In the following section, an integrated approach of ADAS and BBS is described. Both of the approaches have shortcomings that may lead to an overestimation of the safety effects that individual ADAS or BBS approaches have. Although both approaches have been proven to be effective, they lack important aspects for more stable effects and maximum traffic safety potential.

The main shortcoming of the ADAS approach as an attention enhancement measure is that, although the behavioural feedback warnings provide immediate accident prevention, they do not target long-term effects on behavioural change. For example, drivers might forget about the warning that previously identified a behavioural event. Studies have found that an estimated 80% of near-accidents are forgotten after two weeks [73]. This suggests that a driver's memory of his/her glance behaviour and driving performance needs to be reviewed retrospectively. Current ADAS have the potential to signal drivers to direct their attention towards a hazard once a safety-critical event occurs. However, traffic changes can occur very rapidly in the driving environment and, although ADAS are technologically capable of tracking these changes, the driver may fail to react accordingly.

The main shortcoming of the traditional BBS approach is that feedback is provided a relatively long time after the behavioural event occurs. Until now, almost all BBS techniques that are based on driver monitoring systems focus on pre- or post-trip feedback. For these purposes on-board monitoring devices to observe driver behaviours are often used in fleets. In one study, 35.4% of the companies involved used on-board monitoring devices but they did not provide immediate feedback on visual distraction to the driver [58]. According to previous studies, feedback is most effective when given immediately, or as soon as possible, after the occurrence of the behaviour [1]. Research shows that when drivers only receive pre- or post-trip feedback at the office, at-risk behaviour does not continuously decrease. As stated above, Toledo and Lotan [64] showed that drivers only reduced their at-risk behaviour for the first month after cumulative performance feedback (on a personal webpage) was provided. Over a longer time period, this effect diminished and performance remained low because drivers accessed their webpages less frequently. Attention enhancement efforts need to be maintained over a longer period of time.

It has been found that if drivers only receive post-trip feedback they may not be aware of at-risk behaviour once it occurs. Poor immediate feedback about how distraction affects safe driving leads drivers to believe that their driving performance is better than it really is [25]. That people in general are poor judges of their own behaviour was discussed by Bandura [71] who stated, 'Since people's attentiveness to their on-going behaviour fluctuates widely, they are not always all that self-observant' (p. 337). Furthermore, almost all prior BBS research has been applied in work settings where employees can systematically observe safe versus at-risk behaviours of themselves and/or of their co-workers. As Hickman *et al.* [58] argue, truck and bus drivers work alone in relative isolation and thus may require alternative BBS processes.

Previous research suggests that the combination of on-board driver monitoring systems with other safety-management techniques like BBS management is likely to be one of the most powerful approaches in reducing crashes [74]. In a study evaluating the safety benefits of the DriveCam system with driver feedback and offline coaching, results show a 52.2% reduction in safety-related events and a 59.1% reduction in severe safety-related events [63]. Therefore, driver monitoring approaches with *combined feedback* about risky behaviours *at different timescales* are needed. A new approach is to combine ADAS and BBS accident prevention strategies by providing both immediate/delayed feedback and aggregated feedback to achieve a long-term effective BBS-ADAS.

In this chapter, the BBS-ADAS approach as applied to attention enhancement is defined as 'on-board driver monitoring for both immediate and aggregated inattention feedback'. With BBS-ADAS methods, the driver receives feedback about a behavioural event targeting a safety-critical behaviour both immediately and after driving. With the introduction of new technologies, objective real-time measures and feedback systems for driver visual behaviour, with the possibility to save the recorded events, will become available. The type of feedback is informative by notifying the driver about risky behaviour both during a trip and after it. The feedback timing of BBS-ADAS is *a combination* of 'concurrent feedback' within milliseconds, 'delayed feedback' within seconds after the occurrence of the behavioural event [37] and long-term aggregated feedback. The aim of the combination of different feedback timings allows the driver to train for preventing events in future trips.

The feedback goals of BSS-ADAS are to:

1. raise driver alertness to safety-critical situations (as with ADAS)
2. speed up driver responses in a safety-critical situation (as with ADAS)
3. raise driver awareness of safety-critical behaviour (as with BBS)
4. coach the driver about own target behaviour (as with BBS)

The intended BBS-ADAS feedback effects are:

1. driver attention on-road (as with ADAS)
2. quick driver responses (as with ADAS)
3. driver self-awareness (as with BBS)
4. driver calibration (as with BBS)

The BBS-ADAS approach intends to provoke driver behavioural change over different timescales. The unintended BBS-ADAS feedback effects range from the driver feels too safe (as with ADAS) to driver mistrust and scepticism (as with BBS).

The general target user of BBS-ADAS with regard to road safety is, unlike BBS approaches, the individual driver both immediately when a behaviour occurs, and retrospectively over a longer period of time exhibiting multiple instances of this at-risk behaviour. The advantage of the BBS-ADAS approach is that warnings can be recorded and logged to file for a post-trip summary report. This report can be either assessed individually or can be forwarded to the fleet safety manager for further coaching discussions. In summary, inherent in the BBS-ADAS approach is multiple feedback timescales, continuous check-ups, self-awareness enhancement at multiple levels and with various involved personnel. Figure 11.1 gives an overview of the properties of BBS-ADAS feedback and puts them into perspective with existing approaches.

It is important to mention that whilst the BBS-ADAS approach may be an improved approach, if the driver does not appreciate the persistent intervention, or has privacy concerns, he/she may choose to deactivate the ADAS.

11.4.1 Game tactics

In order to achieve sustained change regarding safe glance performance, a combination of immediate feedback *and* post-trip feedback is suggested. A combination of both visual behaviour feedback on a continuum in the pre- and post-trip dimension is the *DO-IT BEST Feedback Model* [75]. In the DO-IT BEST Feedback Model the BBS techniques are coupled to different feedback timescales, adapted from [37], and feedback sources, including the vehicle, aggregated performance analysis, the manufacturer (automotive and telematics devices industry) and the society. The DO-IT BEST Feedback approach is a further development of the four-step DO-IT process on continuous behavioural improvement first introduced by Geller [1]. As described above, DO IT is an acronym for **D**efine, **O**bserve, **I**ntervene and **T**est targeted at-risk and/or safe behaviours. It was originally conceptualised to target immediate behaviour but in an offline setting. Wege and Victor [75] interpret Geller's original four-step process for an example of an immediate/delayed *and* cumulative/aggregated attention enhancement strategy for preventing traffic accidents. The original DO-IT process is not only translated into an online, immediate visual behaviour feedback approach, but also developed further into the DO-IT BEST Feedback approach. BEST is an acronym for **B**ehavioural check-ups, **E**ducation, **S**afety benefit analysis and **T**raining for targeted at-risk and/or safe behaviours. For the DO-IT BEST Feedback approach to be effective, it is necessary to provide immediate feedback to drivers during a safety-critical event (for example looking off the road for an extended period of time) *combined with* post-trip feedback regarding previously recorded safety-critical events.

11.4.1.1 Team ADAS perspective

For the immediate feedback to be effective, vehicles need to be equipped with reliable real-time distraction monitoring technology. Technology can track behaviour and provide an immediate warning when specific behaviours occur. The ADAS warning can be recorded and later used in a BBS-offline setting for further discussion. Recently, research on real-time driver feedback technologies has overcome the problem of capturing and documenting key safety-critical behaviours. Eye-tracking systems [76] and distraction detection algorithms make continuous observation and detection of drivers' glances possible. The main challenge for implementing BBS-ADAS for attention enhancement is accurate, low-cost, mass-market, unobtrusive eye-tracker systems with sufficient tracking uptime. ADAS of these kinds are being developed and real-time distraction detection algorithms have been tested in controlled settings [16, 17, 77, 78]. A review of existing functions can be found in [37, 72, 77]. One example is a Visual-Distraction-Alert System (VDAS), a real-time distraction prevention system based on head/eye movement recordings. It provides informative feedback (a warning) regarding a distracting event (looking away from the road for too long or too frequently) immediately when it occurs [16, 59, 78]. The main purpose of the visual distraction alert is to help distracted drivers realise that they are being 'tricked' into glancing away from the road for too long and/or too often. The alerts should train them to recognise a limit of inattentive glance behaviour. As such, it is a preventative warning system without direct coupling to near-crash situations, but coupled to risky glance behaviours.

In recent years, the focus for technological solutions to the visual distraction prevention problem has intensified. Some first-generation products are already available, but there is little consensus regarding which real-time distraction countermeasure functions are the most effective and useful. Wege and Victor [78] recently found that a reliable warning strategy, taking into account a driver's glance history along with single long glances off the road, increases the percentage of time the eyes are on the road centre as well as driving performance. The findings are in line with Lee *et al.* [16], who evaluated four vision-based algorithms for detecting driver distraction, with the warning algorithm used in Wege and Victor [78] being the most effective. Unlike other ADAS, which target a dangerous situation (e.g. FCW, LDW), the VDAS aims to target dangerous behaviour (e.g. writing a text message while driving). By using technological features that only ADAS are capable of providing (e.g. detection and warning of off-road glances), drivers have the possibility to detect and learn immediately about their risky behaviour. The drivers self-observation component makes it an ideal 'team player' to enhance attention in a BBS-ADAS-style.

Victor *et al.* [59] asked participants about their subjective beliefs as to how efficient a VDAS is compared with other systems. It was found that 63% of the participants stated that a VDAS could reduce crashes. Current scientific research with a similar concurrent driver feedback system, the Driver State Sensor System (DSS), showed a 78% reduction in the distraction event frequency [79]. Another example of an ADAS for enhancing attention is Volvo's Driver Alert Support System (DAS). The DAS alerts the driver when path control deteriorates. It is based on detecting changes in attentional state (due to both drowsiness/fatigue and distraction) from lateral control performance and provides immediate feedback to the driver.

Not only organisations and fleet companies can benefit from the data and feedback gathered from BBS-ADAS devices. Although the BBS-ADAS approach may be most applicable for occupational drivers as retrospective feedback can be provided in a controlled manner by workplaces, the approach can also be applied in non-occupational vehicles. The BBS-ADAS principles may also augment private cars, for example through self-learning tools (post-trip awareness scores and/or web-based self-learning tools). Takeda *et al.* [80] introduced a self-coaching system to improve driving behaviour by allowing drivers to review a record of their own driving performance data (not of their visual behaviour activity).

11.4.1.2 Team BBS perspective

The warnings generated by the ADAS and recorded while driving can be used in a BBS setting for further discussion. During the BBS discussion (e.g. face-to-face meeting among fleet drivers and fleet safety managers) specific behaviours can be identified that will enhance the feedback strategy of ADAS. The ADAS can provide specific warnings according to a set goal (e.g. target behaviour) previously agreed upon. That way, the ADAS feedback can become richer and more tailored to an individual. The warning settings could be adjusted and/or specific warnings could be emphasised according to the BBS goals. This type of *BBS-generated ADAS* is a completely new approach. It is expected that these kind of BBS-ADAS will provide drivers with an individualised performance score (e.g. attention score), which will in turn translate to an *individual safety status* e.g. attention status. The personalised warnings can create a more meaningful and coherent training effect than standardised ADAS warnings would provide. One example of personalised warnings is that the top prioritised risky behaviours identified by BBS (e.g. following too close or drowsy driving) can be given as input to ADAS. ADAS could then emphasise providing more feedback on the prioritised problematic behaviours while driving, for example by providing headway coaching or more drowsiness related feedback. Personalised warnings can create an even more meaningful training effect for an individual driver and can be tied to the current goals and incentives in the BBS program.

11.5 Players debriefing

In conclusion, it is unlikely that distraction will ever be eradicated as a road safety problem – it therefore needs to be effectively managed. Reducing the effects of distracted driving is possible if the right approaches are applied to the problem. Today, there are two separate approaches for improving traffic safety, ADAS and BBS, where development has been carried on largely independently from one another. There is a need to further develop current ADAS and BBS approaches independently, and also in an integrated fashion. Both approaches apply feedback at different timescales to achieve *the same goal:* accident prevention.

Research on HMI and automated ADAS has suggested the theory of 'team play' [2]. Adapting the theory of 'team play' to the field of accident prevention it can be argued that both teams (team ADAS and team BBS) have the same goal, but because of applying different strategies they have yet not been united into one winning team. As stated above, although ADAS and BBS programmes have been proven to be effective, both lack important aspects for more stable effects.

Rudin-Brown [81] addressed one important question: 'Are our "intelligent" transport systems really that intelligent?' Without *adaptive* design, 'intelligent' in-vehicle transport systems have limitations as stated above. Feedback timescales do not have to be mutually exclusive. For a joint BBS-ADAS concept to operate successfully, immediate *and* aggregated feedback should be conceived and designed as 'team players'. Therefore, research focus should not be on either making one or the other 'more intelligent', but rather on how they can *cooperate* as members of the same team on the accident prevention playing field. Adapting the theory of 'team play' to the field of accident prevention, the question that arises is not 'which accident prevention approach has more influence?' The question that should arise is 'How do both team players get along together to play in the same team – united together?'

In the future, ADAS designers and developers, as well as safety management developers, need guidance on how to support the coordination between ADAS and BBS; the mechanisms and 'game rules' need to be defined in order for both team players to cooperate. According to 'team play' theory, good and effective team players make their activities observable for fellow team players, communicate together and are easy to direct.

The combination of BBS and ADAS, a holistic BBS-ADAS approach, can improve safety over and above each approach on its own. On-board driver monitoring devices that provide only immediate feedback may not result in sustained behaviour improvement when implemented in the absence of a BBS program. The combination of an ADAS with a BBS program (including rewards) is expected to result in sustained behavioural improvement when correctly implemented [82]. Hickman *et al.* [58] suggested that weekly feedback in combination with other BBS techniques can result in a very powerful safety improvement. Vice versa, previously discussed BBS techniques can enhance ADAS feedback by providing training sessions or personalised warnings. The BBS-ADAS approach focuses on how to harmonise the ADAS and BBS feedback into one holistic feedback process. The DO-IT BEST feedback approach was presented as one example of such a harmonised process [75].

The mechanisms of cooperation and the effects of different BBS techniques combined in one program need to be further evaluated. Adaptation effects over a longer time period (e.g. several months) have only been tested in two studies; by Krause, Seymour and Sloat [57] and Toledo and Lotan [64]. Further assessment of the effects on safety of BBS-ADAS over a longer period of time is therefore necessary. Further work is also needed to validate that accident prevention strategies work as intended, are wanted by customers and are effective (e.g. in terms of return on investment). As Victor *et al.* [59] stated, recent safety impact results are showing great potential, and no doubt the future driver will be assisted by attention technology.

In this chapter, we presented a reason why the future driver will be assisted by attention technology: when both players are united (BBS-ADAS) they form a team that wins more easily. Due to the use of more precise safety metrics directly representing distraction, the safety benefits of BBS-ADAS are expected to exceed the benefits of ADAS and BBS alone. This can be translated as BBS-ADAS being more than the sum of its parts. ADAS and BBS, united into a BBS-ADAS team with connected strength, may have the best possible impact on accidents rates. With the BBS-ADAS team, traffic safety has the opportunity to advance into an entirely new league.

Acknowledgements

This research received funding from the European Community's Seventh Framework Programme (FP7/2007-2013) under grant agreement no. 238833/ADAPTATION project (www.adaptation-itn.eu). We would like to thank Nick Reed (TRL) and Jeff Hickman (Virginia Tech) for valuable comments on a previous version of this chapter.

References

1. Geller E. S. *The Psychology of Safety Handbook.* Boca Raton, FL: CRC Press; 2001
2. Christoffersen K., Woods D. D. 'How to make automated systems team players' in Salas E. (ed.). *Advances in Human Performance and Cognitive Engineering Research Volume 2.* JAI Press/Elsevier; 2000
3. World Health Organization (Switzerland). *Mobile Phone Use: A Growing Problem of Driver Distraction*. Geneva: WHO Press; Wiley & Sons, Inc.; 2011
4. European Commission (EU). Directorate General for Mobility and Transport. *EU Road Fatalities*; 2011. Available from http://ec.europa.eu/transport/road_safety/pdf/observatory/trends_figures.pdf [Accessed 16 April 2013]
5. World Health Organization (Switzerland). *Global Status Report on Road Safety 2013: Supporting a Decade of Action.* Geneva: WHO Press; Wiley & Sons, Inc.; 2013
6. Bortkiewicz L. *Das Gesetz der kleinen Zahlen.* Leipzig: B. G. Teubner; 1898
7. Dingus T. A., Klauer S. G., Neale V. L., Petersen A., Lee S. E., Sudweeks J. *The 100-Car Naturalistic Driving Study, Phase II – Results of the 100-Car Field Experiment.* Technical Report No. DOT HS 810 593, Washington, DC: National Highway Traffic Safety Administration; 2006
8. Olson R. L., Hanowski R. J., Hickman J. S., Bocanegra J. *Driver Distraction in Commercial Vehicle Operations.* USDOT No. FMCSA-RRR-09-042, Washington, DC: Federal Motor Carrier Safety Administration; 2009
9. Hendricks D. L., Fell J. C., Freedman M. *The Relative Frequency of Unsafe Driving Acts in Serious Traffic Crashes.* NHTSA-DTNH22-94-C-05020, Washington, DC: National Highway Traffic Safety Administration; 2006
10. Angell L., Auflick J., Austria P. A., Kochhar D., Tijerina L., Biever W., et al. *Driver Workload Metrics Task 2 Final Report*. NHTSA Report DOT HS 810 635. Washington, DC: National Highway Traffic Safety Administration; 2006
11. Carsten O., Brookhuis K. 'The relationship between distraction and driving performance: towards a test regime for in-vehicle information systems'. *Transportation Research Part F.* 2005;8:75–77

12. Wierwille W. W., Tijerina L. 'An analysis of driving narratives as a means of determining problems caused by in-vehicle visual allocation and visual workload' in Gale A. G. (ed.). *Vision in Vehicles.* Elsevier Science Publishers; 1996. pp. 79–86
13. National Highway Traffic Safety Administration (USA). *Distracted Driving 2009. Traffic Safety Facts. Research Note.* U.S. Department of Transport HS 811 379. Washington, DC: NHTSA; 2010
14. Cohen J. T., Graham J. D. 'A revised economic analysis of restrictions on the use of cell'. *Risk Analysis* 2003;23(1):5–17
15. Klauer S. G., Dingus T. A., Neale V. L., Sudweeks J. D., Ramsey D. J. *The Impact of Driver Inattention on Near-Crash/Crash Risk: An Analysis Using the 100-Car Naturalistic Driving Study Data.* Technical Report No. DOT HS 810 594, Washington, DC: National Highway Traffic Safety Administration; 2006
16. Lee J. D., Moeckl J., Brown T., Roberts S., Victor T., Marshall D., et al. 'Detection of driver distraction using vision-based algorithms'. *Proceedings of the ESV Conference, Paper Number: 13-0348*; 2013
17. National Highway Traffic Safety Administration (USA). *Distraction Detection Algorithm Evaluation. TRAFFIC SAFETY FACTS Vehicle Safety Research Notes*. DOT HS 811 548. Washington, DC: NHTSA; 2013
18. Liang Y., Lee J. D., Yekhshatyan L. 'How dangerous is looking away from the road? Algorithms predict crash risk from glance patterns in naturalistic driving'. *Human Factors.* 2012;54(6):1104–1116
19. Regan M., Lee J. D., Young K. L. (eds.). *Driver Distraction: Theory, Effects, and Mitigation.* Boca Raton, FL: CRC Press; 2008
20. Regan M., Lee J. D., Victor T. (eds.). *Driver Distraction and Inattention. Advances in Research and Countermeasures. Volume 1.* Human Factors in Road and Rail Transport. Ashgate: SAE International; 2013
21. Engström J., Monk C., Stevens A., Lee J., Regan M., Hanowski R., et al. *A Conceptual Framework and Taxonomy for Understanding and Categorizing Driver Inattention.* US-EU Driver Distraction and HMI Working Group; 2013
22. Lee J. D., Young K. L., Rega M. 'Defining driver distraction' in Regan M. A., Lee J. D., Young K. L. (eds.). *Driver Distraction: Theory, Effects, and Mitigation*. Boca Raton, FL: CRC Press Taylor & Francis Group; 2009. pp. 31–40
23. Young K. L., Regan M. A., Hammer M. *Driver Distraction: A Review of the Literature. Monash University.* Report; 2003
24. Forbes N. L. *Behavioural Adaptation to In-Vehicle Navigation Systems.* Doctoral thesis. University of Nottingham; 2009
25. Horrey W., Lesch M., Garabet A. 'Assessing the awareness of performance decrements in distracted drivers'. *Accident Analysis and Prevention* 2008; 40(2):675–682
26. Elvik R. 'Laws of accident causation'. *Accident Analysis and Prevention* 2006;38(4):742–747

27. Röglinger S. 'A methodology for testing intersection related Vehicle-2-X applications'. *Computer Networks.* 2011;55:3154–3168
28. Röglinger S., Facchi C. 'How can car-2-x-communication improve road safety'. *Arbeitsberichte. Working Paper.* University of Applied Science Ingolstadt; 2009
29. Wege C., Will S., Victor T. 'Eye movement and brake reactions to real world brake-capacity forward collision warnings – a naturalistic driving study'. *Accident Analysis and Prevention.* 2013;58(9):259–270
30. European Commission (EU). *Commission Recommendation of 22 December 2006 on Safe and Efficient In-vehicle Information and Communication Systems: Update of the European Statement of Principles on Human Machine Interface*. Official Journal of the European Union, L 32/200-241; 2007
31. Hickman J. S., Hanowski R. J. 'An assessment of Commercial Motor Vehicle Driver Distraction Using Naturalistic Driving Data'. *Traffic Injury Prevention*. 2012;13:612–619
32. Thalen J. *ADAS for the Car of the Future. Interface Concepts for Advanced Driver Assistant Systems in a Sustainable Mobility Concept of 2020.* Design Report. Faculty of Engineering Technology, University of Twente; 2006
33. Golias J., Yannis G., Antoniou C. 'Classification of driver-assistance systems according to their impact on road safety and traffic efficiency'. *Transport Reviews*. 2002;22(2):179–196
34. Riener A. *Sensor-Actuator Supported Implicit Interaction in Driver Assistance Systems.* Dissertation, University Linz Wiesbaden: Vieweg & Teubner; 2009
35. Wege C., Victor T., Bueno-Garcia M., Beggiato M., Berthon-Donk V., Brusque C., et al. *ADAS within the Adaptation Project – Function Selection, Benchmark, Behavioural Adaptation Effects and Conceptual Framework Development.* Deliverable D4, Public deliverable, Marie Curie ITN, Volvo Technology Corporation (VTEC); 2010
36. Gietelink O. J., Ploeg J., De Schutter B., Verhaegen M. 'Development of a driver information and warning system with vehicle hardware-in-the-loop simulations'. *Mechatronic.* 2009;19:1091–1104
37. Donmez B., Boyle L., Lee J. D. 'Designing feedback to mitigate distraction' in Regan M., Lee J. D., Young K. (eds.). *Driver Distraction: Theory, Effects and Mitigation.* Boca Raton, FL: CRC Press; 2008
38. Piamonte P., Deregibus E., Romera M. *HMI Components (Prototypes: Displays, HUDs, Speech I/O, Sound etc.) – Accompanying Report.* Report associated to Deliverable D3.4.2, AIDE project, IST-1-507674-IP; 2004
39. University of Michigan Transportation Research Institute. *Driver Interface Group, Telematics Guidelines.* [online]. 2013. Available from http://www.umich.edu/~driving/safety/guidelines.html [Accessed 25 March 2013]
40. SAE-J-2400:2003. *Human Factors in Forward Collision Warning Systems: Operating Characteristics and User Interface Requirements.* Safety and Human Factors Standards Steering Committee; 2003
41. European Commission (EU). *Guidelines on Establishing Requirements for High-Priority Warning Signals.* World Forum for Harmonization of

Vehicle Regulations, Geneva. United Nations Economic and Social Council; 2011

42. Campbell J. L., Richard C. M., Brown J. L., McCallum M. *Crash Warning System Interfaces: Human Factors Insights and Lessons Learned.* NHTSA Final Report No. HS 810 697, Washington, DC: National Highway Traffic Safety Administration; 2007. p. 184
43. Alliance of Automobile Manufacturers (US). *Statement of Principles, Criteria and Verification Procedures on Driver Interactions with Advanced In-Vehicle Information and Communication Systems.* Washington, DC: AAM; 2006
44. Japanese Automobile Manufacturers Association (Japan). *Guideline for In-Vehicle Display Systems—Version 3.0, Vol. 1.* Japan: JAMA; 2004. pp. 1–15
45. Swedish Standards Institute (Sweden). *Road Vehicles – Ergonomic Aspects of In-vehicle Presentation for Transport Information and Control Systems – Warning Systems.* Technical Report. ICS 13.180;43.040.15, SIS-ISO/TR 16352; 2006
46. Hesse T., Johansson E., Brockmann M., Rambaldini A., Allgaier A., Esberg I., et al. *interactIVe. Deliverable D3.2.* IWI Strategies. Project Report; 2012
47. Skinner B. F. *Science and Human Behaviour.* New York, NY: Free Press; 1953
48. Ludwig T. D., Geller E. S. 'Intervening to improve the safety of delivery drivers'. *Journal of Organizational Behaviour Management.* 2000;19(4):5–43
49. Fellner D. J., Sulzer-Azaroff B. 'Increasing industrial safety practices and conditions through posted feedback'. *Journal of Safety Research.* 1984;15:17–21
50. Hickman J. S., Geller E. S. 'A safety self-management intervention in mining operations'. *Journal of Safety Research.* 2003;34(3):299–308
51. McSween T. E. *The Values-Based Safety Process: Improving Your Safety Culture with a Behavioural Approach.* New York, NY: John Wiley & Sons, Inc; 1995
52. Komaki J., Barwick K. D., Scott L. R. 'A behavioural approach to occupational safety: pinpointing and reinforcing safe performance in a food manufacturing plant'. *Journal of Applied Psychology.* 1978;63:434–445
53. Krause T. R., Robin J. L., Knipling R. R. *The Potential Application of Behaviour-Based Safety in the Trucking Industry.* FHWAMC-99-071. Washington, DC: Federal Highway Administration; 1999
54. Ranney J., Nelson C., Mozenter J., Coplen M. *The Efficacy of Behaviour Based Safety in the US Railroad Industry: Evidence from Amtrak-Chicago.* Volpe National Transportation Systems Center; 2005
55. Sulzer-Azaroff B., Austin J. 'Behaviour-based safety and injury reduction: a survey of the evidence'. *Professional Safety.* 2000;45(7):19–24
56. Guastello S. J. 'Do we really know how well our occupational accident prevention programmes work?'. *Safety Science.* 1993;16:445–463
57. Krause T. R., Seymour K. J., Sloat K. C. M. 'Long-term evaluation of a behaviour-based method for improving safety performance: a meta-analysis of 73 interrupted time-series replications'. *Safety Science.* 1999;32:1–18

58. Hickman J. S., Knipling R. R., Hanowski R. J., Wiegand D. M., Inderbitzen R. E., Bergoffen G. *CTBSSP Synthesis 11: Impact of Behaviour-based Safety Techniques on Commercial Motor Vehicle Drivers*. Washington, DC: Transportation Research Board of the National Academies; 2007
59. Victor T., Hickman J., Camden M., Jarlengrip J., Larsson C., Morgan J., Tidwell S., Toole L. *Driver Distraction: An Inattention-Mitigation Component for Behaviour-Based Safety Programs in Commercial Vehicle Operations (IM-BBS)*. Final Report. NTRCI University Transportation Center; 2011
60. Krause T. R. *The Behaviour-Based Safety Process: Managing Involvement.* New York, NY: Van Nostrand-Reinhold; 1997
61. Olson R., Austin J. 'Behaviour-based safety and working alone: the effects of a self-monitoring package on the safe performance of bus operators'. *Journal of Organizational Behaviour Management.* 2001;21(3):5–43
62. Horrey W. J., Lesch M. F., Dainoff M. J., Robertson M. M., Noy Y. I. 'On-board safety monitoring systems for driving: review, knowledge gaps, and framework'. *Journal of Safety Research*. 2012;43:49–58
63. Hickman J. S., Hanowski R. J. *Evaluating the Safety Benefits of a Low-cost Driving Behaviour Management System in Commercial Vehicle Operations* (No. FMCSA-RRR-10-033). Washington, DC: Federal Motor Carrier Safety Administration, USDOT; 2010
64. Toledo T., Lotan T. 'In-vehicle data recorder for evaluation of driving behavior and safety'. *Transportation Research Record*. 2006:112–119
65. Horrey W. J., Lesch M. F. 'Factors related to drivers' self-reported willingness to engage in distracting in-vehicle activities'. *Proceedings of the Human Factors and Ergonomics Society 52nd Annual meeting*; 2008
66. Roelofs E., Vissers J., van Onna M., Kern G. 'Coaching young drivers in a second phase training programme' in Dorn L. (ed.). *Driver Behaviour and Training, Volume V*. London: Ashgate; 2012
67. Robin J. L., Knipling R. R., Derrickson M. L, Antonik C., Tidwell S. A., McFann J. 'Truck simulator validation ('SimVal') training effectiveness study'. *Proceedings of the 2005 Truck & Bus Safety & Security Symposium*; Alexandria, VA, November 14–16; 2005
68. Gordetsky M. 'Simulator Gives Better Sense of the Road'. *Transportation Research Board.* Transport Topics, No. 3402; 2000. pp. 37–39
69. Pradhan K., Divekar G., Masserang K., Romoser M., Zafian T., Blomberg R. D., Thomas F. D. 'The effects of focused attention training on the duration of novice drivers' glances inside the vehicle'. *Ergonomics*. 2011;54(10):917–931
70. Barton R., Tardif L. P., Wilde G., Bergeron J. 'Incentive programmes for enhancing truck safety and productivity: a Canadian perspective'. *Transportation Research Board Report* TP 13256E; 1998. p. 85
71. Bandura A. *Social Foundations of Thought and Action: A Social Cognitive Theory. Reproduction*. Prentice-Hall, Englewood Cliffs, NJ. doi:10.2832/54930; 1986

72. Engström J., Victor T. 'Real-Time distraction Countermeasures' in Regan M., Lee J. D., Young K. L. *Driver Distraction. Theory, Effects, and Mitigation*. Boca Raton, FL: Taylor & Francis; 2008
73. Chapman P., Underwood G. 'Forgetting near-accidents: the role of severity, culpability and experience in the poor recall of dangerous driving situations'. *Applied Cognitive Psychology*. 2000;14:31–44
74. Knipling R. R., Hickman J. S., Bergoffen G. *CTBSSP Synthesis 1: Effective Commercial Truck and Bus Safety Management Techniques*. Washington, DC: Transportation Research Board of the National Academies; 2003
75. Wege C., Victor T. 'The DO-IT BEST Feedback Model – distracted driver behaviour management and prevention before, while and after driving'. *Proceedings of the 3rd International Conference on Driver Distraction and Inattention*; Lindholmen Science Park, Göteborg, Sweden, September 2013
76. Ahlström C., Victor T., Wege C., Steinmetz E. 'Processing of eye/head-tracking data in large-scale naturalistic driving data sets'. *IEEE Transactions on Intelligent Transportation Systems*. 2001;13(2):553–564
77. Kircher K., Ahlström C. 'The driver distraction algorithm AttenD' in Regan M., Lee J. D., Victor T. *Driver Distraction and Inattention. Advances in Research and Countermeasures. Volume 1*. Human Factors in Road and Rail Transport. Ashgate: SAE International; 2013
78. Wege C. 'Adaptive eyes: Driver distraction and inattention prevention through advanced driver assistance systems and behaviour-based safety'. Doctoral thesis. Chemnik: University of Technology and Volvo GTT; 2013
79. Croke D., Cerneaz N. 'Real time distraction detection and warning system improves safety on public roads. A case study'. *Proceedings of the first International Conference on Driver Distraction and Inattention*; Lindholmen Science Park, Göteborg, Sweden, September 2009
80. Takeda K., Miyajima C., Suzuki T., Angkititrakul P., Kurumida K., Kuroyanagi Y., et al. 'Self-coaching system based on recorded driving data: learning from one's experiences'. *Intelligent Transportation Systems, IEEE*. 2012;13(4):1821–1831
81. Rudin-Brown C. ''Intelligent' in-vehicle intelligent transport systems: limiting behavioral adaptation through adaptive design'. *IET Intelligent Transport Systems*. 2010;4(4):252–261
82. Knipling R. R., Olsgard P. J. 'Prospectus: the behavioural power of on-board safety monitoring feedback'. *Proceedings of the 10th Annual Meeting of the Intelligent Transportation Society of America (ITS America)*; Boston, May 2000

Chapter 12

Effectiveness of Forward Collision Warning Systems: A contribution from the cognitive analysis combining behavioural and electrophysiological measurements

Mercedes Bueno[1,2,3], Colette Fabrigoule[3] and Alexandra Fort[1,2]

Abstract

This chapter discusses Forward Collision Warning Systems (FCWSs), describing the characteristics and the functions of some of the systems currently on the market and presents an overview of behavioural studies evaluating the effectiveness of these systems on road and in simulators. Results are also presented from recent studies using electroencephalography and the associated event related potentials (ERPs) allowing, through the analysis of brain activity, a more in-depth understanding of the nature of the cognitive processes in the context of FCWSs. These studies address three important questions:

1. Are FCWSs as effective as they are expected to be when drivers are distracted?
2. What are the consequences of driving with a system that is not completely reliable?
3. Is there any behavioural adaptation to the FCWSs over their use in time? Are the consequences of this adaptation positive or negative?

The chapter shows that FCWSs provide potential benefits for road safety, but certain factors such as the attentional state of drivers and the reliability level of the system can mitigate its effectiveness.

12.1 Introduction

Advanced driver assistance systems (ADAS) have experienced a rapid development in recent years to optimise their potential benefits on road safety. It has been

[1] IFSTTAR, Lyon (France)
[2] University of Lyon (France)
[3] University Bordeaux Segalen (France)

estimated that these systems contribute to reducing the number of casualties on the roads. Because of this, the European Commission has recently proposed the mandatory installation of some ADAS such as the lane departure warning system and the collision warning system in heavy vehicles from 1 November 2013 [1]. Nevertheless, light vehicles will be exempted at present because of the lower benefit-to-cost ratio compared to heavy vehicles.

Since ADAS have appeared on the market, research in the field of human factors has been promoted to investigate the impact of both the benefits of these systems on drivers' behaviour and their potential failures or limitations. Some of these systems warn drivers by an audible, visual and/or haptic signal when inappropriate behaviour is detected, either because an action is required but is not carried out or because the action carried out is not the right one. This chapter presents a review of the literature focused on a particular ADAS, the Forward Collision Warning System (FCWS). Nevertheless, some of the questions and discussions considered here could be applied to other warning systems.

FCWSs are designed to reduce the number and/or to avoid rear-end collisions by providing warning to drivers. Traditional behavioural measures obtained in driving simulators and field operational tests are useful tools for evaluating the effectiveness of such systems, but they are not the only ones. Physiological data obtained through the event related potential (ERP) technique can complement behavioural data. Reporting on data from these different methodological approaches, we discuss whether the system reaches one of its main objectives, i.e. to assist distracted drivers, and then how the system reliability affects the efficiency of the system and driver's acceptance. Finally, in the last section of this chapter, certain studies examining the behavioural adaptation to the system are presented. Specifically, the effects on behaviour due to driving with the system over the time (from the introduction of the system to long-term effects of ADAS use) and the consequences of driving without the system after a period of adaptation to the system are discussed.

12.2 The utility of FCWSs

As mentioned above, FCWSs have been developed to warn drivers of potential rear-end collisions. Although the percentage of these collisions resulting in fatalities is relatively low compared to the percentage of injuries [2], these accidents are one of the most prevalent types of collision, disturbing traffic flow and representing an important economic cost for society. The environment (e.g. poor visibility, road type) and the vehicle (e.g. defective brakes) are some of the contributing factors to the rear-end collisions in 5–11% and 12–20% of these accidents, respectively. However, the factor that most frequently contributes to rear-end collisions is the driver, accounting for between 75 and 93% of the cases. More precisely, inside the driver category, distraction has been involved in about 60% of the rear-end collisions [3, 4].

FCWSs are based on sensors which continuously monitor certain parameters such as the relative speed, lateral position and distance between two vehicles. Although a wide variety of algorithms have been designed to calculate the moment for triggering the warning, most of them are based either on the time of collision between the obstacle and the following vehicle or on the minimum distance required to stop the vehicle safely [5]. When a certain threshold value is reached, predicting a potential risk of collision, the system provides a warning signal (e.g. visual, auditory and/or tactile) to alert the driver.

The system, as well as providing assistance by warning the driver (Forward Collision Warning Systems), can also take an active part in the braking process by preparing and/or by applying partial or full braking automatically (Forward Collision Avoidance Systems). This active avoidance module could prevent or reduce the severity of the collision should the driver not react in time or not react at all. Moreover, the active avoidance system has become more relevant in preventing collisions since it has been demonstrated that most drivers do not apply enough force on the brake pedal [6] and that a high percentage of drivers only release the accelerator pedal or even do not react at all [7].

Driving simulators have been one of the most frequent tools used in the evaluation of FCWSs and most research has shown positive effects of the system when drivers are undistracted. This benefit has been demonstrated, for example, by a reduction of the number of collisions [8], by faster braking reaction times (RTs) following the detection of critical situations [9], and by the adoption of longer and safer headways [10]. Other studies such as Georgi *et al.* [11] have estimated the benefits of the system by developing models which apply the algorithm used by the system in real-life accidents. They categorised three different types of drivers: lethargic drivers with the longest RTs and lowest decelerations, realistic drivers as an intermediate level and best drivers with the fastest RTs and highest decelerations. According to their predictions, FCWSs prevented 74% of collisions for best drivers, 38% for realistic drivers and 1% for lethargic drivers. In a more ecological context, using a field operational test, Najm *et al.* [7] assessed the impact of FCWSs and advance cruise control (ACC) systems for four weeks. Sixty-six drivers drove without the system during the first week to obtain the baseline and with the system during the remaining three weeks. The results indicated that the exposure to conflicts (brake or steer at the last second at a comfortable deceleration), near-crashes (brake or steer at the last second at a hard deceleration) and severe near-crashes (minimum time to collision less or equal to 3 s and maximum deceleration greater than 0.3 g) in the last period of the test were reduced by about 20% compared to the first period. Based on different rear-end collision databases, these authors estimated that the system could prevent about 10% of these accidents. Taken together, these studies show a range of benefits from these systems which fluctuates from 10% to 70% approximately [12]. This wide range could be explained by the different variables considered in these studies such as the type of system used and its limitations, the methodological approach chosen, the characteristics of drivers or the weather conditions. Moreover, it is necessary to consider other factors which can have a direct impact on the effectiveness of the system. In this chapter, the impact of the system according to the attentional state of drivers as well as the impact of the system as a function of its reliability level will be analysed. In order to shed light on this aspect, we are going to consider an alternative research method to the classical behavioural data such as the electroencephalography (EEG) and the associated ERPs. Furthermore, the behavioural adaptation to an FCWS as well as the possible consequences of driving without the system will be discussed.

12.3 A new approach for evaluating the effectiveness of FCWSs

Different tools are available for investigating the effect of this kind of assistance system, from the most fundamental techniques such as simplified simulator experiments conducted in a laboratory, to more realistic contexts using driving simulators and to almost or real driving conditions such as field operational tests and naturalistic driving experiments. Among these studies we can highlight the classical trade-off between experimental control and ecological validity.

Given the difficulty in simulating rear-end collisions in real-life situations, most studies are employing driving simulators where different scenarios as well as the severity of collisions can easily be manipulated. Behavioural measures (e.g. RT, steering wheel angle, speed) are used most frequently by researchers in these studies. But not only behavioural data can be recorded in driving simulators. Often, these measures can be enhanced by other complementary techniques such as the analysis of information given by drivers (i.e. questionnaires, in-depth interviews) or the analysis of physiological measures (i.e. electrooculography, galvanic skin response, heart rate, EEG). Although physiological measures are still not very frequent in driving simulators, they provide complementary and additional information that in some cases is not observable in drivers' behavioural performance. In this part, we therefore will focus on EEG and the associated ERP techniques.

EEG is the study of the spontaneous electrical activity of the brain that can be recorded by electrodes placed on the scalp. After repeating a great number of times the stimulus of interest, ERPs are obtained from the electroencephalogram by filtering and averaging the activity time-locked to the stimulus of interest. The filtering and averaging processes allow reducing the random noise (i.e. movement artefacts), and so extracting the specific response to the target stimulus. As a result, a different succession of waves or components can be identified, depending on the type of stimulus presented and the nature of the task carried out. These components are usually designated by a letter (N or P) corresponding to the polarity of the component (Negative or Positive) and a number corresponding to its position in the chronology (i.e. P1, N1, N2, P2, P3) or the classical latency of the peak (i.e. N185 corresponds to a negative component peaking around 185 ms following the stimulus of interest) (see Figure 12.1).

This technique presents a high temporal resolution (on the order of milliseconds) which enables the identification of the different stages of information processing. The first stage of the neural processing chain can be identified even

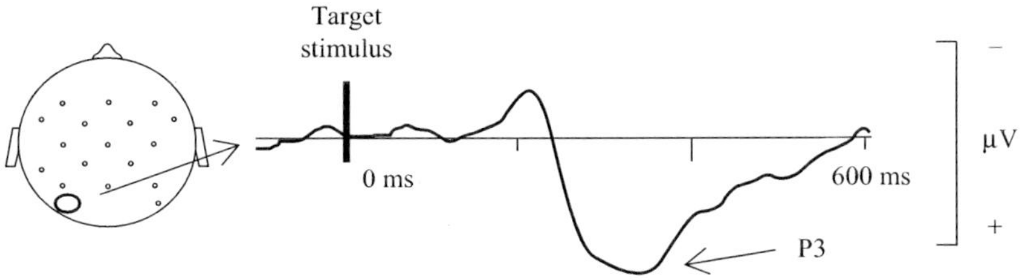

Figure 12.1 Example of a grand average ERP showing a typical P3 component at parietal areas

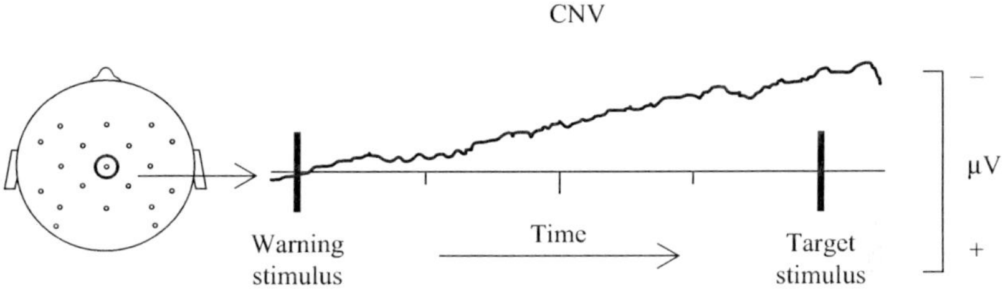

Figure 12.2 Example of a grand average ERP showing a typical CNV component at central areas

before the stimulus appears, reflecting the anticipation of a stimulus. For example, the contingent negative variation (CNV) is a slow negative wave appearing typically by the association of two successive stimuli (Figure 12.2). Thus, when the first stimulus or warning always precedes the second or target stimulus, participants can expect the target stimulus and prepare their responses due to the appearance of the warning stimulus [13].

In the 200 ms following the stimulus presentation, ERP components such as P1 and N1 have been linked to sensory processes as well as to the discriminative processing and are mainly modulated by physical attributes of the stimuli [14]. After about 200 ms, the later ERP components such as N2/P3 are thought to reflect higher cognitive processes. The N2 component is elicited by tasks involving cognitive control, novelty stimuli, perceptual matching and response conflict and the attention required for the processing [15]. The P3 component can be used as an indicator of the attentional resources allocation and working memory updating [16, 17]. In general, it is possible to identify differences between experimental conditions by analysing the differences in ERP in terms of peak amplitude and/or peak latency of the ERP components of interest. In a broader sense, differences in peak amplitude of the wave could indicate the different degree of attentional resources engaged in the processing of the stimulus, and differences in peak latency of the wave (earlier or later occurrence in time) could be related to the speed of stimulus information processing.

The implementation of the ERP technique in driving studies is not widespread due to some constraints such as the high saccadic and motor movement sensitivity and the high number of trial repetitions required in order to reach an acceptable signal-to-noise ratio. Despite these constraints, there are a few ERP simulator studies which mainly investigate the negative effect of performing a concurrent cognitive task while driving [18, 19]. Nevertheless, this technique has been implemented only recently in the study of ADAS and, specifically, in the study of FCWSs [20].

12.4 Are FCWSs effective for distracted drivers?

The major contributing factor to rear-end collisions is driver distraction or inattention. Therefore, if FCWSs are expected to help distracted drivers in particular, the effectiveness of these systems should be evaluated according to the attentional state of the drivers, distracted or not.

Regan *et al.* [21] define driver inattention as ‘insufficient or no attention to activities critical for safe driving’ (p. 5). Driver distraction, or driver-diverted attention, is a type of driver inattention where the attention to activities critical for safe driving is threatened by the diversion towards another competing activity. These competing activities can require cognitive, perceptual (i.e. visual, auditory) and/or motor resources, identifying the different distraction categories. Secondary visuomotor tasks seem to produce the higher degree of interference with the driving task, although this could be mitigated by compensatory strategies. For example, visual concurrent tasks not only tend to impair lane keeping performance and increase the number of off-road glances but also induce speed reductions, whereas cognitive tasks generally provoke the opposite effect of visual concurrent tasks [22, 23].

Several reports have focused on the impact of FCWSs on distracted drivers, but few studies have compared the distraction condition with the control situation (without any distraction); that is, assessing the impact of FCWSs on distracted and also undistracted drivers. Data obtained in simulators show that distracted drivers also benefit from the warning systems by the reduction of the number of collisions [8], by faster braking RTs in detecting critical situations [24] and/or by longer and safer headways [25]. Moreover, in some cases, the system even completely dissipates the negative effect of the secondary task [8, 26].

Electrophysiological studies have been recently conducted in driving contexts in order to assess the impact of secondary tasks. For example, Strayer and Drews [18] found that in a task where participants had to react to intermittent lead vehicle decelerations, the P3 amplitude was reduced and the P3 latency was delayed when they were engaged in phone conversations, suggesting a negative impact on the driving task when the memory load increased. Similarly, Rakauskas *et al.* [19] studied the impact of secondary tasks while driving and detecting unexpected sounds (oddball paradigm). They found that the P3 amplitude to novel sounds was especially reduced by a cognitive secondary task presented by cell-phone, showing that the processing of novel information is deteriorated by the dual task. In the context of FCWSs, the ERP technique was introduced only recently by Bueno *et al.* [20]. These authors conducted an ERP study in a simplified driving simulator to evaluate the impact of a surrogate FCWS according to the attentional state of the participants, distracted or not by a secondary cognitive task. Participants were instructed to follow a lead motorcycle and they had to react by decelerating when the brake light (target) of the lead motorcycle was lit up. An auditory warning could forewarn participants that the motorcycle was going to brake soon. The results showed that the warning system reduced the RT when participants were undistracted but not when they were distracted. ERP data showed a benefit from the warning system at higher cognitive level (N2) suggesting that the warning could enhance the temporal expectancy regarding the target. However, and in parallel to behavioural data, this effect was only observed in simple task condition and not when participants were distracted. A possible explanation for this unexpected result could come from the experimental design. Indeed, an initial deceleration of the motorcycle occurred systematically before braking in all trials, whether or not the braking was preceded by the warning signal. Thus, participants may have used this motorcycle deceleration as an additional and better predictor of the brake light occurrence instead of the warning signal which was not always reliable. Therefore, in a recent study, Bueno *et al.* [27] eliminated the predictive value of this motorcycle deceleration for the forthcoming brake light in order to increase the effectiveness of a FCWS under dual task conditions. The results indicated that participants were faster when they had available the warning signal compared to when no warning signal was given. At the electrophysiological level, this warning effect occurred at preparatory (increase of the amplitude of the CNV) and higher cognitive level (reduction of the peak latency of the P3). In addition, a strong negative impact of the secondary task was observed at behavioural and electrophysiological level with or without the system. This result suggests that the presence of the warning was not enough to compensate the negative effect of the dual task.

These last two studies have contributed to an increase in the knowledge about the processing of warning signals in driving. According to these results, it seems that the warning signal intervenes in motor preparation process (CNV) and higher cognitive processing (P3). These findings are also consistent with the findings of earlier non-driving studies showing warning signal operates in the stage of response selection [28] and that it could reduce the peak latency of the P3 [29].

Contrary to the positive effects on distracted drivers presented at the beginning of this section, the last two studies analysed here [20, 27] showed that the warning system was not always effective when drivers were distracted. Therefore, at present it is not clear whether these systems achieve their main purpose of mitigating the negative effects of the distraction. Although conclusions could not be drawn due to the limited number of studies evaluating the impact of FCWSs in distracted and undistracted drivers and their diverse methodology employed, there are some relevant questions to be considered.

First, given that secondary tasks can differ in modality (visual, auditory, cognitive, etc.) and complexity level, the effectiveness of the system could vary depending on the type and difficulty of the secondary task. It could be possible that the dual task interferes with the warning signal when both warning and secondary task share the same modality. It is known that in dual task paradigms, the likelihood of interference produced when performing two tasks from the same modality (e.g. both tasks are auditory) is higher than when these tasks belong to different modalities (e.g. visual and auditory) [30]. As a result, the effectiveness of an auditory warning could be reduced if, for example, drivers are distracted by a secondary auditory task. Thus, auditory and visual–auditory warnings seem to be effective in helping drivers to detect potential collisions when they are distracted by a visual secondary task [8, 24, 26]. However, when the nature of the secondary task is auditory (i.e. phone conversations) and also when the complexity of the task increases (i.e. mental mathematics and categorisation questions), an auditory warning seems less effective than a tactile warning [31].

Second, from research using driving simulators [32] or ERP approach (but in non-driving related contexts [33]), multimodal warnings seem to be more effective than unimodal warnings. Indeed, receiving redundant warnings from two different modalities could be advantageous in the case of distraction or in the case of warning failures.

Finally, the difficulty level of the secondary task could also undermine the effectiveness of the warning system. It is known that the amount of attentional resources available is considerably reduced by the presence of another competitive task [30, 34]. When secondary tasks are highly demanding on a cognitive level, it might be possible that participants need to invest a lot of attentional resources to deal with both tasks at the same time. Consequently, it is possible that the warning signal cannot be completely processed given the lack of attentional resources available, as could have occurred in the study of Bueno *et al.* [20]. In another study conducted in our laboratory, we investigated the impact of the difficulty of the secondary task on the effectiveness of the warning system. In this study in a driving simulator, only behavioural measures were recorded. The results showed an increase in RT when participants were distracted by a high difficulty cognitive secondary task, and also when they were distracted by a low difficulty secondary task. However, when the warning system was available, RTs were shorter than the baseline (no warning, no distraction) only in the condition of low difficulty distraction, suggesting that the benefit from the system, at least partly, depends on the drivers' attentional resources available (manuscript in preparation). This result could support the hypothesis that warning signals are not treated automatically and need attentional resources to be processed.

12.5 Does the reliability of the system affect its effectiveness?

Besides drivers' attentional state, the effectiveness and acceptance of the system and driver behaviour may also be affected by the reliability level of the system. Indeed, depending on the type of algorithm and scan sensor, the system may malfunction, producing false alerts and/or misses of critical events. False alerts refer to the situations in which an alert is issued in the absence of any potential collision, for example, when there are highly reflective objects. Nuisance alerts can be defined as alerts triggered in an appropriate situation but perceived by drivers as inappropriate because of their frequency, timing, intensity or modality [35].

Nuisance alerts can also be considered as alerts triggered in an inappropriate situation but whose origin could be justified by the presence of a potential threat (i.e. an alert triggered by a stationary vehicle placed out of or near the driver's line) [36]. Misses occur when an alert is not triggered despite the situation requiring it. Such failures are due to the fact that the capacity to detect obstacles could be limited in situations such as bad weather, detecting two-wheeled vehicles, pedestrians, stationary obstacles or very slow-moving vehicles, swerving vehicles or vehicles in bends, abrupt accelerating or decelerating. Because of these limitations, no FCWS currently works perfectly.

Real data about the false alerts and miss rates are not frequent in the literature, but the existing data suggests that this rate is still far from acceptable. As already mentioned, Najm *et al.* [7] collated data from drivers who drove with a vehicle fitted with ACC and FCWS during one month. Their results showed that 44% of the alerts were false alerts and only 56% of all warnings were triggered by in-path targets (whether all of these alerts were necessary or not is unknown). McLaughlin *et al.* [37] evaluated the algorithm used in the previous study using real-crash data and compared it with two other algorithms. Their results showed a benefit from the first system of about 60%. However, this algorithm also accounted with the higher warning frequency, 87 alerts per 161 km driven, compared to the two other algorithms, 83 and 8 alerts per 161 km driven, respectively. Taking into account the recommendation that no more than one nuisance alert per 322 km would be an acceptable rate [38], the data presented above is not very encouraging.

Drivers' trust in the warning system could be considerably undermined if the number of nuisance or false alerts is too high, because these alerts are irritating and distractive [39]. As a result, the acceptance of the warning could be reduced and, therefore, it is likely that users ignore or react inadequately to further warnings, whether or not they are valid [40]. Moreover, concerning behavioural changes in drivers' performance, false alerts seem to lengthen the braking RT [41] and induce unnecessary deceleration responses [39]; however, a higher false alerts rate could be associated with increases in the speed since drivers would tend to ignore the warning [42]. Misses are also critical for the acceptance and trust in the system and drivers' performance could be affected by longer RTs, even longer than driving with no system at all [41]. In addition, the speed could decrease with the increase of the misses which has been interpreted as a reduction of drivers' trust on the system [43].

Besides the rate of false alerts and misses, hit alerts (alerts delivered when a potential collision is imminent) could also be considered as ineffective depending on the timing in which the alert is triggered. In general, early alerts are more effective and account with a higher acceptance level by drivers than late alerts, but if they are prompted too early, they can be considered as nuisance or false alerts [44].

However not all false alerts and misses have negative effects. For example, nuisance alerts can improve the knowledge that drivers have about the way the system functions [36]. In addition, if a system never triggers a false alert, drivers could fail to react adequately when the first hit alert occurs [45]. On the other hand, if a system almost never fails to detect a potential collision, drivers may over-rely on it and become vulnerable or not react adequately, when, for example, driving another unequipped vehicle [45]. It is complicated to assess which level of false alerts/misses is acceptable, given that the probability of experiencing a rear-end collision is quite low, and the efforts for finding out the cut-off point have not been very conclusive as yet [38, 40].

Other studies have investigated the reliability of the system focusing on the percentage of true alerts without specifying the rate of false alerts or misses. In general, higher levels of reliability result in better performance, but the point where the system becomes useless remains uncertain. Bliss and Acton [46] noted that participants responded more frequently to the warning and manoeuvred more appropriately in avoiding collisions when the warning was 100% reliable. Nevertheless, Maltz and Shinar [39] and Ben-Yaacov *et al.* [10] did not find any differences between systems reliable at 60, 80–85 and 90–95%. Subsequently, Wickens and Dixon [47] carried out a meta-analysis of 22 studies and concluded that a reliability level of 70% could be the criterion from which a system could enhance performance.

Finally, it should be noted that failures of the system could have different effects depending on certain driver characteristics such as age, driving experience [36] or attentional state. For example, an alert triggered too early could be considered as a nuisance alert for an undistracted driver but it could be a hit alert for a distracted driver. Abe *et al.* [25] discovered that drivers rated alerts as less unnecessary when they were distracted by a secondary visual task than when they were undistracted. Thus, it seems that distracted drivers try to compensate their lack of attentional resources by the aid of the system. Maltz and Shinar [48] did not find any differences in distracted groups with regard to the percentage of the time spent in a danger or safe zone according to the reliability of the system (high, medium, and low reliability or one, four, and eight errors per minute, respectively). Nevertheless, the analysis of drivers' performance in case the system failed to deliver an alert showed a higher rate of deceleration responses in the danger zone by the low reliability group compared to the medium and high reliability groups and control group. This result could suggest that even when distracted, drivers can adopt a safe behaviour when the system frequently fails in detecting the hazard; however, when the system is highly reliable, drivers did not compensate the error of the system. As in the Maltz and Shinar study [48], Bueno *et al.*[20] analysed the reliability level of an FCWS (100 or 70% reliable) in drivers distracted by a secondary cognitive task but they also analysed this in undistracted drivers. Behavioural data showed that reliable and unreliable systems only reduced RT when drivers were undistracted but not when they were distracted. However, the analysis of the electrical activity of the brain revealed a slightly different pattern of results. ERP data showed a reduction in the amplitude and the latency of the N2 by the reliable system in simple task condition, whereas the unreliable system reduced the latency of the P3 in dual task condition. These results could be interpreted as an increase of the expectancy of the target in the first case and a diminution of the time required for processing the target at higher cognitive level, in the second case. In parallel to Maltz and Shinar's study [48] but with all due caution, it could be speculated that less reliable systems are more effective than more reliable systems in dual task situations. Nevertheless, further research is needed to shed light on these phenomena.

As has been discussed here, the cut-off between a reliable and an unreliable system does not seem to be not only determined exclusively by the characteristics of the system (level of reliability, type of warning signal, etc.) but also by some driver characteristics such as the attentional state.

12.6 Behavioural adaptation to the system

The last factor addressed in this chapter is the behavioural adaptation defined as 'those behaviours which may occur following the introduction of changes to the road-vehicle-user system and which were not intended by the initiators of the change' (p. 23) [49]. Given its repercussion for road safety, the behavioural adaptation to the assistance system has a market negative connotation, although a wider vision could be adopted by considering all relevant changes associated to the introduction of the assistance system [50].

To our knowledge, there are no studies evaluating the behavioural adaptation to a FCWS using ERP data; therefore, only behavioural studies will be considered in this section.

As previously discussed, the introduction of an FCWS can produce different effects on drivers' behaviour. Some of them are positive, such as faster RTs and the reduction of collisions. However, undesirable behavioural changes can also appear, sometimes as a consequence of the interaction with other factors such as the attentional state of the driver or the reliability level of the system. For example, drivers could engage in more secondary tasks or even assume more risky behaviours by delegating the danger monitoring task to the system (compensatory effects). This could have unintended consequences such as an increase in the brake RT in the case of a real collision or in the case of the system failing to warn drivers. Concerning the failures of the system, false alerts could increase the driving speed and RTs and a reduced miss rate could lead to over-relying on the system.

In general, FCWSs have been analysed in short-term studies and, in most of the cases, the effect of introducing the system could not have been assessed in experimental studies because the order effects were eliminated by the counter-balancing. Therefore, at present it is not fully understood how drivers adapt their behaviour to the system in the first-time uses (learning and appropriation phase), nor after a prolonged use during a long period of time (integration phase) [51]. Moreover, other factors could have an influence on the adaptation to the system, such as traffic conditions, driver characteristics and some attitudes and personality traits like driving style, sensation seeking and the locus of control (internal or external responsibility attribution for the outcome of events). Jamson *et al.* [51] conducted a driver simulator study to analyse the benefits of an FCWS taking into account the driver style, with higher or lower sensation seekers. Their results showed a positive effect of the system compared to the baseline; however, the driver style had little or no impact.

To our knowledge, only one study has been carried out showing the behavioural effect of an FCWS in the long term in a field operation test [10]. In this study, participants had to drive in real traffic conditions with the system (second trial) and without it (first and third trials, and fourth trial six months later). Safer time headway was observed immediately after exposure to the system (third trial). Interestingly, this effect was also observed six months later when participants were exposed again to the same drive without the warning assistance. The behavioural effect of an FCWS in the short term has been also analysed in a recent driving simulator study [52]. We have also analysed the behavioural effect of an FCWS in the short term in a driving simulator study. Participants were required to drive without the system during the first session (baseline). Then, the system was introduced in the second, third, and fourth sessions in order to evaluate the immediate effect of introducing an FCWS and the behavioural adaptation to the system in the short term. Finally, participants drove without the system during the fifth session to analyse the consequences of switching off the system after a short period of adaptation. The results showed that the introduction of the system drastically reduced the RT during the second session compared with the control group who drove without any assistance. During the integration phase (sessions three and four), RTs were also faster for the system group than for the control group. Finally, results showed longer RTs during the last session (with the system switched off), with an increase especially noticeable for the system groups. This last result may provide evidence for the disruptive effect of driving without the warning system, for example, by a system breakdown, once the driver is accustomed to driving with the system. Nevertheless, the control group showed a progressive increment of the RT along the five sessions which could be interpreted as a practice and/or fatigue effect. Therefore, it was not possible to conclude whether the longer RTs for the system groups were completely explained by the absence of the system. As there are not many studies in this respect, further research would be necessary to investigate the effects of driving with or without the system after the integration phase.

Although the number of studies and the situations evaluated are not sufficient to draw final conclusions, it is interesting to observe that among these studies no negative behavioural adaptation effects were evidenced. It could be possible that compensatory effects when driving with an FCWS are more difficult to reproduce than when driving with other systems. Rear-end collisions are relatively infrequent situations compared to, for example, monitoring excessive speed or lane departure. Moreover, given the safety risk involved in these collisions, it could be more difficult for drivers to rely completely on the system. Certainly the reliability level can have a strong impact on the behavioural adaptation to the system. However it has been demonstrated that unreliable systems can also be useful in terms of assistance for drivers.

Psychology of learning could enable interpretation of why unreliable systems can be effective. The association between the warning (stimulus 1) and the hazard (stimulus 2) is produced in very different contexts (i.e. high traffic flow, good or bad weather conditions, secondary tasks, etc.). Moreover, drivers learn that the causal relation between the two stimuli is not perfect. Indeed, as discussed earlier, the perfect system does not exist. Therefore, the presence of false alerts and misses produces an intermittent reinforcement of the warning and, consequently, more resistance to extinction [53]. Nevertheless, given the limited number of studies in this regard, further research would be necessary in order to determine the positive and negative effects of FCWSs used in the short and long term.

12.7 Conclusions

This chapter aimed at reviewing literature assessing the effectiveness of FCWSs. Although it has been shown that the system provides potential benefits for road safety, certain factors such as the attentional state of drivers and the reliability level of the system can mitigate its effectiveness. Further research seems necessary to clarify whether the system helps distracted drivers by manipulating the difficulty level of the secondary tasks and the type of warnings. If, as it has been hypothesised here, the processing of the warning information is not an automated task and needs some attentional resources, it is possible that a training programme could optimise drivers' performance when driving with the system.

The reliability of the system is likely to be one of the most critical factors for the acceptance of the system by drivers. Although failures of the system can have a positive impact on road safety, it is necessary to find a good balance to avoid drivers deciding not to use it. Probably one of the most negative effects of behavioural adaptation could be over-reliance. The studies presented in this chapter did not show evidence of this phenomenon in drivers; however, currently there are only a few studies and, therefore, more research assessing behavioural adaptation after short- and long-term use needs to be conducted.

Finally, ERP studies have shown to be promising in the analysis of the effectiveness of FCWSs. Until now, FCWSs had been mainly studied by recording measures of driver performance and, eventually, by analysing the visual behaviour of drivers. The ERP technique permits to elucidation and dissociation of the different stages of the information processing, some of which are not observable in open behaviour. Thus, the results showed in the studies presented above revealed that the warning signal could operate at preparatory (CNV) and higher cognitive stages (P3).

Throughout this chapter, it has been shown that the effectiveness of a FCWS and its acceptance by drivers depend on numerous factors. On the one hand, factors that directly affect the effectiveness of the system can come from the system itself. For instance, the moment for triggering the warning, the reliability level of the system or the modality of the warning (visual, auditory) are critical elements to take into consideration. But even the 'best' of the systems in terms of technical parameters could not be appropriated for certain drivers. It has been demonstrated that the effectiveness of the system could vary according to the attentional state of the driver. The different driving style, age, experience or health status of drivers should also be considered in order to insure that specific populations, for example, older drivers, could benefit as expected from these systems. Moreover, one of the issues evidenced in some of the studies presented above has been the large inter-individual variability observed among the same type of drivers. Therefore, we strongly recommend developing adaptive FCWSs which take such characteristics into account.

Acknowledgement

This research received funding from the European Community's Seventh Framework Program (FP7/2007-2013) under grant agreement no. 238833/ADAPTATION project (www.adaptation-itn.eu).

References

1. 'Regulation (EC) No 661/2009 of the European Parliament and of the Council of 13 July 2009 concerning type-approval requirements for the general safety of motor vehicles, their trailers and systems, components and separate technical units intended therefor'. 2009
2. United Nations. *Statistics of Road Traffic Accidents in Europe and North America*. Geneva; 2007
3. Knipling R. R., Wang J. S., Yin H. M. *Rear-end Crashes: Problem Size Assessment and Statistical Description*. Washington, DC: National Highway Traffic Safety Administration, DOT HS 807-994; 1993
4. Vogel L., Bester C. J. 'A relationship between accident types and causes'. *24th Southern African Transport Conference*, South Africa; 2005
5. Bella F, Russo R. 'A Collision Warning System for rear-end collision: a driving simulator study'. *Procedia – Social and Behavioral Sciences*. 2011; 20(0):676–686
6. Breuer J. J., Faulhaber A., Frank P., Gleissner S. 'Real world safety benefits of brake assistance systems'. *20th International Technical Conference on the Enhanced Safety of Vehicles (ESV)*, Lyon; 2007
7. Najm W. G., Stearns M. D., Howarth H., Koopmann J., Hitz J. '*Evaluation of an Automotive Rear End Collision Avoidance System*'. Cambridge: National Highway Traffic Safety Administration; 2006
8. Lee J. D., McGehee D. V., Brown T. L., Reyes M. L. 'Collision warning timing, driver distraction, and driver response to imminent rear-end collisions in a high-fidelity driving simulator'. *Human Factors: The Journal of the Human Factors and Ergonomics Society*. 2002;44(2):314–334
9. Abe G., Richardson J. 'Alarm timing, trust and driver expectation for forward collision warning systems'. *Applied Ergonomics*. 2006;37(5):577–586
10. Ben-Yaacov A., Maltz M., Shinar D. 'Effects of an in-vehicle collision avoidance warning system on short- and long-term driving performance'. *Human Factors: The Journal of the Human Factors and Ergonomics Society*. 2002;44(2):335–342
11. Georgi A., Zimmermann M., Lich T., Blank L., Kickler N., Marchthaler R. 'New approach of accident benefit analysis for rear end collision avoidance

and mitigation systems'. *21st International Technical Conference on the Enhanced Safety of Vehicles*, Germany; 2009

12. Anderson R. W. G., Doecke S. D., Mackenzie J. R. R., Ponte G., Paine D., Paine M. *Potential Benefits of Forward Collision Avoidance Technology*. South Australia: Centre for Automotive Safety Research, CASR0106; 2012
13. Loveless N. E. 'The contingent negative variation related to preparatory set in a reaction time situation with variable foreperiod'. *Electroencephalography and Clinical Neurophysiology*. 1973;35(4):369–374
14. Vogel E. K., Luck S. J. 'The visual N1 component as an index of a discrimination process'. *Psychophysiology*. 2000;37(2):190–203
15. Folstein J. R., Van Petten C. 'Influence of cognitive control and mismatch on the N2 component of the ERP: A review'. *Psychophysiology*. 2008;45(1): 152–170
16. Polich J. 'Updating P300: an integrative theory of P3a and P3b'. *Clinical Neurophysiology*. 2007;118(10):2128–2148
17. Kok A. 'On the utility of P3 amplitude as a measure of processing capacity'. *Psychophysiology*. 2001;38(3):557–577
18. Strayer D. L., Drews F. A. 'Multi-tasking in the automobile'. In Kramer A, Wiegmann D, Kirlik A. (eds.). *Applied Attention: From Theory to Practice*. New York, NY: Oxford University Press; 2007. pp. 121–133
19. Rakauskas M. E., Ward N. J., Bernat E. M., Cadwallader M., Patrick C. J., de Waard D. 'Psychophysiological measures of driver distraction and workload while intoxicated'. *3rd International Symposium on Human Factors in Driver Assessment, Training, and Vehicle Design*, Rockport, Maine; 2005
20. Bueno M., Fabrigoule C., Deleurence P., Ndiaye D., Fort A. 'An electrophysiological study of the impact of a Forward Collision Warning System in a simulator driving task'. *Brain Research*. 2012;1470:69–79
21. Regan M. A., Hallett C., Gordon G. P. 'Driver distraction and driver inattention: definition, relationship and taxonomy'. *Accident Analysis and Prevention*. 2011;43:1771–1781
22. Engström J., Johansson E., Östlund J. 'Effects of visual and cognitive load in real and simulated motorway driving'. *Transportation Research Part F: Traffic Psychology and Behaviour*. 2005;8(2):97–120
23. Kaber D. B., Liang Y., Zhang Y., Rogers M. L., Gangakhedkar S. 'Driver performance effects of simultaneous visual and cognitive distraction and adaptation behavior'. *Transportation Research Part F: Traffic Psychology and Behaviour*. 2012;15(5):491–501
24. Kramer A. F., Cassavaugh N., Horrey W. J., Becic E., Mayhugh J. L. 'Influence of age and proximity warning devices on collision avoidance in simulated driving'. *Human Factors: The Journal of the Human Factors and Ergonomics Society*. 2007;49(5):935–949
25. Abe G., Itoh M., Yamamura T. 'Visual distraction when driving and its influence on driver response to a forward collision warning system'. *SICE Annual Conference* 2011, Tokyo, Japan; 2011

26. Ho C., Spence C. 'Using peripersonal warning signals to orient a driver's gaze'. *Human Factors: The Journal of the Human Factors and Ergonomics Society*. 2009;51(4):539–556
27. Bueno M., Fort A., Francois M., Ndiaye D., Deleurence P., Fabrigoule C. 'Effectiveness of a Forward Collision Warning System in simple and in dual task from an electrophysiological perspective'. *Neuroscience Letters*. 2013;541:219–223
28. Hackley S. A., Valle-Inclán F. 'Which stages of processing are speeded by a warning signal?'. *Biological Psychology*. 2003;64(1–2):27–45
29. Rudell A. P., Hu B. 'Does a warning signal accelerate the processing of sensory information? Evidence from recognition potential responses to high and low frequency words'. *International Journal of Psychophysiology*. 2001;41(1):31–42
30. Wickens C. 'The structure of attentional resources'. In Nickerson R. (ed.). *Attention and Performance*. Hillsdale, NJ: Lawrence Erlbaum; 1980. pp. 239–257
31. Mohebbi R., Gray R., Tan H. Z. 'Driver reaction time to tactile and auditory rear-end collision warnings while talking on a cell phone'. *Human Factors: The Journal of the Human Factors and Ergonomics Society*. 2009;51(1): 102–110
32. Ho C., Reed N., Spence C. 'Multisensory in-car warning signals for collision avoidance'. *Human Factors: The Journal of the Human Factors and Ergonomics Society*. 2007;49(6):1107–1114
33. Santangelo V., Van der Lubbe R., Olivetti Belardinelli M., Postma A. 'Multisensory integration affects ERP components elicited by exogenous cues'. *Experimental Brain Research*. 2008;185(2):269–277
34. Baddeley A. D. 'Exploring the central executive'. *The Quarterly Journal of Experimental Psychology*. 1996;49A(1):5–28
35. Campbell J. L., Richard C. M., Brown J. L., McCallum M. *Crash Warning System Interfaces: Human Factors Insights and Lessons Learned*. Washington, DC: National Highway Traffic Safety Administration; 2007
36. Lees M. *Context Conditions Drivers' Disposition Towards Alarms*. PhD thesis. Iowa: University of Iowa; 2010
37. McLaughlin S. B. *Analytic Assessment of Collision Avoidance Systems and Driver Dynamic Performance in Rear-End Crashes and Near-Crashes*. Virginia: Virginia Polytechnic Institute and State University; 2007
38. Kiefer R. J., LeBlanc D., Palmer M., Salinger J., Deering R., Shulman M. *Forward Collision Warning Systems: Development and Validation of Functional Definitions and Evaluation Procedures for Collision Warning/Avoidance Systems*. Washington, DC: National Highway Traffic Safety Administration, DOT HS 808-964; 1999
39. Maltz M., Shinar D. 'Imperfect in-vehicle collision avoidance warning systems can aid drivers'. *Human Factors*. 2004;46(2):357–366

40. Lerner N. D., Dekker D. K., Steinberg G. V., Huey R. W. *Inappropriate Alarm Rates and Driver Annoyance*. Washington, DC: National Highway Traffic Safety Administration, DOT HS 808-533; 1996
41. Abe G., Itoh M., Tanaka K. 'Dynamics of drivers' trust in warning systems'. *15th IFAC World Congress*, Barcelona, Spain; 2002
42. Cotté N., Meyer J., Coughlin J. F. 'Older and younger drivers' reliance on collision warning systems'. *Human Factors and Ergonomics Society Annual Meeting Proceedings*. 2001;45:277–280
43. Yamada K., Kuchar J. K. 'Preliminary study of behavioral and safety effects of driver dependence on a warning system in a driving simulator'. *Systems, Man and Cybernetics, Part A: Systems and Humans, IEEE Transactions on*. 2006;36(3):602–610
44. Abe G., Richardson J. 'The effect of alarm timing on driver behaviour: an investigation of differences in driver trust and response to alarms according to alarm timing'. *Transportation Research Part F: Traffic Psychology and Behaviour*. 2004;7(4–5):307–322
45. Parasuraman R., Hancock P. A., Olofinboba O. 'Alarm effectiveness in driver-centred collision-warning systems'. *Ergonomics*. 1997;40(3):390–399
46. Bliss J. P., Acton S. A. 'Alarm mistrust in automobiles: how collision alarm reliability affects driving'. *Applied Ergonomics*. 2003;34(6):499–509
47. Wickens C. D., Dixon S. R. 'The benefits of imperfect diagnostic automation: a synthesis of the literature'. *TTIE*. 2007;8(3):201–212
48. Maltz M., Shinar D. 'Imperfect in-vehicle collision avoidance warning systems can aid distracted drivers'. *Transportation Research Part F: Traffic Psychology and Behaviour*. 2007;10(4):345–357
49. OECD. *Behavioural Adaptations to Changes in the Road Transport System*. Paris: Organisation for Economic Co-operation and Development; 1990
50. Saad F., Hjälmdahl M., Cañas J., Alonso M., Garayo P., Macchi L., *et al. Literature Review of Behavioural Effects*. Brussels: Information Society Technologies; 2004
51. Jamson A. H., Lai F. C. H., Carsten O. M. J. 'Potential benefits of an adaptive forward collision warning system'. *Transportation Research Part C: Emerging Technologies*. 2008;16(4):471–484
52. Bueno M. 'Impact of a collision warning system on information processing and driver behaviour'. Doctoral Thesis, University of Bordeaux and IFSTTAR, France, 2013
53. Mackintosh N. J. *The Psychology of Animal Learning*. New York, NY: Academic Press; 1974

Chapter 13

Introduction of automated platooning in traffic: What is the impact on non-automated drivers? – An analysis of the influence of the short time headways held by automated platoons

Magali Gouy[1], Nick Reed[1] and Alan Stevens[1]

Abstract

Automated driving is a realistic future scenario that will arise from a combination of different Advanced Driver Assistance Systems (ADAS) that are gradually being introduced in vehicles to support aspects of the driving task. As the traffic fleet becomes more varied and equipped vehicle drivers (EVDs) are mixed with unequipped vehicle drivers (UVDs), it becomes important to ensure that this evolution is safe for EVDs and remains safe for UVDs. The aim of this chapter is therefore to introduce a new research field investigating the impact of EVDs on the UVDs with a particular emphasis on one scenario: the behavioural adaptation of the UVDs to the short time headway (THW) kept by automated vehicles. The chapter commences with an overview of the development of automated systems and their increasing role within the driving task towards the implementation of fully automated driving. Finally, some guidelines are provided for conducting research in this new area.

13.1 From ADAS to automated driving

Significant technical progress over the last 30 years has increased the capability for a vehicle to gather information about the driving environment, to support the driver in the execution of manoeuvres and to communicate with other vehicles or infrastructure. Several milestones have demonstrated that the grouping of all these capabilities makes fully automated driving a possibility. The idea of an automated vehicle was launched by General Motors at the 1939 World's Fair [1] and in 1980, Ernst Dickmanns and his team at the Bundeswehr University and Mercedes-Benz equipped a van with cameras and sensors, able to control autonomously steering

[1] TRL, Crowthorne House, Nine Mile Ride, Workingham, Berkshire, RG40 3GA, United Kingdom

wheel, throttle and brakes through a computer. In 1994, at the EUREKA-PROMETHEUS project's final demonstration, Ernst Dickmanns and his team demonstrated that their driverless cars were able to drive more than thousand kilometres in the traffic on a three-lane motorway [2]. Further progress in automated driving was inspired by the 2005 DARPA (the US Defense Advanced Research Projects Agency) Grand Challenge in which a variety of research and commercial organisations developed autonomous vehicles that were required to complete an off-road route [3]. In 2007, the DARPA Urban Challenge required autonomous vehicles to manage behaviour amongst other vehicles and obey traffic rules [4]. The 2010 VisLab Intercontinental Autonomous Challenge [5] exposed supervised fully autonomous vehicles to general traffic, requiring the vehicles to drive 13,000 km from Parma, Italy to the World Expo in Shanghai, without human intervention. In 2010, Google announced that their fleet of Toyota Prius hybrids and Lexus RX hybrids had driven more than 140,000 miles on public roads with only a few human interventions [6].

These milestones show that automated driving is technically achievable. The implementation of automated driving is progressively taking place as Advanced Driver Assistance Systems (ADAS) taking over an increasing part of the driving task are developed and fitted in cars. ADAS taking over either the lateral or longitudinal control of the vehicle are already on the market. ADAS combining the two are on the brink of the market such as the Autopilot system announced by Volkswagen [7].

To understand the evolution of automation within the driving task, the advantages of automated systems are described followed by a description of the scope of actions of ADAS which provide the building blocks for automated driving. Then, the technical aspects and deployment of automated driving are addressed. Finally, the chapter focuses on behavioural adaptation of UVDs that could possibly arise in mixed traffic.

13.1.1 Advantages of automated systems

The influence of human error in road accidents was revealed by a famous study from Treat and colleagues [8]. They investigated 2,258 road accidents and drew the conclusion that in 93% of the accidents in their sample, human errors was a contributory factor (against 34% of accidents where environmental factors were a contributor and 13% of accidents where vehicle factors contributed). Fuller argued that road accidents occur when task demands exceed drivers' capabilities [9]. Hence, in a driving situation containing a large amount of relevant information, drivers' capability to attend to all the relevant information may reach its limit. In such a situation of driving demand, drivers can fail to detect critical changes in the environment, leading to loss of control and/or a collision. For instance, drivers may maintain that they did not see the object with which they collided, even though this object was in an apparently clearly visible position within their field of view. This type of error has been classified as 'looked-but-failed-to-see' [10] because it seems clear that the object passed within the driver's field of view but was not consciously detected. ADAS are designed to address this discrepancy; by sensing the driving environment and by providing the driver with information, warnings and/or direct assistance, it is envisaged that task demand is reduced, thereby decreasing the likelihood that it will exceed drivers' capability and the chance of an accident is diminished.

Beyond improving safety within the existing driving situation, the benefits of automated control can be exploited to increase road capacity. By automating longitudinal control of vehicles, gaps between vehicles can be reduced, increasing traffic capacity of the existing road infrastructure. Besides saving space and increasing the traffic flow [11], tightly spaced vehicles have a positive effect on energy consumption induced by the slipstream effect [12].

13.1.2 Different categories of ADAS

Driving is a complex task that encompasses several sub-tasks. The scope of action for ADAS is therefore manifold. ADAS can be divided into categories; the most popular taxonomies to classify the large variety of different ADAS are based on: the level of intervention, the type of intervention and the level of automation.

- The *level of intervention* classifies ADAS depending on which of the three hierarchical levels of the driving task they support: the strategic, the tactical or the operational level [13]. At the lowest level (operational level), control-based behaviours take place such as steering, braking and speed control; ADAS corresponding to this level would be, for example, ESP (electronic stability program) or ABS (anti-lock braking system). On the intermediate level (tactical level) manoeuvre control is exercised by the drivers, which includes interacting with traffic signs, other road users, merging and lane changing. An example for this category is the intersection assistance system that warns drivers if there are other road users on their trajectory if they want to accomplish a turning manoeuvre or change lane. Finally, decisions are made at the strategic level regarding means of transportation and the route to drive. An example within this level is the navigation system, which assists drivers in choosing an appropriate route.
- The *type of intervention* also provides a distinction between systems and three different categories can be distinguished. First, driver information systems provide information either related to the driving task (e.g. navigation, traffic or weather information) or unrelated to the driving task (e.g. emails). The second category represents driver warning systems that alert drivers of potential dangers, such as lane departure and, as a third category, intervening systems provide an active support to drivers in taking over parts of the driving task.
- One of the most cited taxonomies of *levels of automation* is that proposed by Sheridan and Verplank [14]. It divides automation into ten different levels, where the first level is full manual control and the tenth level represents full automation of a system. These automation levels also can be related to a simple four-stages view of human information processing (Information Acquisition, Information Analysis, Decision Selection, Action Implementation) [15].

Each of the classifications presented in this section sheds light on one particular aspect of ADAS. Automated driving is derived from developments with ADAS and will thus encompass different aspects of ADAS as defined by the different categories.

13.1.3 Automated driving

The availability of reliable new technology and the expected benefits of automated driving have brought an increasing interest in its development and implementation. Two different lines of development are emerging: individual self-driving cars and cooperative automated systems. Information about the development of individual self-driving cars is generally led by private companies (such as Google, Continental or BMW) and to some extend supported by public institutions (e.g. European Commission). As a result, knowledge on development of self-driving cars is scarcely available as private companies have an interest in retaining intellectual property in relation to their technical work. The development of cooperative automated systems is generally undertaken by public institutions (e.g. U.S. Department of Transportation, European Commission) that are more predisposed to disseminate information. Cooperative automated systems can, again, be divided into two different categories depending on whether they are operating directly on a standard motorway or on a separate, specialised route such as an automated highway system (AHS) [16].

Platoons of electronically coupled vehicles represent a form of cooperative automated system currently in development in Europe and designed to operate on a normal unmodified motorway. Within a platoon, the first vehicle is driven manually while the others follow automatically in single file at relatively short following distances. Many successive research projects have been aimed at developing and implementing the system. First, the PROMETHEUS project (Program for European Traffic with Highest Efficiency and Unprecedented Safety; 1988–1995) achieved the technical development of an autonomous vehicle [17]. The first step towards the development of platoons in Europe was made within the projects CHAUFFEUR I and II that achieved a platoon of three vehicles electronically coupled [18, 19]. Thereafter, research on platooning was conducted in Germany with different projects financed by the Federal Ministry of Education and Research (BMBF): EFAS (2001–2002) [20], MFG (2003–2004) [21] and KONVOI (2005–2008) [22]. Technical feasibility has been proven throughout the consecutive projects and research increasingly focuses on how to make the system useable and acceptable. As well as work on the technical development of platoons, aims of the European Commission funded project SARTRE (Safe Road Trains for the Environment) (2009–2012) included solutions to acceptability issues, defining a set of 'platooning' strategies to operate on public motorways and determining business models to encourage the use of platoons [23].

Lank and colleagues [22] explained how the KONVOI project imagined the function of a platoon (illustrated in Figure 13.1). Drivers that wish to cover a

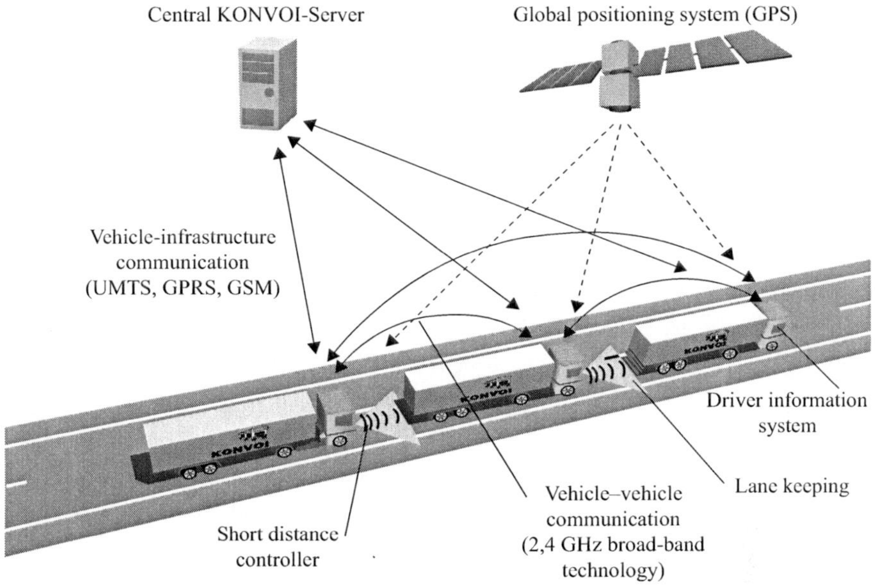

Figure 13.1 Overview of a truck platoon as developed in KONVOI (reproduced with permission from [22])

certain distance in a platoon communicate this intention to the system via a human–machine interface (HMI). A connection to the central server is established by means of mobile communication (UMTS, GSM, GPRS). The KONVOI system transmits the vehicle's current GPS location, destination and desire to join a platoon to the central server. The central server scans for a platoon heading in a compatible direction and able to accept an additional vehicle. If a suitable platoon is found, the server informs the driver and subsequently monitors the merging process. Once drivers are sufficiently close to the platoon (60 m), a request is sent to the leading vehicle. If the request is accepted, the automated system takes over the lateral and longitudinal control of the vehicle. The distance between the vehicle that joins a platoon and the lead one is reduced to a target distance of 10 m. At a speed of 90 km/h, this represents time headway (THW) of 0.4 s. The safety indicator THW is commonly used to estimate the criticality of a driving situation; this is the elapsed time between the back of the lead vehicle passing a point on the roadway and the front of the front vehicle passing the same point [24]. The THW is a timescale (s) measured by dividing the distance (m) through the speed (m/s): THW (s) = distance (m)/speed (m/s).

In the United Kingdom, the Driving Standard Agency recommends that drivers should maintain a THW of at least two seconds when following a vehicle [25]. In Germany, a rule of thumb recommends that drivers keep a distance in metres of 'half the speedometer' [26]. At a speed of 100 km/h, this results in a distance of 50 m and thus a THW of 1.8 s.

The system enabling platooning, as developed in KONVOI, combines several ADAS of which the main components are [27]:

- Actuators (steering and powertrain)
- Sensors (RADAR-sensor for object recognition in close-up, LIDAR-sensor for far-range and CMOS-camera for lane recognition)
- Vehicle–vehicle communication (V2V) over a wireless local area network (WLAN)
- Automation units (coordination of the different components)
- Control unit (adaptive cruise control and automatic lane keeping guidance)
- Driver information system: human–machine-interface, central server (or organisation assistant), GPS and vehicle-infrastructure communication

Longitudinal control is taken over by an adaptive cruise control module based on distance sensor systems that capture the distance to the vehicle ahead, supported by vehicle–vehicle communication. These data enable the ACC to keep a tight vehicle–vehicle distance safely. Lateral control is achieved by an automatic guidance system (or lane keeping system) based on image data processing (to identify current lane position) and a distance sensor (to determine lateral offset of the driven vehicle in relation to the lead vehicle). A steering actuator delivers the necessary steering movement to correct the position of the vehicle in lane.

13.2 Behavioural adaptation of EVDs to the short THWs held by automated platoons

The introduction of automated platoons will be accompanied by a reduction of the THW as it results in increased network and energy efficiency. In the framework of the KONVOI project on truck platooning, Skottke [28] investigated whether the small THW adopted when driving within a platoon cause carry-over effects in the form of reduced THWs in the subsequent driving period after having left a platoon. The results of a first experimental trial using a car simulator showed that participants adapted their behaviour to the small THW maintained in a platoon by choosing shorter THW after the platoon drive. The results of a second experimental trial showed that behavioural adaptation occurred even after a short time spent in a platoon. Skottke [28] subsequently investigated possible underlying factors causing this behavioural adaptation in two experimental trials investigating whether this behavioural adaptation is perception or action based. These options are discussed below.

13.2.1 The perception based component of behavioural adaptation

The idea is that behavioural adaptation occurs through a change in the driver's frame of reference. Theories about frames of reference influencing perception originate in psychophysics [29, 30] and consider the influence of context in judging a stimulus.

Table 13.1 Serial stimuli for the two subject groups

Group 1 (low series) (s)	0.3	0.6	0.9	1.2		
Group 2 (high series) (s)			0.9	1.2	1.5	1.8
			Overlapping stimuli, rated by subjects			

Source: Reproduced with permission from [31].

Participants drove within a platoon and were exposed to different sets of THWs within a platoon. One group were exposed to sets of four 'small distances' (0.3 s/0.6 s/0.9 s/1.2 s) and the other group were exposed to sets of four 'large distances' (0.9 s/1.2 s/1.5 s/1.8 s) (see Table 13.1).

Each group had to give their opinion about each of the different distance afterwards (close/far and comfortable/uncomfortable). In addition, drivers' following behaviour was assessed after each drive in platoon with different distance.

Two of the four distances were similar between the two groups (0.9 s and 1.2 s) forming an overlapping condition that enabled a comparison between the two groups. The expectation was that the judgement of the overlapping stimuli would be evaluated as more comfortable by the group exposed to 'small distances' than by the group exposed to 'large distances'. Of interest was also the THW behaviour after the experience with the different distances in the platoon. The expectation here was that manual driving behaviour after leaving the platoon will be commensurate with the judgement of distance in the platoon. It was expected that participants judging the overlapping stimuli in a more aversive way would keep a larger distance and that the participants having a less aversive judgement would keep a smaller distance.

Results showed that the group with large distances rated similar conditions as more aversive than the small distances group but the difference was statistically not significant. A second expectation was that driving behaviour changed as a consequence of the drivers' judgements. The large distances group showed no pre–post effects; the small distances group significantly reduced their THW after having been coupled with distances from 0.3 s to 1.2 s. The conclusions were drawn that the shift in the frame of reference changed drivers' behaviour but not their judgement and consequently; this suggests that behavioural adaptation occurs unconsciously.

These results are in line with other evidence in the literature showing behavioural adaptation as the effect of a shift in drivers' frame of reference, notably in studies on speed adaptation. Studies on speed adaptation have shown that after travelling at high speed, drivers will underestimate the velocity at which they are travelling [32–34]. In a simulator study, participants were required to decelerate a vehicle from 110 km/h to 64 km/h and subsequently underestimated their speed [32]. In a field study, vehicles' speed on a highway was assessed by means of radar deployed in a 80 km/h speed limit area [33]. The data from the northbound traffic that previously experienced a speed limit of 105 km/h were compared to the southbound traffic previously exposed to a speed limit of 64 km/h. Results show that northbound traffic velocities exceeded those of southbound traffic by an average of 6.9 km/h. In another field study, the speed of two groups were compared after each of the group had to drive at one specific speed, either low or high. The average speed of the group driving at higher speed was higher than the speed of the group driving at lower speed [34].

13.2.2 The action based component of behavioural adaptation

The action based approach is based on task switching theories [35]. In a task switching paradigm, performance on repeated tasks is compared with performance when there is a requirement to switch between different tasks. Typically, reaction time and error rate are greater on task switch than on repetition trials [36]. This suggests that there is a performance cost in shifting from one task set to another. Furthermore, Mayr and Keele [37] demonstrated that the implementation of a new task requires the inhibition of the previous one (backward inhibition). As an expression of backward inhibition, Mayr and Keele [37] predicted increased response time when shifting to a task set that had to be abandoned recently and thus, suffers residual inhibition. Skottke [28, 31] transfers this effect from the micro level of reaction time in milliseconds to the macro level of traffic behaviour, theorising that drivers who experienced small THWs in a platoon and need to change THW when leaving the platoon would be backward inhibited by the THW held during the platoon drive. Thus, drivers would not adopt an appropriate THW when driving manually after leaving a platoon but keep a similar THW to that within the platoon drive.

To test the hypothesis of backward inhibition, Skottke [28] tested how behavioural adaptation changed after leaving a platoon. One of the scenarios tested whether drivers would jump to appropriate THWs, whereas the aim of the second scenario was to test if THW would linearly increase to an appropriate one. A jump could be interpreted as a task switch, delayed by the backward inhibition responsible for the behavioural adaptation. The results show neither evidence for a jump in the choice of the THW during the manual drive following the platoon drive nor for a linear change; drivers constantly keep the same THW as during the platoon drive. The study showed no evidence for backward inhibition but demonstrated that behavioural adaptation resulting from a drive with small THW had a long lasting effect.

In summary, Skottke's [28] results support the idea that a shift in the frame of reference leads to behavioural adaptation and that although the shifted frame of reference changed behaviour, it had no significant impact on the judgement, suggesting that the behavioural adaptation occurred unconsciously.

13.3 The possibility for behavioural adaptation of UVDs to the short THW held by automated platoons in the light of social psychological models

The progressive introduction of ADAS and ultimately of automated vehicles will lead to a mixed traffic scenario where unequipped vehicle drivers (UVDs) will have to interact with equipped vehicle drivers (EVDs). In a mixed traffic scenario, changes in the behaviour of EVDs (i.e. reduction of distance between vehicles) induced by automated system will become visible. As a typical manifestation of human intelligence, drivers adapt their behaviour to perceived changes. The research literature contains many examples of behavioural adaptation occurring with different types of ADAS such as ACC [38], Intelligent Speed Adaptation (ISA) [39], Lane-Keeping Assistance [40], ABS [41], night vision systems [42], navigation systems [43] and platoons [28, 31] and concerns have been raised that drivers may adapt to these systems in an unintended manner, potentially jeopardising the expected safety benefits [44]. The most well-known theories to explain the origin of behavioural adaptation are motivational models such as risk [45] and workload models [46]. Trust theories developed in the area of automation are also suitable to explain behavioural adaptation [47].

Thus far, research has mainly focused on the behavioural adaptation of the driver using ADAS and has largely neglected to investigate potential behavioural adaptation on behalf of UVDs encountering EVDs. This focus is justified as long as the penetration rate (or number of equipped vehicles) is negligible. Over recent years, however, an increasing number of ADAS have entered the market and ADAS penetration rates can be expected to steadily increase over time, which may result in the number of equipped vehicles outnumbering unequipped vehicles in the future.

Often overlooked, but perhaps also a contributing factor in behavioural adaptation, are social psychological processes. Since the driver is not alone on the road, social processes can be expected to influence the driver. A great deal of research within the realm of social psychology of driving has investigated the link between intentions and behaviour using the framework of the theory of reasoned action (TRA) [48, 49] and its extension, the theory of planned behaviour (TPB) [50] (see for example [51]). The TRA states that a person's behaviour is determined by his or her intention to perform this behaviour [48, 49]. Three theoretical constructs jointly determine behavioural intentions: attitude to behaviour, subjective norms and perceived behavioural control. The attitudes towards the behaviour are formed by the product of an individual's beliefs about consequences and the evaluation of outcomes (i.e. the seriousness of consequences). The subjective norm represents the perceived social pressure by significant others and reference groups and the motivation to comply. The TPB adds the construct of perceived behavioural control that refers to the individual's perceived ease or difficulty to perform the behaviour of interest. Figure 13.2 gives an illustration of the model as described above.

In an extended TPB model, additional norms have been specified (normative, descriptive and personal) [52]. The authors of the extended model specify that the 'significant others' affecting the subjective norms in the TPB [50], are rarely present in driving situations and so have negligible influence. However, drivers can be influenced by the behaviour of other road users in general, which the authors refer to as the normative norm. In a number of studies it was demonstrated that displaying to drivers the number of other road users that have not broken the speed limit previously on a particular road leads to a decrease of speed violation on that road [53]. A similar study observed driver behaviour in response to Variable Message Signs (VMS) which carried information about the percentage of drivers who were not speeding and tailgating in the last few days [54]. Results showed that

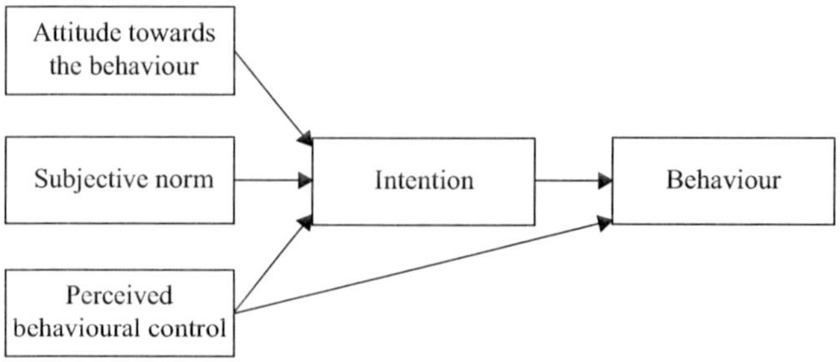

Figure 13.2 Simplified model according to the theory of planned behaviour

the incidence of these behaviours reduced when the VMS messages were displayed. Descriptive norms are characterised by the imitation of other road users in the immediate vicinity. To illustrate this, several models exist which explain the process of the social contagion of speed [55]. Limited but convincing empirical support of the models was provided by observational study of drivers' speed choice in following and free-following situations, showing that vehicles close to each other travel at similar speeds. More empirical evidence show a significant correlation between observed speed and the speed of others as well as between a driver's estimate of own speed and their estimate of other's speed [56]. Finally, personal norms determine what people should do in order to be consistent with their self-image. Before engaging in certain behaviour, people will consider the impact on their self-image. If the behaviour is inconsistent with the personal norm, the anticipated damages to the self-image inhibit the behaviour.

In summary, the extended TPB model [52] stresses the importance of the influence of others on behaviour. The presence is either imagined: in the descriptive norm the others' behaviour serve as a model ('That's what people do') and in the personal norm the judgement of others is instrumental ('What will people think about it?'); or the presence of others is direct, real and visible, as reflected by the normative norm: drivers are inclined to imitate the observed behaviour of others surrounding them.

13.4 Investigations of behavioural adaptation of UVDs to the short THW held by automated platoons

Concerns have been raised in the research community, which have been encapsulated in a prominent report from the Organisation for Economic Co-operation and Development (OECD), that drivers can react in unexpected ways to the introduction of safety measures and ADAS, a phenomenon labelled 'behavioural adaptation' [44]. Thus far, research has focused almost exclusively on the behavioural adaptation of the driver as a result of direct interaction with EVDs. Both the market penetration and functionality of ADAS is increasing; it is therefore conceivable that they may influence traffic norms. The UVDs might become aware of the changes generated through ADAS in two different ways: indirect feedback, e.g. via the media, or direct feedback, e.g. obvious changes in other vehicles' behaviour [57]. For instance, indirect feedback can make UVDs aware that ADAS can improve safety for EVDs and may lead to less cautious behaviour from the UVDs. Since it is difficult to foresee what form indirect feedback will take in the future, it might be challenging to test experimentally how indirect feedback will influence drivers. However, to some extent it is already possible to test experimentally the impact of direct feedback on the UVDs as some obvious changes in vehicles dynamic are already known. By automating longitudinal control of vehicles and supported by vehicle-to-vehicle communication (V2V), gaps or THW between automated vehicles will be reduced, increasing traffic capacity of the existing road infrastructure.

Section 13.2 reviewed research showing that a change in the drivers' frame of reference is a cause for behavioural adaptation. There is evidence that drivers will underestimate the speed at which they are travelling after an exposure to a certain speed. Inversely, drivers will overestimate their distance to the lead vehicle after an extended exposure to a small distance. After a platoon drive, the shorter headways have become 'normal', leading drivers to reduce the THW they would normally keep. In addition, Section 13.3 showed evidence in the literature for influence on the driver of norms generated either by the direct or imagined presence of other drivers. As visual processes seemed responsible for behavioural adaptation of platoon drivers to short THWs in platoons and because the increasing amount of vehicles keeping short THWs in traffic might change the norm related to THW, it is postulated here that UVDs observing smaller THWs within platoons may also experience a shift in their frame of reference. Consequently, they may reduce their own THW, increasing the probability of a rear-end accident. This scenario highlights a potential risk that may emerge from the interaction between UVDs and EVDs in a mixed traffic situation and suggests that this interaction should be investigated prior to the widespread introduction of ADAS that enables platooning.

Recent research has illustrated that UVDs may indeed modify their behaviour as a result of driving in mixed traffic including platoons of vehicles maintaining short THWs. A simulator study conducted at TRL investigated whether drivers are willing to keep a THW smaller than their preferred one to conform to the norm established by the presence of platoons holding short THWs [58]. Two constructs were distinguished to answer this question: the preferred THWs represent a range of THWs drivers feel safe with and the THW that is indeed adopted by drivers in a given situation (adopted THW). Hence, comparing the minimum adopted and minimum preferred THW informs whether drivers would adopt a THW beyond their preferred limit as a result of the influence of platoons.

To measure the influence of short THWs kept in platoons on the THW adopted by drivers, we carried out an experiment in the driving simulator. Forty-two participants were confronted with a car-following task in different traffic conditions: in two conditions, there was a platoon of vehicles in the inside lane, where the THW between the vehicles was either large (THW = 1.0 s) or short (THW = 0.3 s); in a third baseline drive, there was no platoon present in the inside lane.

Drivers' minimum preferred THW was determined using the psychophysical method of limits [59] in the simulator. The participants' car was alternatively

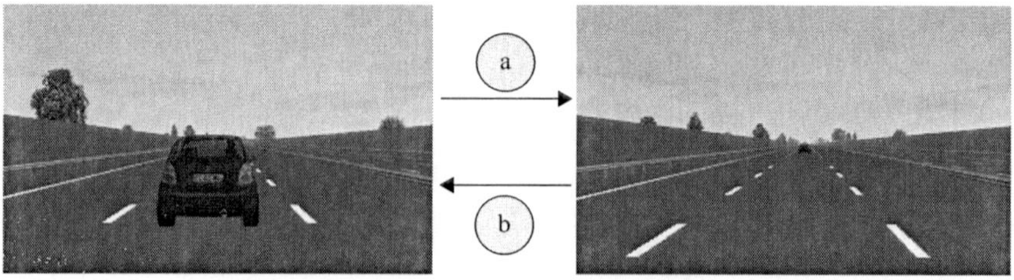

Figure 13.3 The psychophysical method of limits implemented for the assessment of preferred THW includes the presentation of a set of (a) increasing THWs starting with a very small THW (0.1 s) and (b) decreasing THWs starting with a large THW (2.5 s)

gradually either opening (Figure 13.3a) or closing (Figure 13.3b) the gap to a lead vehicle (LV) in steps of THW = 0.1 s. After a THW was presented for 5 s, the screen was blanked and participants were asked to respond 'yes' if they would normally adopt this THW or whether it was 'too short' or 'too large'. In each series, there was a point where participants' judgement changed, thus forming a threshold. The presentation of the set of increasing and the set of decreasing distances was counterbalanced and the results from both sets were averaged. As the result in each set represented a threshold, which was the lowest THW that participants would accept, the output of the THW assessment represents, in fact, a minimum preferred THW.

The study alternated the evaluation of the adopted THW (simulator drive) under three different conditions with the evaluation of the preferred THW (psychophysical method of limits) taking place after each of the simulator drives.

Figure 13.4 depicts the average minimum preferred and adopted THW at the three measuring points. Results showed that the average minimum preferred THW varied significantly [$F(2, 82) = 3.26$, $p = 0.04$, $n_p^2 = 0.07$] but the F value is closed to the significant level and the effect is small. Moreover, pair-wise comparisons using Bonferroni correction showed that these differences were not statistically significant. In addition, Cronbach's alpha was computed for the three measurements of preferred minimum THW and results showed high internal consistency ($\alpha = .83$). It can be concluded that there is a consistency between the measures of preferred THW taken at different points. Results showed that the average minimum adopted THW varied significantly [F (1.42, 58.3) = 9.65, $p < .00$, $n_p^2 = 0.20$]. The results support the idea of two distinct constructs. Comparing the two parameters will inform about drivers' propensity to keep a THW that is shorter than the one they would prefer to keep resulting from a shift in their frame of reference. Dependent t-test has been computed and Bonferroni correction has been applied to compare minimum adopted and preferred THW in each traffic condition. There was a significant difference between the adopted and preferred THW in condition BL [t (41) = −3.44, $p = .001$, $r = -0.23$] but no significant differences were observed in conditions THW03 [t (41) = − .88, $p = .39$, $r = -0.05$]. In THW10, the difference is close to the significant level ($\alpha = 0.0167$) but does not reach it [t (41) = −2.07, $p = .04$, $r = -0.11$]. It thus emerges from the results that

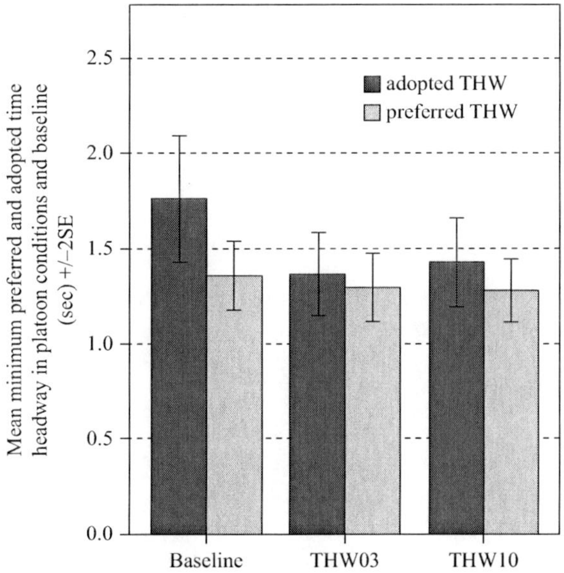

Figure 13.4 Averaged minimum preferred and minimum adopted THW and standard error (+/− 2SE) in the three traffic conditions (BL, THW03, THW10) (N = 42)

participants' minimum adopted THW did not drop under but was very close to the minimum preferred THW in condition THW03. It can be concluded that, in the conditions of the study, platoons can lead drivers to drive closer to their limits.

This study illustrates the impact of short THWs on UVDs in one very specific driving scenario. Behavioural adaptation arises as a result of a complex combination of the elements of the driver–car–environment system. Therefore, each change in variables of one system element might impact the magnitude or direction of behavioural adaptation. Some examples for variables in each of the elements are presented below:

- An important variable of the environment is the traffic. Possible variations operating within traffic not only can be the penetration rate of automated vehicles but also can be the traffic flow and the traffic density.
- Characteristics of the equipped vehicles can vary such as the length of the formed platoon, the THW kept between vehicles and the type of vehicles involved in a platoon (cars or trucks).
- Inter- and intra-variability of drivers' characteristics will influence the way drivers adapt their behaviour to EVDs in traffic. On the one hand, drivers must be aware of the presence of automated vehicles in traffic, which might rely for instance on their visual system or on their available cognitive capacity (changing depending on the driving situation). On the other hand, some more innate characteristics might affect the way drivers adapt their behaviour to certain situation such as their driving skills and their personality.

It is recommended that further studies should also assess whether a driver is still in control of the situation when the adopted THW is close or below their preferred THW. For instance, can the driver react to unexpected braking behaviour of the lead vehicle? Further studies should also investigate the cost for drivers, as reflected in the experienced workload, of keeping a THW that is close or below their preferred one.

13.5 Conclusion

This chapter has highlighted the absence of research on the behavioural adaptation of UVDs to the presence of (semi-) automated systems in traffic. Evidence presented suggests that drivers are influenced by other drivers in the surrounding area. Arguably, if automated systems induce a change in the driver–car–environment system, it will have an impact on the UVDs. This chapter illustrated the issue focusing on the behavioural adaptation of UVDs to automated systems keeping short THWs. Based on the study presented here, it is concluded that behavioural adaptation can occur in certain driving conditions. However, as discussed, a large range of distinct variables might impact on the magnitude and direction of behavioural adaptation. Further studies are therefore required to investigate the impact of these variables on UVDs behavioural adaptation and, particularly, to examine conditions in which the interactions may be unsafe.

Beyond the focus presented here and because other systems' characteristics might affect UVDs, it is of paramount importance to develop good practice in systematically investigating the impact of ADAS and automated driving on the UVDs in order to prevent any undesirable effects. If behavioural adaptation is observable in a consistent way and in a way that is a threat for drivers' safety, then countermeasures should be investigated. Behavioural adaptation can be controlled using a range of traditional methods such as the famous 'three E's': Engineering, Education and Enforcement. The first engineering measures could be implemented on platoons themselves. For instance, rules enumerated by the gestalt psychology [60] could be applied to make platoons look like a separate entity. Rules that could be applied in this context are, for instance, the law of proximity or law of similarity. The law of proximity is put into practice anyway when cars are driving sufficiently close together and the law of similarity could be put into practice by, for instance, using a certain light pattern for vehicles belonging to the same platoon. Still, it is questionable whether according to these laws, platoons are perceived as a distinct entity and whether the perception of platoons as a separate entity would mitigate any negative impact on UVDs. Of course, all vehicles could be fitted with a system warning against unsafe THW [61].

Other countermeasures could highlight education. Informing the driver about the risk either through special training or warning signs at the entrance of the zone dedicated to platoons, could be effective. Finally, enforcement could supplement the two kinds of measures aforementioned. Increasing the control of THW kept by a vehicle and using legal sanction to prevent close-following could urge drivers to be more vigilant towards their THW. Enforcement could also concern use of the system, such as restrictions allowing use only in certain areas or during the night.

Acknowledgement

This research received funding from the European Community's Seventh Framework Program (FP7/2007-2013) under grant agreement no. 238833/ ADAPTATION project (www.adaptation-itn.eu) © Transport Research Laboratory 2013.

References

1. Urmson C., Duggins D., Jochem T., Pomerleau D., Thorpe C. (eds.). 'From automated highways to urban challenges'. *IEEE International Conference on Vehicular Electronics and Safety*; Columbus, OH, September 2008. New York, NY: Institute of Electrical and Electronics Engineers (IEEE); 2008. pp. 6–10
2. Dickmanns E. D., Behringer R., Dickmanns D., Hildebrandt T., Maurer M., Thomanek F., Schiehlen J. (eds.). 'The seeing passenger car "VaMoRs-P"'. *The Intelligent Vehicles '94 Symposium*; Paris, France, 1994. New York, NY: Institute of Electrical and Electronics Engineers (IEEE); 1994. pp. 68–73
3. Buehler M., Iagnemma K., Singh S. *The 2005 DARPA Grand Challenge: The Great Robot Race*. Berlin Heidelberg: Springer Verlag; 2007
4. Buehler M., Iagnemma K., Singh S. *The DARPA Urban Challenge: Autonomous Vehicles in City Traffic*. Berlin Heidelberg: Springer; 2009
5. Broggi A., Medici P., Cardarelli E., Cerri P., Giacomazzo A., Finardi N. 'Development of the control system for the VisLab Intercontinental Autonomous Challenge'. Presented at *13th International IEEE Annual Conference on Intelligent Transportation Systems*; Portugal, 2010
6. 'Google cars drive themselves, in traffic', *The New York Times*, 2010, October 10, A1
7. Bartels A., Karrenberg S., Weiser A. 'Hochautomatisches Fahren auf der Autobahn/Highly automated driving on the motorway'. *Verkehrsunfall und Fahrzeugtechnik*. 2011;49(11):2–6
8. Treat J., Tumbas N., McDonald S., Shinar D., Hume R., Mayer R., Stansifer R., Catellan N. Tri-level study of the causes of traffic accidents. *Final Report, Vol. 1: Causal Factor Tabulations and Assessments*, 1979
9. Fuller R. 'Towards a general theory of driver behaviour'. *Accident Analysis & Prevention*. 2005;37(3):461–472
10. Brown I. D. *A review of the 'looked but failed to see' accident causation factor*. 1851125272, 2002
11. Van Arem B., Van Driel C. J. G., Visser R. 'The impact of cooperative adaptive cruise control on traffic-flow characteristics'. *IEEE Transactions on Intelligent Transportation Systems*. 2006;7(4):429–436

12. Zabat M., Stabile N., Farascaroli S., Browand F. *The Aerodynamic Performance of Platoons: A Final Report.* UCB-ITS-PRR-95-35, 1995
13. Michon J. A. 'A critical view of driver behavior models: What do we know, what should we do'. In L. Evans and R. C. Schwing (eds.). *Human Behavior and Traffic Safety*. New York, NY: Plenum Publishing Corporation; 1985. pp. 485–520
14. Sheridan T. B., Verplank W. *Human and Computer Control of Undersea Teleoperators, Man-Machine System Lab*. Cambridge: MIT; 1978
15. Parasuraman R., Sheridan T., Wickens C. 'A model for types and levels of human interaction with automation'. *IEEE Transactions on Systems, Man and Cybernetics, Part A: Systems and Humans*. 2000;30(3):286–297
16. Horowitz R., Varaiya P. 'Control design of an automated highway system'. *Proceedings of the IEEE*. 2002;88(7):913–925
17. Williams M. (ed.). *PROMETHEUS – The European Research Programme for Optimising the Road Transport System in Europe*. London. New York, NY: IEE; 1988. pp. 1–8
18. Friedrichs A., Henning K. 'Entwicklung und Untersuchung von Fahrerassistenzsystemen für elektronisch gekoppelte Lkw-Konvois'. In VDI_issensforum (ed.). *Fahrzeugs- und Verkehrstechnik*. Duesseldorf: VDI Verlag; 2006. pp. 223–237
19. Harker B. J. (ed.). *PROMOTE-CHAUFFEUR II & 5.8 GHz Vehicle to Vehicle Communications System*. Birmingham, September 2001. New York, NY: IET; 2001. pp. 81–85
20. Preuschoff E., Henning K., Happe J. 'Einsatzszenarien fuer Fahrerassistenzsysteme im Gueterverkehr und deren Bewertung'. In K. Henning (eds.). *Fortschritt-Berichte VDI: Verkehrstechnik/Fahrzeugtechnik*. Düsseldorf: VDI-Verlag; 2003. pp. 164–183
21. Savelsberg E. 'Lastenheft für elektronisch gekoppelte Lkw-Konvois'. *VDI Fortschrittsberichte, Reihe 22*. Düsseldorf: VDI Verlag; 2005
22. Lank C., Haberstroh M., Wille M. 'Interaction of human, machine, and environment in automated driving systems'. *Transportation Research Record: Journal of the Transportation Research Board*. 2011;2243(-1): 138–145
23. *SARTRE Project website* [online]. 2013. Available from www.sartre-project.eu [Accessed 1 August 2013]
24. Evans L. *Traffic Safety and the Driver*. New York, NY: Van Nostrand Reinhold Company; 1991
25. DSA 'Official Highway Code'. Norwich: TSO; 2011
26. Bouska W., Leue A. *StVO-Straßenverkehrs-Ordnung*. Heidelberg: Auflage Heidelberg; 2009
27. Kunze R., Ramakers R., Henning K., Jeschke S. 'Organization and operation of electronically coupled truck platoons on German motorways'. *Intelligent Robotics and Applications*. 2009;135–146
28. Skottke E.-M. *Automatisierter Kolonnenverkehr und adaptiertes Fahrverhalten. Untersuchung des Abstandsverhaltens zur Bewertung möglicher künftiger Verkehrsszenarien*. Hamburg: Dr. Kovac; 2007

29. Helson H. 'Adaptation-level as frame of reference for prediction of psycho-physical data'. *The American Journal of Psychology*. 1947;60(1):1–29
30. Helson H. 'Current trends and issues in adaptation-level theory'. *American Psychologist*. 1964;19(1):26–38
31. Eick E. M., Debus G. 'Adaptation effects in an automated car-following scenario'. In G. Underwood (eds.). *Traffic and Transport Psychology: Theory and Application*. Oxford: Elsevier Ltd.; 2005. pp. 243–255
32. Denton G. 'The influence of adaptation on subjective velocity for an observer in simulated rectilinear motion'. *Ergonomics*. 1976;19(4):409–430
33. Matthews M. L. 'A field study of the effects of drivers' adaptation to automobile velocity'. *Human Factors: The Journal of the Human Factors and Ergonomics Society*. 1978;20(6):709–716
34. Casey S. M., Lund A. K. 'Changes in speed and speed adaptation following increase in national maximum speed limit'. *Journal of Safety Research*. 1992;23(3):135–146
35. Allport D. A., Styles E. A., Hsieh S. 'Shifting intentional set: Exploring the dynamic control of tasks'. In C. Umiltà and M. Moscovitch (eds.). *Attention and Performance 15: Conscious and Nonconscious Information Processing*. Cambridge, MA: The MIT Press; 1994. pp. 421–452
36. Monsell S. 'Task switching'. *Trends in Cognitive Sciences*. 2003;7(3):134–140
37. Mayr U., Keele S. W. 'Changing internal constraints on action: The role of backward inhibition'. *Journal of Experimental Psychology: General*. 2000;129(1):4–26
38. Rudin-Brown C. M., & Parker H. A. 'Behavioural adaptation to adaptive cruise control (ACC): implications for preventive strategies'. *Transportation Research Part F*. 2004;7(2):59–76
39. Comte S. 'New systems: new behaviour?'. *Transportation Research Part F: Traffic Psychology and Behaviour*. 2000;3(2):95–111
40. Breyer F., Blaschke C., Farber B., Freyer J., Limbacher R. 'Negative behavioral adaptation to lane-keeping assistance systems'. *Intelligent Transportation Systems Magazine, IEEE*. 2010;2(2):21–32
41. Sagberg F., Fosser S., & Saetermol I.-A. F. 'An investigation of behavioural adaptation to airbags and antilock brakes among taxi drivers'. *Accident Analysis & Prevention*. 1997;29(3):293–302
42. Stanton N. A., Pinto M. 'Behavioural compensation by drivers of a simulator when using a vision enhancement system'. *Ergonomics*. 2000;43(9):1359–1370
43. Forbes N. L. *Behavioural Adaptation to In-Vehicle Navigation Systems*. Nottingham: University of Nottingham; 2009
44. OECD. *Behavioural Adaptations to Changes in the Road Transport System*. Paris: OECD; 1990
45. Wilde G. 'Critical issues in risk homeostasis theory'. *Risk analysis*. 1982;2(4):249–258
46. Fuller R. (ed.). *Driver Control Theory: From Task Difficulty Homeostasis to Risk Allostasis*. Waltham, MA: Academy Press; 2011. pp. 13–26

47. Rudin-Brown C., Ian Noy Y. 'Investigation of behavioral adaptation to lane departure warnings'. *Transportation Research Record: Journal of the Transportation Research Board.* 2002;1803(2002):30–37
48. Ajzen I., Fishbein M. *Belief, Attitude, Intention and Behavior: An Introduction to Theory and Research.* Reading, MA: Addision-Wesley; 1975
49. Ajzen I., Fishbein M. *Understanding Attitudes and Predicting Social Behaviour.* Prentice Hall, NY: Englewood Cliffs Prentice Hall; 1980
50. Ajzen I., Madden T. J. 'Prediction of goal-directed behavior: Attitudes, intentions, and perceived behavioral control'. *Journal of Experimental Social Psychology.* 1986;22(5):453–474
51. Parker D., Stradling S. G., Manstead A. S. R. 'Modifying beliefs and attitudes to exceeding the speed limit: An intervention study based on the theory of planned behavior'. *Journal of Applied Social Psychology.* 1996;26(1):1–19
52. De Pelsmacker P., Janssens W. 'The effect of norms, attitudes and habits on speeding behavior: Scale development and model building and estimation'. *Accident Analysis & Prevention.* 2007;39(1):6–15
53. Van Houten R., Nau P. A. 'A comparison of the effects of posted feedback and increased police surveillance on highway speeding'. *Journal of Applied Behavior Analysis.* 1981;14(3):261–271
54. Groeger J., Chapman P. 'Normative influences on decisions to offend'. *Applied Psychology.* 1997;46(3):265–285
55. Connolly T., Åberg L. 'Some contagion models of speeding'. *Accident Analysis & Prevention.* 1993;25(1):57–66
56. Fildes B., Rumbold G., Leening A. *Speed behaviour and drivers' attitude to speeding* Report No. 16, 1991
57. Weller G., Schlag B. 'Verhaltensadaptation nach Einführung von Fahrerassistenzsystemen: Vorstellung eines Modells und Ergebnisse einer Expertenbefragung'. In B. Schlag (eds.). *Verkehrspsychologie: Mobilität – Sicherheit – Fahrerassistanz.* Lengerich: Pabst Science Publisher; 2004. pp. 351–370
58. Gouy M., Diels C., Reed N., Stevens A., Burnett G. 'Do drivers reduce their headway to a lead vehicle due to the presence of platoons in traffic? A conformity study conducted within a simulator'. *IET Intelligent Transport Systems Journals.* 2013;7(2):230–235
59. Fechner G. T. *Elemente der Psychophysik* (Vol. 1). Leipzig, Germany: Breitkopf and Härte; 1889
60. Madsen K. B. 'Gestalt psychology'. In *Advances in Psychology.* Elsevier; 1988
61. Fairclough S., May A. J., Carter C. 'The effect of time headway feedback on following behaviour'. *Accident Analysis & Prevention.* 1997;29(3):387–397

Chapter 14

User-centred design approach to model scenarios on driving simulators

Ghasan Bhatti[1,2,3], Roland Bremond[2], Jean-Pierre Jessel[3], Fabrice Vienne[2] and Guillaume Millet[1]

Abstract

Modelling scenarios on driving simulators is a critical and complex task for behavioural researchers. It requires specific technical and programming skills, for which researchers are typically not formally trained. The main reason for this complexity is the lack of User-Centred Design (UCD) in scenario authoring tools, including an unintuitive and user-unfriendly interaction environment. This could account for some of the challenges faced by behavioural researchers in achieving their objectives with driving simulators. In this chapter, we discuss the problem in detail and propose a user-centred solution, which has been evaluated by users. We also propose a meta-language and an interoperability framework to generalise our solution, so that our approach could be used on most driving simulators.

14.1 Introduction

A driving simulator is a useful tool for researchers to study driver behaviour, to analyse road safety features and to study and evaluate Advance Driving Assistance Systems (ADAS) without any safety risk. Designing and implementing an experiment on a driving simulator is quite a critical and complex task, which requires careful considerations and a controlled environment. It is often carried out by behavioural researchers with an educational and research background in psychology or ergonomics. However, scenario creation may require technical and programming skills for which these researchers typically do not have sufficient training. Thus, they depend on technical personnel or the scenario developers in their respective organisations who programme scenarios for them. This can be a time-consuming task because of the dependency on other people and the need for

[1] OKTAL SA, Meudon, France
[2] IFSTTAR/LEPSiS, Paris-Est University, Champs Sur Marne, France
[3] VORTEX Group, IRIT, Paul Sabatier University, Toulouse, France

the researcher to create a precise specification of the exact scenario they wish to be created (perhaps with limited awareness of the capabilities of the simulator).

Driving simulation platforms are quite heterogeneous, and it is not possible at the moment to port scenarios from one driving simulation platform to another. Also, there is no common and standard way of describing the scenarios on driving simulators, though it is a common practice to execute the same scenarios, especially in European projects, where similar scenarios are executed on different partner sites for different purposes. So there is a need for a common methodology and a framework that could provide interoperability of driving scenarios on different driving simulators.

14.2 What is a driving simulator scenario?

The term scenario in general means a sequence of actions or the events; however, this term is used with some variance in the driving simulation literature. A driving simulator scenario is the specification of the road and traffic situations along the road in the driving simulator [1]. According to Papelis et al. [2], scenarios are termed as, 'everything that happens in the driving simulator', which include specifying and controlling ambient traffic and its attributes, ambient environment and simulation conditions, route of the participants and their position and the series of real-world situations which are to be manipulated on driving simulators. Woleffelaar defines the scenario as a constellation of consecutive traffic situations [3]. It is sometimes used to specify both the layout and the activities during an experimental trial. Other authors use the term 'Scene' to specify the layout (road network, terrain, driving environment, e.g., city, motorway etc.) and the term 'scenario' to specify what is going to happen during the simulation, that is, critical situations and events [4]. So the scenario is usually regarded as the sequence of activities that happens during the experimental trial. These activities include different situations and manoeuvres to be tested; for example, crash situations, and traffic jams. In this chapter, scenario modelling includes specifying the ambient traffic, environment, simulation conditions and the manipulation of real-world traffic situations.

14.2.1 Scenario modelling systems

In this section, we discuss briefly different types of interaction environments for scenario modelling systems and the methodology used by the driving simulators for modelling scenarios. We do not go into details of these systems or their software components, because this is not the focus of our chapter. We discuss their interaction environment, that is, how to place scenario objects in the scenario and how to manipulate the scenario events.

SCANeR software [5], developed by OKTAL SA and Renault, has a Graphical User Interface (GUI) for implementing experimental protocol and to model scenarios. The scenario objects (vehicles, pedestrians, traffic signals) are placed on a map using a mouse. The ambient traffic is generated by placing the 'sources' (traffic generation source) on the map or by placing individual vehicle on the map. To build critical situations, it uses a GUI for defining condition–action pairs (if–else statements).

ARCHISIM [6], developed by IFSTTAR, has a textual interface to define an experimental protocol and to model scenarios. The scenario objects are specified in the text files by specifying the locations on the road. Ambient traffic is created manually using text files. To build critical situations, it uses a simple text editor like a 'notepad'. In the notepad, the user can specify the condition–action pairs for various events.

STISIM developed by System technology Inc. has a textual interface for the scenario modelling process. Scenario Definition Language (SDL) is a scripting language developed to define the scenario events [7]. The scenario objects are placed in the scenario according to the route travelled by the driver using the following statement of SDL:

On Distance, Event, Appear Distance, Parameter1, parameter2, . . . ParameterN

where 'On Distance' is the distance travelled by the driver during the simulation. 'Event' refers to the specific object (i.e. buildings, vehicle or pedestrian). 'Appear Distance' is the longitudinal distance relative to 'On Distance', when the event will appear in the scene. Ambient traffic is generated by defining events in the text file using the textual statement as described above.

Besides these three different ways of specifying the scenario objects and scenarios (by placing objects directly on a map like in SCANeR, by specifying the position on the object using textual statements like in ARCHISIM, and by the route travelled by the driver during the simulation trial), some other simulators use different metaphors and ways of describing the traffic scenarios. For example, sometimes it is required to generate a script dynamically during the simulation. Some driving simulators provide this functionality, and some use other means to fulfil this requirement. Wasshink et al. [8] proposed a movie set metaphor to generate scenarios dynamically based on Green Dino Virtual Realities' Dutch Driving Simulator. They have proposed the movie set as a driving simulator, where actors (vehicles, pedestrians etc.) come into the scene and play certain sets of roles, which are assigned to them in the script. They have also emphasised the problem for users of model scenarios using a scripting language. A tile-based approach is also used to specify scene and scenario elements in the driving simulator. The whole world is divided into different tiles, which are configured, assembled and then loaded into the driving simulator during the experimental trial. A 'Tile' is a section of the route which contains the elements such as roads, traffic signals, buildings, trees and scenario objects. These tiles are then grouped together and loaded using an interface or by specifying the tile sequence. The scenario events are then created on the map. In some systems, tiles are static and may not be altered or moved during the experimental run [2], while there are some systems in which tiles as well as data on the tiles (scene objects, scenario objects) can be altered dynamically [9].

14.2.2 Challenges with programming and authoring tools

One of the reasons why programming is difficult is that programmes are abstract [10]. It is difficult for people who do not have a solid programming background to think in an abstract manner to implement the same situation, as compared to thinking about the situation in the real world. So abstraction is one of the barriers, especially for novice users. Also, it is a common observation that, the more complex the situation, the more abstraction is required in the programme. Also, programming languages have been designed without careful attention to human–computer interaction issues [11], which makes programming more difficult than necessary [12]. Different design techniques and solutions [13–15] have been proposed in the past two decades, which has made significant improvements for the non-programmers to undertake programming.

In the case of scenario programming on driving simulators, two main factors present the key challenges to making events happen at the right time and at the right place in a natural way [4]. First, driving behaviour is complicated and not all aspects are well-understood; so it is difficult to create realistic as well as controllable traffic. The second factor is the variability of driving behaviour, as people change their speed, lane position and tactical decisions with the time during the simulation trials, which leads to variance in driver behaviour. As a simple but challenging example, imagine that a researcher wants to create an accident situation, where a car overtakes the participant and then stops in front of their vehicle. To implement this using a modelling language, there are many parameters to control and consider including speed and position of the overtaking vehicle, speed of the participant vehicle and the distance to stop from the participant vehicle. This needs to be accomplished through the scripting environment provided with the driving simulator, by using triggers and actions or some event-driven mechanism.

14.2.3 User-centred design in scenario authoring tools

'User-Centred Design (UCD) is a multidisciplinary design approach based on the active involvement of users to improve the understanding of user and task requirement, and the iteration of design and evaluation' [16]. As the name UCD suggests, the outcome of the product using UCD approach is developed by placing the user at the centre of the design. The international standard ISO 9241-210 (formerly known as ISO-13407) provides the basis for UCD processes [17].

Many UCD methods (e.g. User analysis, prototyping, heuristic evaluation, cognitive walkthroughs, etc) are employed by usability practitioners. Surveys [16, 21] have been conducted to find out which UCD methods are being practiced in the industry, and which methods are effective in different contexts. UCD is a common practice in the development of software which follows different phases, such as analysis (user and task analysis), design, prototype and evaluation (design iteration), implementation of the design and finally product deployment. UCD methods claim to provide an end-product which satisfies the users and enables them to achieve their goal while taking their profile into account. However, many claims have been found in the literature about the UCD methods being impractical and inefficient for different reasons [16, 18]. Many developers do not use basic usability principles, because they find them very expensive, time-consuming and intimidating in their complexity [19, 20].

Modelling scenarios on a driving simulator is a complex and difficult task for the behavioural researchers, because of their complex and technical nature. Though recent scenario modelling systems are powerful enough to manipulate many kinds of traffic manoeuvres and situation, they do not take into account the profile of the researchers, who are the primary users of the driving simulators. We think that the needs and profile of the end-users of driving simulators should be reflected in the design of scenario authoring tools. That's why we have used the UCD approach to model scenarios in a driving simulator. In the next section we explain our method in detail.

14.3 Method

As explained in the previous section, we have used a UCD approach to involve the user at the start of design process and we have used interview techniques as part of the USD practice. The use of interviews has many advantages. By asking open questions of simulator users, we gathered information about needs and requirements and learnt about user profiles. After the interviews, we analysed the user profiles and their requirements and expectations for developing their experimental protocols. Based on analysis of the user survey (explained in the next section), we proposed a solution. Feedback was then collected from the users on this proposed solution and, based on the proposed solution, we have also proposed an interoperability framework.

14.4 User survey

We interviewed 19 driving simulator users with diverse background and experience levels on different driving simulators, as shown in Tables 14.1 and 14.2 [22]. Six out of the 19 users were interviewed in person, 1 was interviewed using Skype and the remaining 12 users were interviewed on the phone. The interviews were not recorded but relevant details were captured by preparing formats on note-taking

Table 14.1 Driving simulators

Driving simulator	Number of user(s)
ARCHISIM	7
SCANeR studio	6
ST SIM	2
STI SIM drive	1
VTEC driving simulator	1
Honda simulator	1
DriS	1

Table 14.2 User profile

Users	Number of user(s)
PhD researchers	7
Behavioural researchers	8
Software engineer	2
Research engineer	1
Post-doctorate	1

sheet. The average duration of the interviews in person was 40–45 minutes, the average duration of telephone interviews was 50–60 minutes and the Skype interview took about 83 minutes. The participants were not given any incentive to participate in the interviews.

Behavioural researchers had an average experience of three years using driving simulators, while software engineers, research engineer and PhD researchers had an average of two years' experience. PhD and post-doctoral researchers used a driving simulator during their PhD and were using it currently.

14.4.1 Results of the user survey

The user survey gave us a good insight of the users' experience and their requirements to model scenarios on driving simulators. Table 14.3 shows the programming knowledge of the end-users.

The software engineers and the research engineer had a good level of programming skills. They usually model scenarios for the researchers with no or very low programming skills. Seven users had neither programming knowledge nor practice; they always asked scenario developers or software engineers to model scenarios for their experiments. The rest of the users had a very basic knowledge of programming and usually receive help from technical persons.

The focus of the interviews was on four different topics. First, the general problems that the users face while modelling scenarios. Second, the concept-to-script translation problem, that is, when the users think about traffic situations in

Table 14.3 Programming language knowledge of the users

Programming language	Number of users
C++	3
Visual basic	3
Programming in Matlab-Simulink	3
HTML	2
Java	1
Action script	1
Python	1
No experience of programming language	7

the real world and have to transform them into a script. Third, they were asked what would be an ideal approach for them to model scenarios. Fourth, what improvements would they suggest for the driving simulator they use. Apart from simulator-specific problems, the users provided details about the problems and requirements concerning the scenario modelling systems of their driving simulators. The summary of the problems and ideas are described below.

14.4.1.1 General problems

General problems refer to those faced by the user while using software to model scenarios. The common problems identified by the users were:

1. Finding/selecting the relevant functions in order to perform a specific task.
2. The difficulty of testing the scenarios before running the simulation (e.g. a specific situation has to occur after 25 minutes of simulation time; it can only be tested by driving for 25 minutes through the scenario and not before).
3. Debugging the script in order to find errors.
4. The syntax of the scenario modelling language which is difficult for the users to follow.
5. The difficulty of controlling and programming the scenarios at a quite low and detailed level.
6. When the script is long and complex, it is difficult to go back and forth between different testing conditions (mainly the IF-ELSE condition).

14.4.1.2 Concept to script translation problems

The ‘Concept to script’ problems refer to the problems the users face when they want to manipulate real-world situations on driving simulators using the simulator’s programming language. According to the users, this is the most difficult step for them as they cannot script the situation in the same way as they imagine. So the main problems identified by the users were:

1. Controlling the ambient traffic using the script.
2. Tuning the events; for example, timing a vehicle approaching an intersection or when a pedestrian is crossing.
3. Selection of triggers (speed, distance etc.) in order to tune the events.
4. Reproducibility of situations (needs extensive scripting and is difficult to manage).

14.4.1.3 Ideal approach to model scenarios

After the users had described their problems, they were asked about their ideal approach to model scenarios and how their problems could be solved. The main ideas were:

1. Drag and drop situation for a scenario. Ten users out of 19 proposed the idea of dragging and dropping the scenario situations. They think that, this way, they would just have to specify the specific parameters for a specific situation, and not the complete code.

2. High-level scripts for scenarios. This idea is the same as 1, but when users were told that every user has different requirements, and it would therefore be difficult to manage libraries for dragging and dropping, they then specified that in order to use high-level script for complex scenarios, there should be a functionality of editing the high-level script at a low level.
3. Interaction with the world. The users wanted to interact with a world/map while modelling the scenarios.
4. There should be an image/preview of the work users are doing, which would help them in taking decisions.
5. It would be more comfortable for the users if they could click on autonomous vehicles participating in the scenario and then set their rules and behaviours.
6. It would be easier to specify and follow the tasks/events based on a time-line or distance-line of the experiment protocol.

14.4.1.4 Improvements of the driving simulator in use

Users were asked how driving simulator software could be improved to model scenarios. The users' responses to this question are listed below.

1. Searching of the relevant functions should be made easier.
2. Interface of the scenario modelling software. All users have used textual interface for scenario modelling, except the user of SCANeR software, which has a GUI. Most of the users who are using textual user interface (UI) would have preferred to use a GUI for the scenario modelling.
3. There should be a very user-friendly documentation for how to use the relevant functions, with some examples.
4. In some simulators, users have to configure different files in order to run a simulation trial. According to users, there should be only one interface in order to configure different files.

From the results, we have also identified seven different steps that users undertake while developing an experimental protocol on a driving simulator. The steps are:

1. Terrain selection.
2. Configure participant/subject vehicle (initial position, ADAS, speed and other platform-dependent parameters).
3. Configure the ambient traffic (number of vehicles per hour, configuration of vehicles, i.e., speed, itinerary) and traffic signals.
4. Configure the environment (i.e. daylight/night, weather).
5. Set the independent variables and experimental design.
6. Construct critical events using scripting.
7. Experiment execution and data collection (dependent variables).

During the survey, users were asked which software they felt comfortable with, which they thought was the ideal software for end-users, or the software they thought had interaction techniques that could be helpful in designing an experimental protocol. It was not possible to generalise across all the different software packages specified by the users. However, features of the software packages were named, examples of which include Matlab-Simulink, E-Prime and SuperLab, and we have tried to adopt some principles from these packages.

We have summarised the requirement of one class of the users (behavioural researchers). As one can see in Table 14.1, we have interviewed the users of different driving simulators and the requirements we have presented are those which are identified by most of these users. Although it is difficult to generalise these requirements for all driving simulators, if we explore these requirements in a contemplative manner, then these problems are quite obvious. Although users have mentioned their problems, they have also proposed solution in order to solve them. For example, if we adopt ideas 1 and 2 given in the subsection 14.4.1.3, then we can solve problems such as 1, 3, 4, 7, 8 and 9 outlined above. We think that, by following the problems identified by the users, and the ideas proposed by users, and applying the appropriate interaction techniques to these ideas, we can provide an intuitive, user-friendly interaction environment appropriate for their skills.

14.5 Proposed solution

We propose a new UI and experiment development approach based on the survey we conducted and the steps that we have identified. Traditionally, in order to implement an experimental protocol, a user (technical person or researcher) uses the same interface for implementing an experiment protocol, regardless of the level of their technical and programming skills. We now propose a multi-layer programming approach. As the name suggests, there will be different layers of programming, and researchers and technical persons will interact at different layers corresponding to their skills.

14.5.1 Multi-layer programming

In our proposed multi-layer programming solution, end-users (researcher) interact with the higher-layer, which is close to the human language or real world, and the technical persons interact with the middle layer, which are the Application Programming Interfaces (APIs) or the high-level functions of the driving simulator software. API is the set of programming instructions to access software applications. In the case of driving simulators, the APIs can be, for example, Change Lane function (instruction to change the lane of the specific car), or, Change Speed function (instruction to change the speed of the car). Technical persons develop templates for the researchers using the API of the driving simulator.

Then, there is the lower-layer, which is a platform specific layer and close to the machine language, for example, C++ or PHP, depending on the software architecture of the driving simulator. Driving simulator manufactures usually interact with this layer.

The multi-layer programming architecture is shown in the Figure 14.1. The middle layer functions (APIs) are usually developed by driving simulators developers (e.g. SCANeR by OKTAL), or it may be developed by the technical persons,

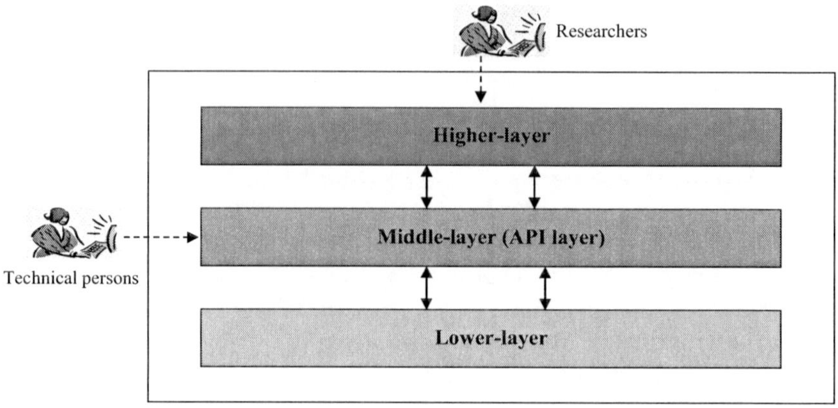

Figure 14.1 Multi-layer programming architecture

particularly if the driving simulator is built in house by a research institution (e.g. ARCHISIM by IFSTTAR. So, in principle, the researcher will interact with the higher-layer and the technical person will interact with the middle layer. Traditionally, researchers and technical persons interact with the API layer to develop an experiment, where they have to follow the programming rules and syntax. Here, we add an extra layer (higher-layer) for the researcher, where they can develop the experimental protocol by exploiting the programming primitives.

14.5.2 User-interface development

During the user-survey, we identified three different kinds of users who actively interact with driving simulators. So in the proposed design, we split the scenario modelling activity into three sub-interfaces depending on the roles they have to perform while modelling scenarios and the set of skills they have. The three roles are 'Technical Person' as 'R1' (good programming/technical skills), 'Researcher' as 'R2' (low or no programming skills) and 'Operator' as 'R3' (no technical skills required). Based on the user roles, we split the scenario modelling activity into three sub-interfaces which are 'Template Builder', 'Experiment Builder' and 'Experiment Interface'. The technical persons will interact with the middle layer of our multi-layer programming architecture (see Figure 14.1) via 'Template Builder' performing role 'R1'. The researcher will interact with the higher-layer via 'Experiment Builder' performing role 'R2' and the operator will also interact with the higher-layer via 'Experiment Interface' performing role 'R3'.

We explain this new approach and interface with the help of an example. Our example scenario contains two events:

- Accident Event: A vehicle overtakes the participant vehicle by increasing its speed and changes its lane to the lane of the participant vehicle and then brakes.
- Pedestrian cross event: A pedestrian walks and then crosses the road as the participant vehicle approaches.

14.5.2.1 Template builder

This sub-interface will be used by technical persons performing role R1 having good programming and technical skills. R1 will design the GUI-based templates of the scenario events. The Template Builder will let R1 use existing functions offered by the scenario-modelling environments of the driving simulator to model scenario events. In our example, for the template 'Accident', the 'Template Builder' will let R1 programme a vehicle around the participant vehicle to accelerate, change position and apply brakes at some distance from the subject vehicle. For the event 'Pedestrian crossing', the 'Template Builder' will let R1 programme a pedestrian to walk and cross the road as the participant vehicle approaches the intersection. At the 'front end' of the template, there would be different text fields to specify the parameters for the events 'Accident' and 'Pedestrian crossing'. These parameters will be filled by the researcher in the 'Experiment Builder' sub-interface, if he or she wants to modify the default values. The templates developed by R1 will be stored in a template library, so that researchers could access the templates in the 'Experiment Builder' interface as described in Figure 14.2.

14.5.2.2 Experiment builder

This sub-interface will be used by researchers/trainers performing role R2 and possibly having low or no programming skills. R2 will define the whole experiment

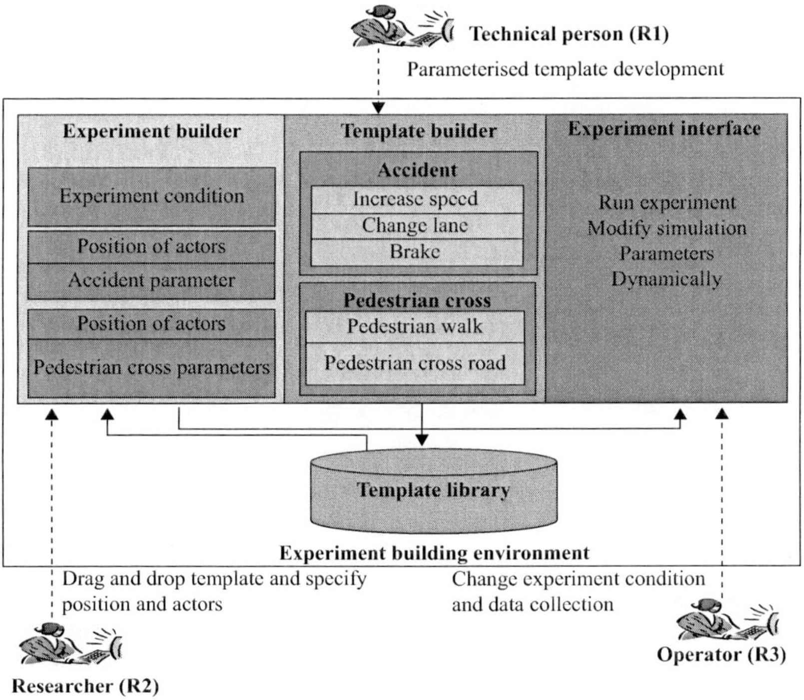

Figure 14.2 Experiment building procedure

using a user-friendly and intuitive GUI which includes specifying experiment condition, environment, ambient traffic and data to be collected (all the steps of an experiment). To specify the critical events or situations to be studied, R2 will access the template library developed by technical persons in the 'Template Builder' and will place them in the scenario editor using drag and drop and proceeded by a user-defined trigger. If needed, R2 will fill the parameters of the template.

In our example, the end-users will specify the position and actors involved in the templates 'Accident' and 'Pedestrian Crossing' besides the template parameters, if needed. There will always be appropriate default values for all the parameters of the template.

Traditionally, end-users have to configure different files to create an experimental protocol, for example, for autonomous vehicles, environment, variables to study and critical events. In some tools, they have to configure everything in the same scenario editor or different text files, which are sometimes difficult to understand, particularly when the experiment is long and complex. So for researcher, we propose a wizard-based interface, in which they can develop the experimental protocol in the same way as they design it.

14.5.2.3 Experiment interface

The 'Experiment Interface' will be used by the user (the researcher or the person who will execute the experiment) performing role R3. Using this sub-interface, R3 will load and execute the scenarios in the driving simulator, developed with the 'Experiment Builder'. R3 can change the parameters of the scenario or template (if needed), during the experiment trial and finally collect the data.

14.6 Prototype building

A prototype based on the proposed approach has been developed. In this chapter, we focus on the Experiment Builder sub-interface, which will be used by researchers. The prototype of our solution is based on the problems identified, user suggestions and the steps that were identified during the user survey. The prototype is built using 'Justinmind Prototyper' [23], a tool to develop prototype for any kind of application. We briefly explain the steps which end-users will have to undertake in order to develop an experimental protocol and these steps correspond to the skills of the end-users. The user is guided using breadcrumb navigation [24] during the experiment development process. The details of the steps are as follows:

Step 1: Experiment description

In this step, end-users provide the name and description of the experiment as shown in Figure 14.3.

Step 2: Terrain selection

In this step, end-users select the terrain/map for the experiment from a library as shown in Figure 14.4.

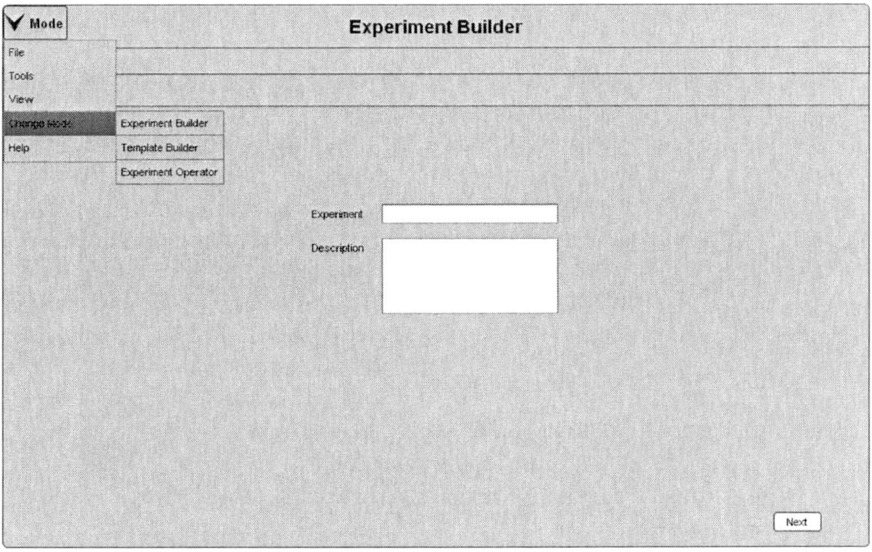

Figure 14.3 Experiment description

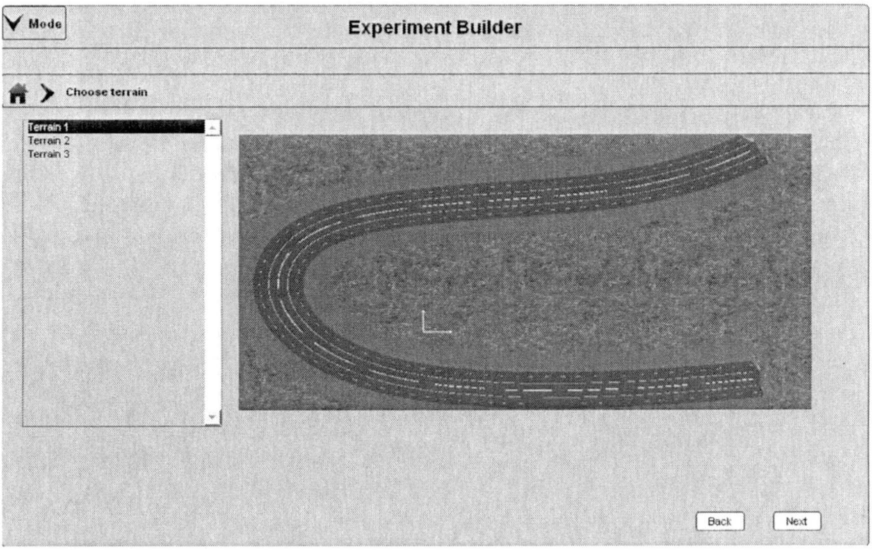

Figure 14.4 Terrain selection

Step 3: Configure subject vehicle

In this step, end-users configure the subject vehicle. They specify the vehicle type, model, its initial position, heading maximum speed etc. on the road. They can also specify the itinerary for the vehicle, if it has to travel specific path, and they

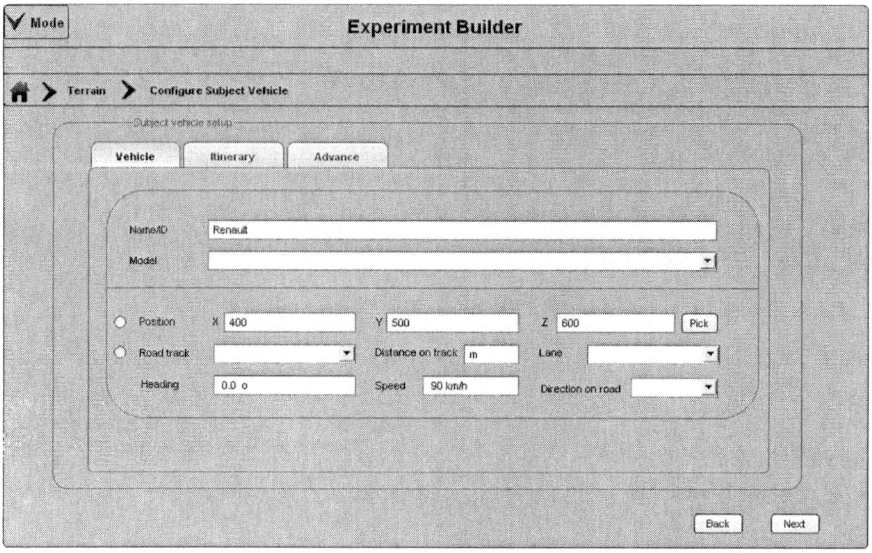

Figure 14.5 Subject vehicle configuration

can also specify the ADAS and other parameters in the 'Advance' tab as shown in Figure 14.5.

Step 4: Configure autonomous traffic

In order to specify the autonomous traffic, the user can define zones for the traffic, as they anticipate while designing the experiment. A traffic zone has a start and an end position in the terrain. End-users can specify either the traffic density or the traffic flow in the traffic zone. They can specify the density or they can accept a default density value for the traffic. They can also specify the traffic distribution and traffic behaviour distribution in the zone as shown in Figure 14.6.

Step 5: Configure environment

In this step, end-users specify the environment (light, rain, fog, snow etc.). They can also specify zones for the environment as well as for traffic. A zone has a start and an end position, and various zone parameters can be configured, for example, rain, fog and snow, as shown in Figures 14.7 and 14.8.

Step 6: Select dependent variables

In this step, end-users can explicitly select the dependent variables that they want to analyse quickly, as shown in the Figure 14.9.

Step 7a: Construct critical events and scenarios

This is a very critical step for the end-users. In this step, end-users have to manipulate traffic situations and vehicle manoeuvres, which they want to study during the experimental trials. End-users drag the template of the traffic situation or the vehicle manoeuvres into the scripting area and configure the parameters of the

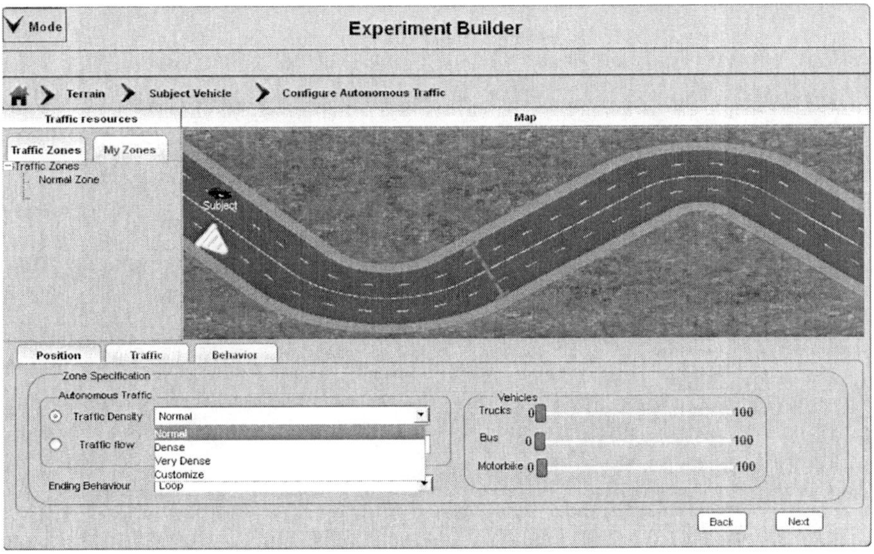

Figure 14.6 Autonomous vehicle configuration

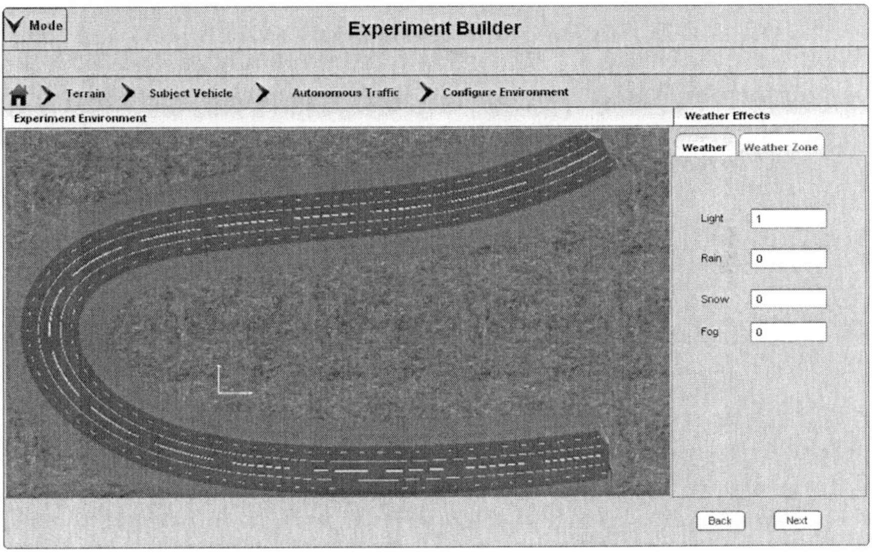

Figure 14.7 Configure environment

template. End-users will drag the template onto the map, followed by configuring the condition (trigger) for this template to execute as shown in the Figure 14.10. The templates that end-users will drag are built by the technical persons using the 'Template Builder'.

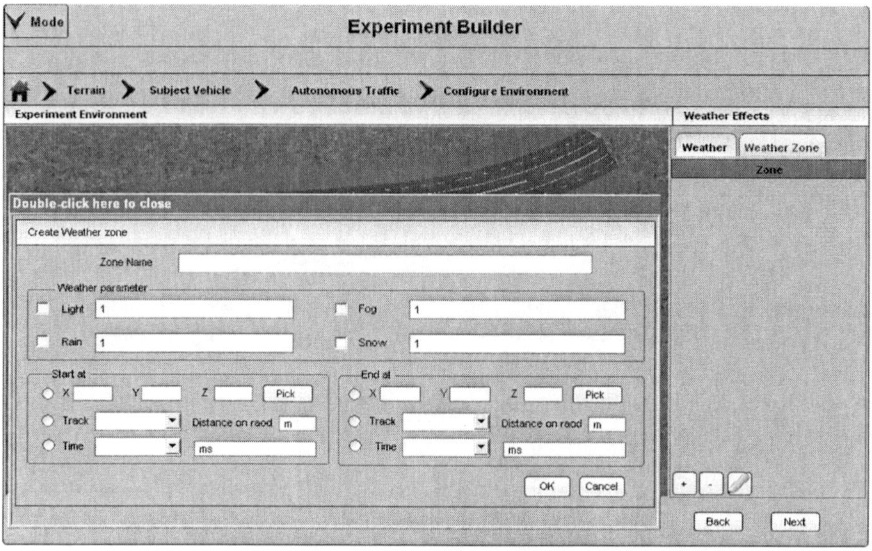

Figure 14.8 Weather zone configuration

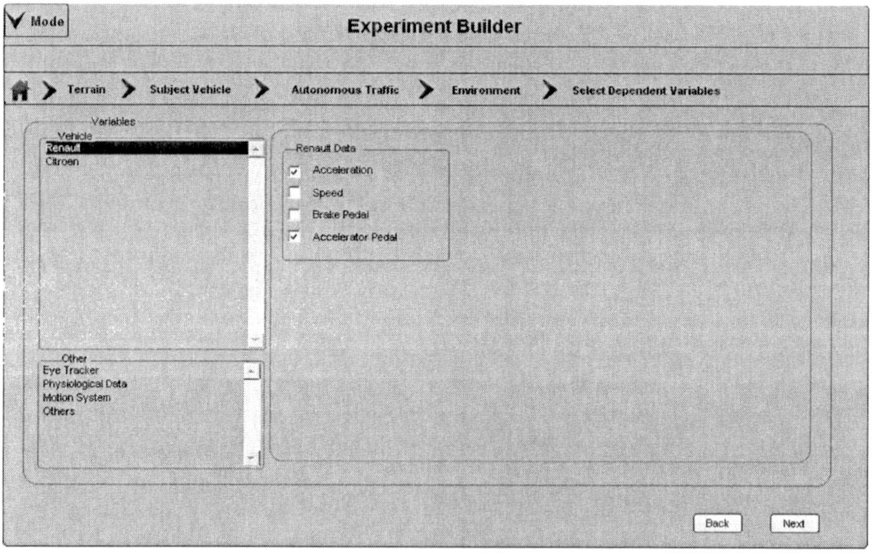

Figure 14.9 Variable selection

Step 7b: Specify template parameters

After the selection of trigger type and value for the template, end-users specify the parameters for the template as shown in Figure 14.11.

Step 7c: Overview of the experiment

After the user has developed the experiment, they can visualise the whole experiment in a spatial-temporal representation. The templates to be triggered

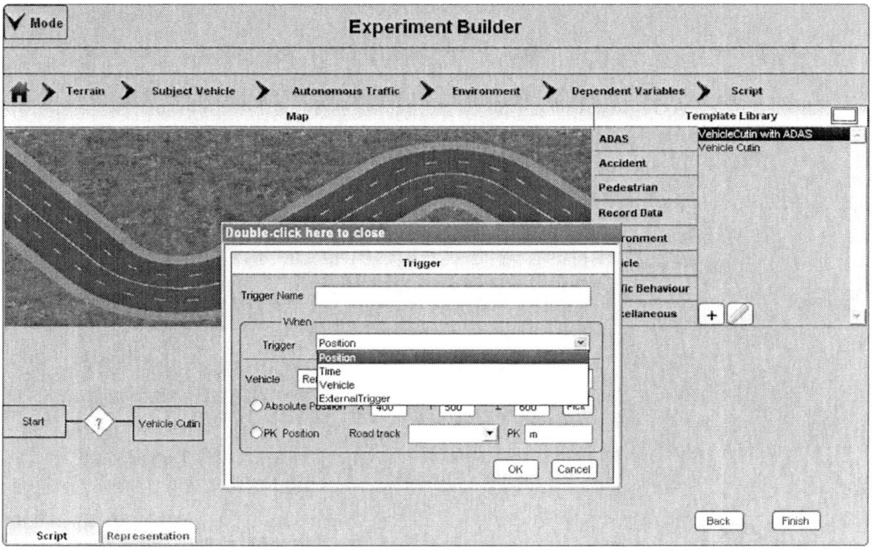

Figure 14.10 Trigger for the template

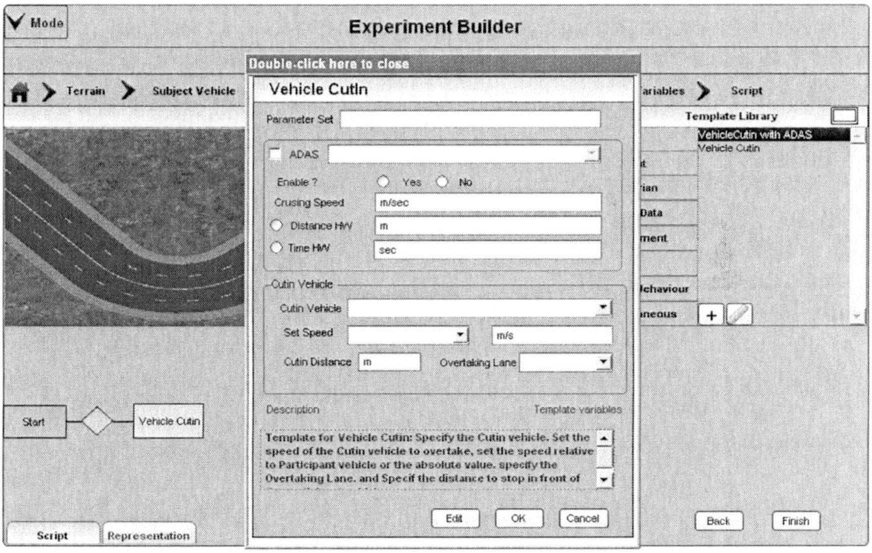

Figure 14.11 End-user specifying template parameters

based on position can be viewed on the map and also on the distance-line. The templates, to be triggered based on time, can be viewed on the timeline and their description in the tool tip, as shown in Figure 14.12. This idea of a global view of the experiment is inspired by the E-prime software package, which is widely used among research psychologists [25].

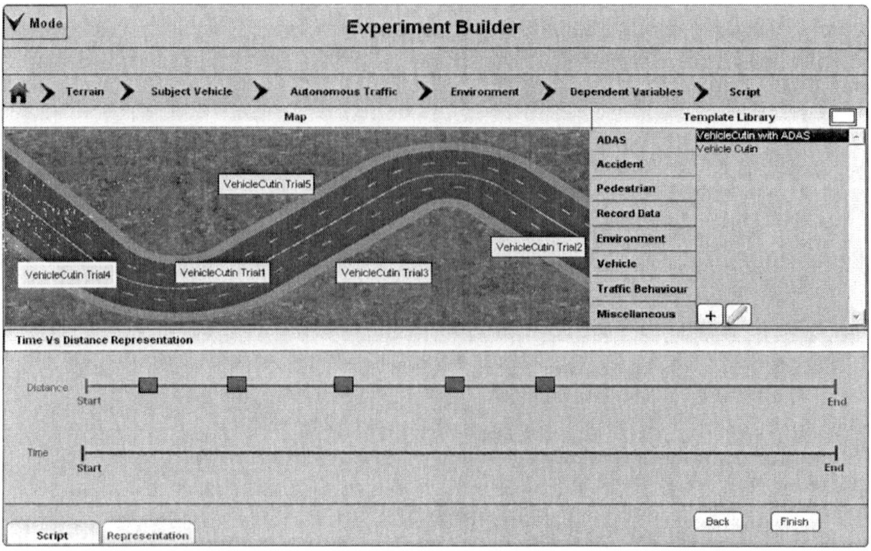

Figure 14.12 Visualise the experiment on the timeline and distance-line

During each step in the experiment development process, support is provided to end-users using different interaction techniques. We keep the end-users away from the technical part of experimental development process, in which they would have to write low-level scripts. In fact, we empower the user by providing an interface with which they can develop scenarios using the skills they have and with minimal help from technical persons.

Traditionally users have to configure and set up each autonomous vehicle, whether it is interacting with the subject vehicle or not. In this way, they have to configure many vehicles, which can be complex, especially when the traffic is dense. Most of the time, the user requirement is to specify traffic in a general way; for example, the terminology they use is 'normal traffic', 'dense traffic' or 'specific number of vehicles' in a specific area (Zone). So we also enable the user to specify the traffic at this general-/high-level. If the users want a vehicle with specific characteristics or a vehicle which will be involved in the scenario, they can configure and create it in Step 7, during the construction of critical events.

In order to reduce complexity, we have also separated the environment configuration from the scenario specification. The users can change the environment (fog, rain, light, snow) by interacting with the map (Figures 14.7 and 14.8). They can set the environment based on the experiment time or the vehicle position. If needed, users can also configure the environment by configuring a template for the environment.

The end-users develop critical events and situations using the drag and drop function. They can interact with the map, while specifying triggers for the templates, which is very important for them (they said in the user survey that without interacting with the map, the process is not clear to them). Time/distance representation of the experiment enables users to have an overview of the experiment, so they can anticipate what is going to happen.

14.7 Prototype evaluation

As we are following the UCD approach, so end-users have evaluated our prototype [26]. It was a qualitative evaluation, as it took place during the initial design phase, and we were interested to discuss the issues in detail during a subjective evaluation. Nine users (behavioural researchers with low or no programming skills) were involved in this evaluation.

The approach was explained to the user and then they performed a small exercise using the prototype. Users were observed during the exercise and later interviewed. The interviews were not recorded. The average time for a user interaction was 30–40 minutes, with 5–7 minutes for explanation, 15–20 minutes for the exercise and 10–20 minutes for the interviews. In the interviews, users were asked about their feedback with respect to the new approach, the division of experiment protocol steps and difficulties during the exercise, and they were also asked if there was anything that needed to be improved. The questions were not limited to what we had specified or prepared and we discussed in detail any issues raised by the users. In the exercise, the users developed a small scenario, in which, they had to create a lead vehicle and a critical situation in which they had to enable a vehicle to cut-in ahead of the participant vehicle. After this, there was a second event, in which the leading vehicle brakes. Finally, there was another vehicle cut-in situation.

The structure of the user interface and the steps to create the experimental protocol were perfectly clear to all the users. They all gave quite positive feedback about the approach for creating scenarios/events using templates. There were some minor problems; for example, four users stated that creating a zone for autonomous traffic was not very intuitive. These and a further two users found the process a bit uncomfortable. Three users also suggested that, in order to create events based on position, they should be able to drop the template on the map. Five users attempted to do that as well. During the interviews, they felt it was easy and intuitive to configure the templates, rather than doing the low-level coding, and they were also interacting with the map, so they were aware of where and what was happening.

14.8 Interoperability of the approach

In order to generalise our solution, so that it could be used on different driving simulators, we propose an interoperability framework. As previously mentioned, there is no common or standardised way of describing scenarios, so it is difficult to port scenarios from one platform to another. Each driving simulator platform has its own means of describing scenarios and the API of each simulator provides different levels of details. In our proposed solution described above, we enable the user to specify at a high-level, which provides enough detail to develop an experimental protocol. So we also propose a meta-language called Scenario Meta-Language (SML)

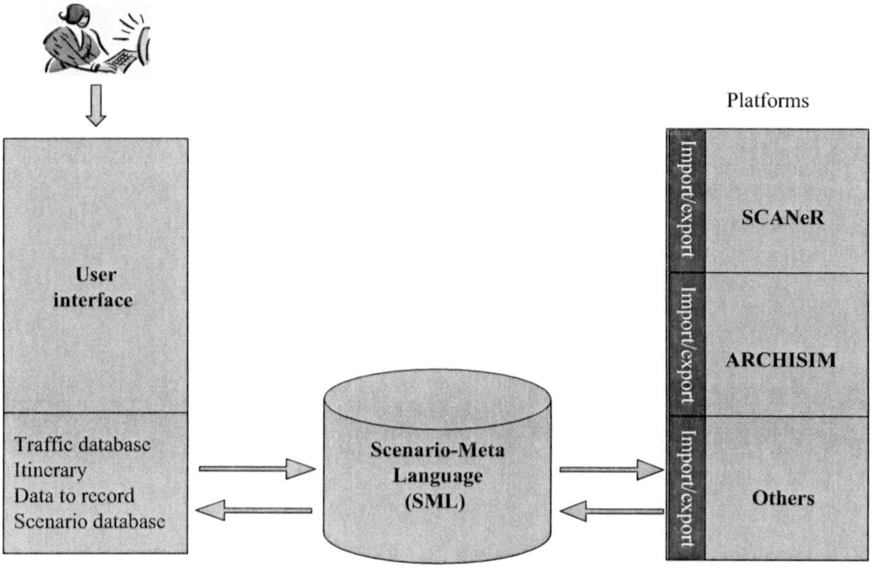

Figure 14.13 Scenario modelling process across different platforms

to integrate our solution within an interoperability framework. Through this, researchers will develop the experimental protocol using the proposed approach which will be converted into SML format, and later it can be executed on different platforms. The SML contains the experiment information at high-level, which will be used by the majority of driving simulation platforms. The scenario modelling process across different platforms can be seen in Figure 14.13.

14.8.1 Development of the meta-language

Extensible Markup Language (XML) is used to define the format of SML. XML is a very commonly used mark-up language that defines a set of rules to encode a document in a format which is understandable by both humans and machines. XML is independent of any execution platform and widely used over the internet, as well as for desktop and simulation application.

Some efforts have been made to develop XML structures for specifying scenarios and for different purposes for modelling and simulation and a scenario mark-up language for authoring in the transport domain has been developed [27]. This language has been proposed for an online virtual world platform called Open-EnergySim. However, the structure of this language is tightly bound to the execution platform. The OneSAF Objective System (OOS) programme has created an XML-based mechanism for specifying scenarios: the Military Scenario Definition language (MSDL) [28]. It is an open specification, designed to allow the decoupling of planning and execution systems to enable a broad range of interoperability and reuse during simulation. Like the concept of MSDL, an attempt to create an

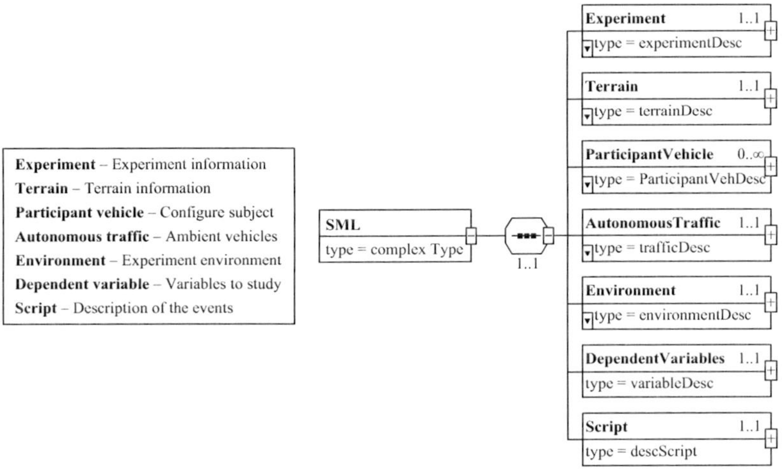

Figure 14.14 SML Schema overview

open source battle management language (BML) has also been carried out [29]. BML controls military operations performed by the forces and provide situational awareness and a shared common operational picture.

There is no common methodology to describe scenarios, but we have pointed out the steps that end-users usually follow during the experiment design process. On the basis of these steps, we have developed the XML structure for SML and all the experiment information specified follows this structure. An overview of the schema for SML is shown in Figure 14.14.

14.8.2 Interoperability framework

The proposed approach (interface) can be integrated with SML using the framework as shown in Figure 14.15. The technical person will develop a template, which has two representations: XML representation and low-level specification. The end-users will fill the template parameters along with the other experiment steps in the interface and the whole scenario will be converted into SML format.

The whole framework can be regarded as a multi-layer programming environment, where end-users interact with the upper layer (high-level). In this way, the typical programming process has been converted into a customisation process. The technical persons interact with the lower-layer and develop the templates to be used at the upper layer, along with the low-level representation. There are two representation of a template, XML and low-level. The XML representation contains the parameters of the template and the low-level specification contains the logic and the programming representation of the template developed with the API of a specific driving simulator.

The end-users will develop the experiment using the GUI and configure the template parameters which will be converted into SML format. The SML will then be imported onto the specific platform and use the low-level specification and

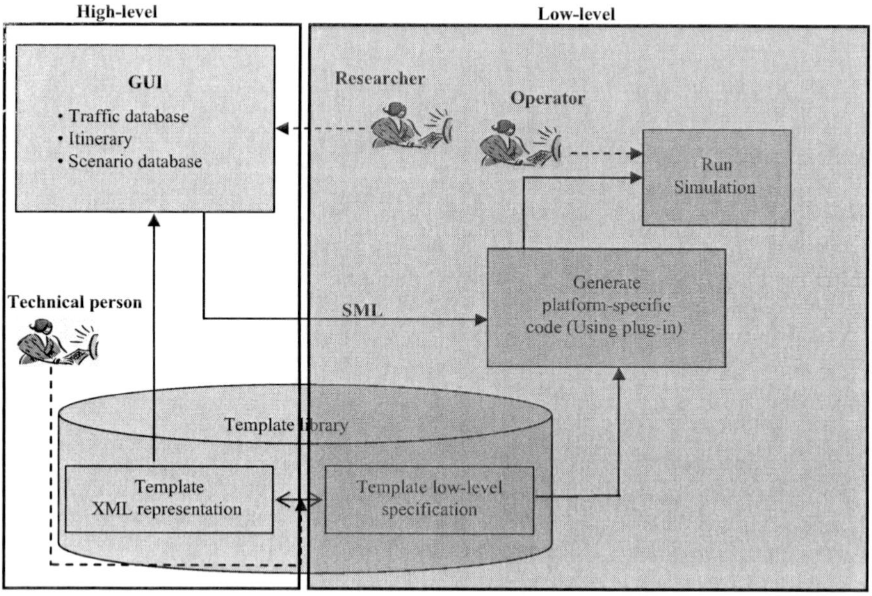

Figure 14.15 Interoperability framework

template parameters in the SML to generate the platform specific code. Finally the experiment can be executed.

This framework is being implemented, and the interoperability of the scenarios between ARCHISIM and SCANeR platform will be validated.

14.9 Conclusion

We have discussed in detail the problems and requirements of researchers as end-users of driving simulators. There were two main objectives; one was to provide a user-friendly and intuitive interaction environment for the end-users to model scenarios, and the other was to provide a framework so that scenarios can be ported from one platform to another.

We used a USD approach to address the first objective, where we focused on the Experiment Builder sub-interface, which is for the researchers who do not have programming skills. The multi-layer programming solution implemented reduces the complexity in developing the scenarios by separating the roles of the users who interact with the driving simulator. One of the difficulties in performing a programming task is to split the task into meaningful sub-tasks. We have split the whole experiment development process into different steps following the user design process for developing an experimental protocol. We have also separated the role of technical persons, who will implement the domain logic of the critical situation and manoeuvres using the low-level languages provided by the driving simulators. In this way, the end-users will not have to go through the typical programming process, yet they will just undertake the process of developing an experimental protocol. They will drag and drop templates from a template library to the scripting area and this will reduce their time in the programming process and reduce their dependency on technical persons.

Another problem in modelling scenarios is that the process is not transparent to the users, as they cannot have a global view of the experiment unless they execute the scenario, and then they have to edit the scenarios following the syntax of the programming language. Our solution provides a better transparency to the users by enabling them to interact with the map during the experiment development process.

The end-users had given a positive feedback on our prototype, which was then improved based on this evaluation. The solution is being implemented and, in order to make our solution effective and efficient for the user, future work is planned to find out the extent to which we have empowered the researcher and how this approach can help technical persons to use driving simulators effectively and efficiently.

To address the second objective of portability we have proposed a high-level solution, so that, whilst the user feels close to the real world, it is also easy to port from one platform to another. The proposed interoperability framework and the SML have been implemented and further work will be undertaken to verify the approach as proposed above.

Overall, we believe that our solution can provide support to the end-users by filling the gap between the user-skills and the goals they want to achieve using driving simulators. Also, our interoperability framework can enable driving simulators to be used efficiently and effectively.

Acknowledgements

This work has received funding from the European Community's Seventh Framework Program (FP7/2007–2013) under grant agreement no. 238833/ADAPTATION project.

Tables 14.1, 14.2 and 14.3 and some text in Section 14.4 are excerpted by permission of the Publishers from 'User requirements to model scenarios on driving simulators', in Driver Behaviour and Training edited by Lisa Dorn (Farnham, UK: Ashgate, 2012), pp. 493–499. Copyright © 2012.

References

1. Olstam J., Espie S. 'Combination of autonomous and controlled vehicles in driving simulator scenarios'. Presented at International Conference on Road Safety and Simulation; Rome, Italy, 2007
2. Papelis Y., Ahmad O., Watson G. 'Developing scenarios to determine effects of driver performance: techniques for authoring and lessons learned'. Presented at Driving Simulation Conference North America; Dearborn, Michigan, 2003

3. van Wolffelaar P.C. *Functional aspects of the driving simulator at the University of Groningen*. Groningen, Netherland: University of Groningen; 1999
4. Kearney J.K., Timofey F.G. 'Scenario authoring' in Fisher D.L., Rizzo M., Caird J.K., Lee J.D. (eds.). *Handbook of Driving Simulation for Engineering, Medicine, and Psychology*. Boca raton, FL: CRC Press/Taylor & Francis; 2011
5. Reymond G., Heidet A., Canry M., Kemeny A. 'Validation of Renault's dynamic simulator for Adaptive Cruise Control experiments'. Presented at Driving Simulation Conference (DSC00); Paris, France, 2000
6. Espié S., Saad F., Schnetzler B., Burlier F. 'Microscopic traffic simulation and driver behavior modeling: the ARCHISIM project'. Presented at Road Safety in Europe and Strategic Highway Research Program; Lille, France, 1994
7. Park G.D., Allen R. W., Rosenthal J. 'Flexible and real-time scenario building for experimental driving simulation studies'. Presented at *CHI 2011*; Palo Alto, CA, 2011
8. Wassink I.H.C., Van Dijk E.M.A.G., Zwiers J., Nijholt A., Kuipers J., Brugman A.O. 'Bringing hollywood to the driving school: dynamic scenario generation in simulations and games'. *Proceedings of the First international Conference on Intelligent Technologies for Interactive Entertainment*; Madonna di Campiglio, Italy, 2005. New York, NY: Springer-Verlag; 2005. pp. 288–292
9. Suresh P., Mourant R.R. 'A tile manager for deploying scenarios in virtual driving environments'. Presented at Driving Simulation Conference North America; Orlando, Florida, 2005
10. Cypher A., Smith D.C. (eds.). 'KidSim: end user programming of simulations'. *Proceedings of the SIGCHI Conference on Human Factors in Computing Systems*; Colorado, USA, ACM Press/Addison-Wesley Publishing Co.; 1995. pp. 27–34
11. Allen N., Stuart K.C. 'The prospects for psychological science in Human-Computer interaction'. *Human–Computer Interaction*. 1985;1(3):209–242
12. Pane J.F., Ratanamahatana C., Myers B.A. 'Studying the language and structure in non-programmers' solutions to programming problems'. *International Journal of Human-Computer Studies*. 2001;54(2):237–264
13. Pane J.F., Myers B.A., Miller L.B. (eds.). 'Using HCI techniques to design a more usable programming system'. *Proceedings of IEEE Symposium on Human Centric Computing Languages and Environments*; Arlington, VA, IEEE Computer Society; Sep 2002. pp. 198–206
14. Cypher A. 'Programming repetitive tasks by example'. *Proceedings of the SIGCHI Conference on Human Factors in Computing Systems: Reaching through technology*; ACM. New Orleans, New York, USA; 1991. pp. 33–39
15. Ioannidou A. 'Programmorphosis: a knowledge-based approach to end-user programming. Bringing the bits together'. Presented at 9th IFIP TC13 International Conference on Human–Computer Interaction; Zurich, Switzerland, 2003
16. Mao J.-Y., Vredenburg K., Smith P.W., Carey T. 'The state of user-centred design practice'. *Communications of the ACM*. 2005;48(3):105–109

17. ISO FDIS 9241-210. Ergonomics of human system interaction – Part 210: Human-centered design for interactive systems, *International Organization for Standardization (ISO)*, 2010
18. Gulliksen J., Lantz A., Boivie I. 'User Centered Design in practice – problems and possibilities'. *SIGCHI Bulletin*. 1999;31(2):25–35
19. Nielsen J. 'Guerilla HCI: Using discount usability engineering to penetrate the intimidation barrier' in: Bias R.G., Mayhew D.J. (eds.). *Cost-Justifying Usability*. Boston, MA: Academic Press; 1994. pp. 245–272
20. Bellotti V. 'Implications of current design practice for the use of HCI techniques'. *Proceedings of the Fourth Conference of the British Computer Society on People and computers IV;* University of Manchester, United Kingdom. 1988. Cambridge University Press; 1988. pp. 13–34
21. Vredenburg K., Mao J.-Y., Smith P.W., Carey T. 'A survey of user-centered design practice'. *Proceedings of the SIGCHI conference on Human factors in computing systems: Changing our world, changing ourselves*; 2002; Minneapolis, Minnesota, USA. ACM; 2002. pp. 471–478
22. Bhatti G., Brémond R., Jessel J.-P., Millet G., Vienne F. 'User requirements to model scenarios on Driving Simulators'. *Proceedings 5th International conference on Driver behaviour and Training (ICDBT)*; Paris, France, November, 2011. Farnham, UK: Ashgate; 2012. pp. 493–499
23. JusInMind. [online]. Available from: http://www.justinmind.com/ [Accessed 6 March 2013]
24. Nielsen J. *Breadcrumb navigation increasingly useful.* [online]. April 10, 2007 [updated April 10, 2007]; Available from: http://www.nngroup.com/articles/breadcrumb-navigation-useful/ [Accessed 6 Mar 2013]
25. Schneider W., Eschman A., Zuccolotto A. *E-Prime reference guide*. Pittsburgh: Psychology Software Tools, Inc. 2002
26. Bhatti G., Brémond R., Jessel J.-P., Dang N.-T., Millet G., Vienne F. 'Filling the user skill gap using HCI techniques to implement experiment protocol on driving simulator'. *Proceedings 6th International conference on Advances in Computer-Human Interactions (ACHI)*; Nice, France. February–March, 2013. New York, NY: Curran Associates, 2013. pp. 401–406
27. Gajananan K., Nakasone A., Prendinger H., Miska M. 'Scenario markup language for authoring behavioral driver studies in 3D virtual worlds'. *Proceedings of the 2011 IEEE Symposium on Visual Languages and Human-Centric Computing (VL/HCC)*; 2011. Pittsburgh, PA; 2011. pp. 43–46
28. Franceschini D., Franceschini R., Burch R., Sherrett R., Abbott J. 'Specifying scenarios using the military scenario definition language'. Presented at Fall Simulation Interoperability Workshop; Orlando, FL, 2004
29. Hieb M.R., Tolk A., Sudnikovich W.P., Pullen J.M. 'Developing extensible battle management language to enable coalition interoperability'. Presented at European Simulation Interoperability Workshop; Edinburgh, Scotland, 2004

Chapter 15

Development of a database for storage and analysis of behavioural data

Marta Pereira[1], Holger Lietz[2] and Matthias Beggiato[1]

Abstract

This chapter presents results of one of the post-doctoral projects conducted in the framework of the ADAPTATION project. This study aimed to develop and implement a database and data analysis protocol enabling the storage, joint access, individual analysis and integration of different types of data that are collected during driving simulator and field studies. Such a tool should allow researchers to gather large amounts of data in a unique place, from which it can be managed and retrieved efficiently. The potential of a database for exchanging data with other research institutes and for conducting secondary analysis is discussed. This is one of the first databases of its kind in Europe for archiving Human Factors in the field of road transportation research.

15.1 Database fundamentals

The number of in-vehicle information systems (IVIS) and advanced driver assistance systems (ADAS) available on the market has been increasing in the last decade. The presence of these systems in vehicles enables novel human–machine interactions and thereby alters the driving task [1]. Studying the consequences of such changes is crucial to understand the real impact of technology on road safety. Behavioural research plays an important role in providing guidelines for the improvement of technological innovations. Apart from striving for accuracy, researchers face other crucial challenges: they should also be efficient in producing results and be aware of technological developments within the road transportation sector. The tools and techniques utilised by researchers should support them in reaching these goals. These instruments need to be functional, promote

[1] Department of Cognitive Psychology and Engineering, Chemnitz University of Technology, 09107 Chemnitz, Germany
[2] Department of Communications Engineering, Chemnitz University of Technology, 09107 Chemnitz, Germany

communication with a broad research network, and be applicable to a specific research context. Databases are capable of fulfilling these needs and supporting researchers in their daily tasks.

A database is a structured collection of datasets hosted by a database management system (DBMS). The first electronic DBMS was created in the 1960s [2]. The pioneer in this field was Charles Bachman, whose research focused particularly on the development of efficient storage and access management of datasets. Since then, database management systems have changed considerably and the requirements of modern systems are manifold. Capabilities such as permanent and secure data storage, transaction control and maintenance of data consistency are required. Atomicity, consistency, isolation and durability of transactions ('ACID') are the four minimum properties required for secure transactions. Database management systems should also include user management and user-rights systems, backup capability and a user interface or a data query language for accessing the data. Many database management systems are designed as client-server systems and allow simultaneous access by multiple users.

There are several types of database models. Such a data model is 'an abstract, self-contained, logical definition of objects, operators and so forth that together constitute the abstract machine with which users interact' ([3], p. 15). Other examples include the hierarchical, object-oriented and object-relational models. The relational model is frequently utilised in several database management systems, such as Microsoft SQL Server, PostgreSQL and MySQL. In the relational model, data are sorted in tables and the information stored in each row is called a record. The conceptual modelling of such a database is usually done with entity-relationship diagrams. These diagrams support the developer in the definition and representation of the database architecture, i.e. the different elements (entities) and the links between them (relationships).

Entities are independent sets of information that are included in the system. They have certain properties that correspond to the groups of information that are recorded about them. The level of complexity of an entity may vary considerably as it depends on the type of information being stored [3]. Relationships connect entities together and describe how the entities and associated information are related. Relationships can involve two or more entities. To establish these relationships, primary keys must be defined. The primary key is a unique identifier for any record in a relational database. Each relation is usually described as a table, which is organised into records (rows of a table) and attributes (columns of a table). The primary key can be a single attribute or a set of interrelated attributes. Additionally, tables can be modelled using a foreign key in order to reflect a more complex data model. The foreign key is a referential link between corresponding attributes in two or more entities, and ensures that data consistency is maintained.

The development of database architecture may be considered on three levels: internal, conceptual and external levels (according to ANSI-SPARC group [4]). First, the internal level is most closely linked with the physical storage of the database. It defines how data is stored inside the system and built in terms of machine-oriented constructs (like bits and bytes). The conceptual level is an intermediary between external and internal levels. The entire content of the database is represented at this level. In contrast to the internal level, the conceptual level is more abstract and shows what the stored data represents. Third, the external level is closest to the user in that it corresponds to how data will be seen by users. Thus, as different users require different types and quantities of information from a database, the external level is represented by unique views depending on user or user group. These external views may also be distinct in terms of content, showing distinct portions of a database [3]. Defining these three levels at the outset of database development helps identify how data can be organised within the system and is a good basis for further steps.

15.1.1 Databases for human factors in road transportation

There are numerous examples of databases that store diverse sorts of information relevant to Human Factors in road transportation research. They include collections of published scientific documents e.g. ScienceDirect [5], compilations of road accident statistics e.g. CARE [6], detailed information about traffic safety-related projects financed by the European Union, e.g. [7], or data about public and private research institutions that conduct field operational tests, e.g. work conducted by FOT-Net [8].

Databases are frequently used to store information collected in empirical studies. They are usually developed in individual research institutes and are often accessed exclusively by members of that organisation. Because the process of collecting empirical data is highly time and resource-consuming, the information stored in such databases is valuable and seldom shared with external researchers [9]. Free access to databases of other research institutes is not common. When access is given, however, restrictive conditions are applied and compensation may be required.

One example of cooperative data sharing is the project conducted jointly by the Chemnitz University of Technology (Technische Universität Chemnitz, TUC) and the German Aerospace Center (DLR) in 2008. To fulfil the project's goals, an extract of the University of Michigan Transportation Research Institute (UMTRI) database was purchased. External researchers from Germany received files containing CAN bus data, video and audio data, as well as data from external sensors. These data were part of a project stored in the UMTRI database (the Road Departure Collision Warning Field Operational Test project) [10]. These data were then stored in TUC's own database.

Rare examples of databases that can be freely accessed by external researchers are those developed by the Virginia Tech Transportation Institute (VTTI). Two repositories can be accessed: one through Virginia Tech's Scientific Data Warehouse environment [11] and the other via a website dedicated to the project SHRP 2 (Strategic Highway Research Program 2) [12]. As reported in the 2011 VTTI annual report [13], the Data Warehouse hosts datasets from the 100-Car Study [14] and the 8-Truck Naturalistic Truck Driving Study (NTDS) [15]. In both databases the user has access to a list of datasets containing several types of information.

These include files with information such as vehicle characteristics from which the data was collected, time stamps, codifications of eye glances or event narratives. Supporting documentation for these datasets is also given (metadata), including data dictionaries, vehicle sensor information and guidelines for secondary analysis [13]. When downloading a file, users are presented with a disclaimer regarding the utilisation of such information for research purposes, and the need for authorisation in case further research is conducted with the downloaded data.

Similarly, the SHRP 2 website allows validated forum members to access data collected in the study (for further information see [16]). To become a member, a free registration form must be completed, and then all features become available to the user. When downloading a file a disclaimer is also presented and more details about the nature of the data are given.

15.1.2 Advantages and disadvantages of working with databases

The need to work with a database can be triggered by several events. It can, for example, start when the amount of data is so high that the capacity of file-based systems and programmes are no longer adequate. It can also derive from the wish to rapidly combine or select certain segments of data stored in one or more independent datasets and last, but not least, the need to share information among several users who want to access data at distinct places and/or points in time may necessitate the establishment of a database. In all of these cases, using a database can help ease data handling and offer several advantages to its users.

The advantages that a database can offer depend on the goals of such a system and type of data stored. For databases that contain information collected in empirical studies conducted in the field of Human Factors in road transportation, these tools provide researchers with the opportunity to:

- store large amounts of data in a unique place
- put together distinct sets of information that can be linked to each other
- have fast access to considerable amounts of data
- select, filter and sort the most adequate data for further analysis
- share several types of data
- access the most updated information
- maintain the consistency of the data after its modification
- manage users' rights, giving different access and utilisation rights to distinct users
- conduct data mining with ease
- achieve immediate access together with other users at the same time and in different parts of the world

To benefit from all of these advantages, the main constraints related to database utilisation have to be overcome. One of the problems is the complexity of database systems, which can lead to difficulties in data management and user interactions. Training in database functionality is usually required for all researchers, and especially those without a background in information technology (IT).

Even standard database interfaces are complex and require advanced knowledge of a particular database language. To this end, software has been developed to facilitate the user–database interaction. This software works as an interface, containing more user-friendly characteristics and simplifying the tasks of users who are not IT specialists. The goal of such software is to enhance the link between the user and the system by associating, for example, a row in a relational database to an element in the design.

An interface can be seen as 'the part of the product that the user interacts with to accomplish the task' ([17], p. 7). Its characteristics are very important as they are capable of defining the quality of the interaction between the user and the system. Thus, if an interface is inadequate for a specific user or user group, then interactions with the database may be compromised.

Another major concern is related to the utilisation of secondary data. The literature highlights that differences between primary and secondary data are related to the purpose of data collection. Hox and Boeije [18] define secondary data as 'the one collected for a different purpose and reused for another research question'. Thus, performing secondary analysis is indeed a process of analysing data originally gathered for another purpose [9].

Though working with secondary data offers several advantages that are strongly related with the economy of resources, its use has several inherent limitations. A major disadvantage lies in the fact that the data were collected for different purposes, in response to different research questions. Thus, particular information needed to perform the secondary analysis might not be available or arranged in an adequate way. A second concern is that the researcher conducting the secondary analysis is not the individual who collected the data in the first place. The planning and execution of the collection process is performed by a researcher who has clearly defined knowledge about the nature of the data and its structure. Researchers who are not involved in this process might have more difficulties in understanding the data and consequently more difficulty analysing it [19]. Therefore adequately detailed documentation of the data is required when a secondary analysis is conducted by a researcher not involved in the data collection process.

Both advantages and disadvantages of utilising such a system were considered in the development of the ADAPTATION database. This process is further explained, describing what motivated the development of this particular database, its main goals and development. Major limitations encountered during the development process and future perspectives on the utilisation of such systems as basis for sharing empirical data are also presented.

15.2 ADAPTATION database

15.2.1 Aims and objectives

The aim of this work was to develop and implement a database and data analysis protocol to enable the storage, joint access, individual analysis and integration of different types of behavioural data, which were collected during driving simulator and field studies performed within the project. To this end, the following first steps were identified:

- define the specification requirements of the database, taking into account the nature of data collected and type of analysis usually conducted in this research domain
- define the legal boundaries to enable data storage and fruitful sharing
- develop and implement a database for the storage of behavioural data

Electronic DBMS have existed for more than four decades and the technical developments and expertise of today's technicians make the creation of such tools a reality. Though considerable research is still underway in database development, the present work was not directly focused on technical aspects of such tools, but on two main challenges: (1) testing the concept of freely sharing empirical data among researchers and (2) developing a system for use by non-database-experts in which users can not only benefit from stored information but also contribute to the database without requiring technical support.

The idea of creating such a tool was motivated by researchers' past experiences. To plan and conduct empirical experiments in this research area, institutes have to invest considerable time, as well as human and material resources, for data collection. After the data are used for defined purposes, the data are frequently archived and not used for further studies. Because such empirical experiments are so expensive (especially those conducted in driving simulators and in on-road environments), researchers usually collect data fitted to several research questions, frequently resulting in copious amounts of data.

A Europe-wide project network was considered the most adequate environment for the development of a system that would allow data sharing. Within this framework, the development of a database was done in parallel with planning and conducting empirical experiments. Thus, the requirements of researchers, while planning an experiment, were easily accessed, as well as the broad scope of tools and work strategies utilised to conduct such work. To achieve the defined goals for the development of the ADAPTATION database, the work was divided into three main parts: (1) legal boundaries: to ascertain the relevant legal aspects regulating the collection, storage and sharing of data of European drivers, (2) development of the database: to define and implement the technical requirements of the database and respective interface and (3) small-scale testing: to identify and prevent the occurrence of major problems. These three stages are described in detail in the following sections.

15.2.2 Legal boundaries

The design and development of the ADAPTATION database began with an assessment of its legal feasibility. This aspect is crucial, as all subsequent phases of the project are dependent on what is possible in legal terms. Strong legal constraints may require considerable revision to the database concept, and influence its main characteristics and functionalities. Thus, individuals responsible for legal aspects were consulted with the intent of clarifying two main aspects: (1) database access rights and (2) the possibility of uploading driver data and making it accessible.

After consultation, it was decided that database access rights would be given to all partners in the consortium and that, initially, institutions outside the consortium would not have access. The conditions under which partners were allowed to share data were framed by the Consortium Agreement and by the Grant Agreement of ADAPTATION.

Researchers normally have to attain approval of their local ethics committees to collect empirical data. Additionally, participants of such studies must authorise researchers to record and store their data. In order for data collected during the project to be stored in the database, it would have to follow the usual procedures and subject consent agreements completed as usual.

The regulations of the country in which the data would be stored (physically) were also analysed. As TUC (Chemnitz, Germany) was the host of the database, two legal documents were consulted: (1) the Federal Data Protection Act (BDSG, promulgated on 14 January 2003 and last amended on 14 August 2009) and (2) the Saxon Data Protection Act (SächsDSG, promulgated in August of 2003, with adjustments in 2007). Both acts clarify the protection of individuals against infringement of their rights to privacy as the result of handling their personal data. These legal instruments can be applied to the collection, processing and use of personal data by certain public and private institutions.

The possibility of storing and sharing data belonging to EU citizens was assured through compliance with requirements defined in both legal documents. One major issue was the storage of personal data without participants' written consent. Informed consent signed by participants of an experiment usually does not specifically account for this authorisation (rather stating that all data is treated by the researchers anonymously). Therefore it was decided that the database would not contain personal data (i.e. data through which a single person could be identified). Data to be omitted included names and contact information, as well as videos recorded during the studies. So, whilst protecting individuals, the absence of video can strongly limit the potential for secondary analysis. This places additional emphasis on comprehensive planning of data capture since, once coding of video has been completed (extraction of data from video), the original video is not then available for subsequent additional coding at a later date.

Another obligation of German law is the need to register all data transfer events performed in the database. The identification of the user, the moment and the nature of each transfer would have to be saved to allow data tracking. Thus, a register of all transfers performed in the database would be kept.

15.2.3 Development process

The development of the database architecture began by considering the three levels of the ANSI-SPARC group [4]. These structural levels were utilised to define and organise the elements composing the system. Figure 15.1 shows a representation of these levels, with examples of behavioural data that would be collected in the project.

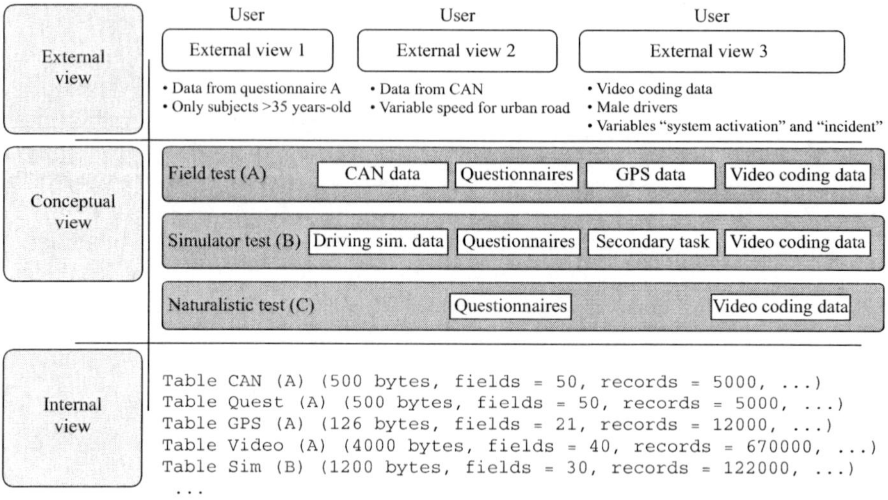

Figure 15.1 Three levels of database structure and examples of behavioural data content

The internal view includes the tables stored in the database. They can contain information from various sources such as questionnaires, GPS, dynamic parameters of a driving simulator or dynamic parameters of an instrumented vehicle. At the intermediate level (the conceptual view), the data is organised differently as the different sets of information belong to a specific study (e.g. Field Test A). Datasets belonging to one study are linked, but datasets from different studies are not. The external view represents what users can see when interacting with the database. This is dependent on the specific requests they make. The external views represent the requests made, either by distinct users or by the same user, on separate occasions.

As a next step, information was collected for the benchmarking process. Aspects regarding database functionalities, main advantages and most appropriate software were identified. To obtain an impression of how the database might appear, similar tools used in road transportation research were analysed (e.g. Virginia Tech Transportation Institute data repositories [11, 13]). While moving forward to the development process, it was considered that a customised interface should be created in parallel to the development of the ADAPTATION database. To achieve elaboration of the database and interface, a user-centred design process was utilised (for further information see [20]), in which the end users' needs and capabilities were taken as the centre of the approach. How users are likely to interact with a product is important both before and during product development [17]; information provided by and about users informs the design process.

15.2.3.1 Understanding and involving users

To better understand user behaviour, their involvement through the design and development processes was requested. The early stage researchers (ESR) of the

ADAPTATION network were selected and involved in the process from the beginning. After being informed about the general aims of the database in one of the first network meetings, their contribution was requested via questionnaire. This questionnaire, sent to all 10 ESR, aimed to document their understanding of the database as well as their needs and expectations. It was also important to assess their level of experience with database use and to invite suggestions regarding functionalities and interface characteristics considered important. Knowledge of their experience in conducting secondary analysis was also important to understand their familiarity with managing data that was collected by others or collected with different research purposes. Similarly, information about the foreseen amount of data collected within the ADAPTATION project, the type/format of these data and the predicted usage of the database within the ADAPTATION framework was gathered.

The answers given by the ESR ($N = 10$) revealed that five respondents had already used database management systems such as MS Access, PostgreSQL, MySQL or Oracle. Databases dedicated to literature research e.g. ScienceDirect, MedLine, ProQuest, administration of research participants and storage of empirical data from the road transportation sector (EuroFOT study; see also [21]) were also mentioned. Eight of the 10 respondents mentioned not having experience with database development and affirmed not knowing SQL language. In addition, eight respondents had no experience conducting secondary analysis.

Seven ESR expected to use the ADAPTATION database in the future to support their work conducted within the project. This means that they considered utilising the content stored in the database. When questioned about their own contributions, six confirmed their willingness to share data collected in their ADAPTATION projects. One of these respondents thought about sharing data from other projects as well, while three partners were not able to contribute data: one was not planning to collect empirical data and the other two mentioned legal constraints that would prevent sharing the data collected in their companies.

Respondents provided suggestions concerning additional functionalities that the database could have, illustrating particular concerns regarding this type of tool:

a. Easy upload process with minimum preparation of data

   'If I have for example 30 different log-files generated by the simulator, I would like to click on just one button in the database and have all the files saved in the database in one document. Finally, I should be able to export them easily in excel or SPSS'.

   'Tools to automatically upload and connect different data types (e.g. video data with driving data)'.

b. Include a search function to find experiments and variables with certain keywords

   'It would be nice if the database could provide the answer to specific queries of the researcher for the specific experiment. For example in an experimental dataset, researchers might be interested in speed, or any other variable of the driver, only at an intersection'.

c. User-friendly interface
'Graphical tools ("Table designer") are better to customise the resulting tables, selecting variables, adding filters'.
d. No need for additional software to have access to database content
'Open access to the database without a lot of additional software'.
e. Possibility of immediate help/support
'Good documentation of the database and a "Hotline" (email, phone) support for urgent problems'.
f. Detailed information about the studies included in the database
'I would worry about having a very precise definition of the design of the study and how the data has been collected exactly: frequency, material (brand, software, version, etc.), documentation of possible confounding variables, etc.'

These results were used to define some of the database functionalities and to choose among certain interface characteristics.

15.2.3.2 Development requirements

The responses to the questionnaire framed the first development stage, both in terms of functionality and design. Accordingly, the tool should offer the following possibilities:

- Provide complete information about the data stored (type of data, how it is structured, how it was collected, who made the collection, what does it mean)
- Allow data download to a common format
- Permit the selection of separate sections of data for download
- Permit data updating without the need to install additional software

The questionnaires revealed that end-users are not database experts. Therefore, the functionalities offered should match users' system knowledge. With respect to design, the interface should be similar to the ADAPTATION website (in terms of colour scheme, basic commands, areas and elements), and the actions analogous to the ones performed in well-known software programs e.g. excel, SPSS.

Database access is an important consideration because the locations of target users are dispersed in several countries. It was decided that the ADAPTATION database would be hosted on a virtual server, integrated in the TUC server structure. Further, database access would be provided via a browser-based interface.

15.2.3.3 Technical development

The ADAPTATION database system consists of two parts: (1) graphical user interface (GUI) and (2) database itself. The GUI is an online web platform that gives participants access to the ADAPTATION project through a login system. The database stores both data uploaded by ADAPTATION users and metadata for data management. The database is exclusively accessible through the web-based GUI.

The GUI is written in PHP and Javascript and runs on an Apache web server. PHP-generated dynamic HTML code is needed for data transfer from the database

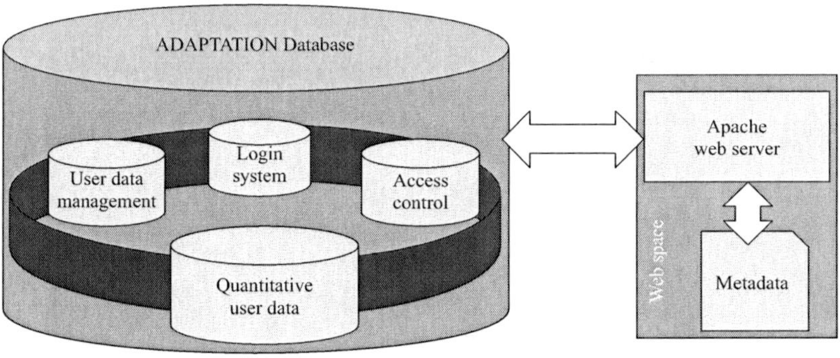

Figure 15.2 Illustration of the management data

to the user's computer and vice versa. Javascript code is executed within the user's web browser and used mostly for displaying information and hiding, or highlighting, GUI elements. Users must have Javascript enabled in their browser settings to access the ADAPTATION database system. Older browsers without Javascript functionality are not supported. The GUI can be linked to one of the following relational database management systems: MySQL or PostgreSQL. Data within the database are used for management of the ADAPTATION database system and for making the metadata available. The management data can be subdivided into data for (1) access control ('AccessCtrl'), (2) login system ('LoginSys') and (3) user data management ('UserDataMgmt'). The interaction of these data parts is illustrated in Figure 15.2.

Data that are uploaded by users may be empirical data (usually placed in tables as a result of experimental measurements) or metadata (text data describing which type of information is contained in the empirical data explaining for each experiment who, when and how the data was collected and also which variables were measured and in which form). When metadata text files are uploaded to the ADAPTATION database system, they are stored in web-space and not in the database itself. Only a link between the file and the related experiment is placed in the database. After having selected empirical data and the respective variable descriptions, the entire dataset is transferred from the web-space to the database, upon selection of the 'Save data' button. A dataset of empirical data is then stored as a table in the database. The name of the table inside the ADAPTATION database is generated by the name of the experiment and the name of the dataset. Both parts are separated by an underscore ('_').

The login system ensures that only authorised users can access the ADAPTATION database system. This system also assures that database users are automatically logged out after being idle for more than one hour. Password recovery functionality is also implemented. Sharing login data with multiple users is ruled out by the login control system. Different users may be logged in at the same time, but not with the same login name. The data related to the access control system are needed for archiving history and for the logbook. Any modifications to the data (upload, download, deletion, etc.) are logged.

Table 15.1 Description of tables used to store the management data

Table name	Subsystem	Data included
adaptation_accesscontrol	AccessCtrl	Log book of modifications, uploads and downloads of data
adaptation_datasets	UserDataMgmt	Descriptions of uploaded datasets
adaptation_experiment_types	UserDataMgmt	Information about supported experiment types
adaptation_experiments	UserDataMgmt	Descriptions of experiments
adaptation_files	UserDataMgmt	Data of uploaded metadata
adaptation_institutions	LoginSys	Information about participating institutions
adaptation_logbook	AccessCtrl	Login/logout time of users including IP addresses
adaptation_login	LoginSys	Login time and time of last action of currently active users
adaptation_researchers	LoginSys	Information about accepted users, including login names
adaptation_variables	UserDataMgmt	Descriptions of variables inside the datasets

The user-based data management system enables linking different datasets or additional files to experiments and also allows the attachment of general information and descriptions to datasets or experiments. Search functionality is available for empirical data stored in the ADAPTATION database. This means that any description, experiment name, dataset or variable can be retrieved. Searching words or phrases inside metadata files, however, is not part of this functionality. Within the ADAPTATION database, several tables are used to store the management data. In Table 15.1 an overview of table names, their subsystems and data is provided.

15.2.3.4 Interface design

For the development of the interface, different areas in the webpage were defined. The elements were characterised in terms of position, physical characteristics (in both active and inactive states) and behaviour (i.e. action produced when the element is activated). Detailed sketches were produced for all different pages of the interface. All had the same header (with the ADAPTATION logo) and a navigation menu placed on the left side. Seven distinct page groups were created: (1) login/password recovery, (2) main menu, (3) view data, (4) create/modify data, (5) search, (6) user profile and (7) help.

The ‘view data’ pages (3) inform users about experiments stored in the database. In the centre of the pages, a list of experiments appears. On the right-hand side, technical details of the selected experiment are summarised (metadata). The respective datasets belonging to an experiment can be accessed and downloaded on subsequent pages. The ‘create/modify data’ (4) pages allow users to upload information. Details about each experiment have to be entered and the data uploaded in datasets. The ‘search’ page (5) allows researchers to look for experiments and variables using specific keywords. The ‘user profile’ pages (6) provide information

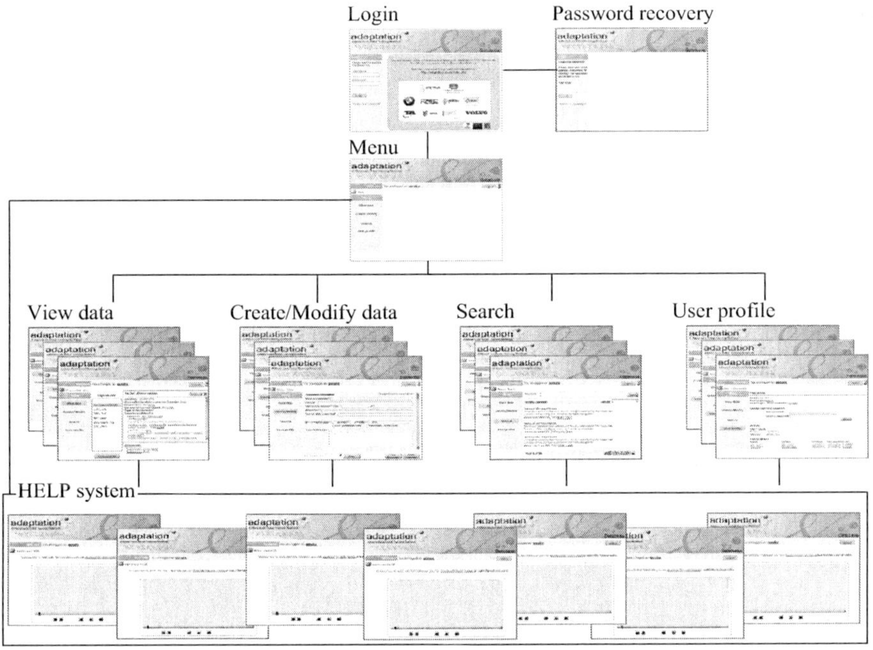

Figure 15.3 Scheme of web platform pages and respective links

about past logins and give users the possibility of changing their password. Further, a help system (7), which can be accessed from every page in the website, provides detailed information on how to proceed. A scheme showing web platform pages and respective links is provided in Figure 15.3.

15.2.4 Small-scale testing

During the development process, two small-scale tests were conducted to identify the main problems of the database both in terms of functionality and interface. The tests were performed at different stages of database development. As the project had a specific timeframe, conducting small-scale tests during the development process helped save time as identifying the main problems earlier gave researchers the opportunity to make the required adjustments before the end of the project.

The main goals of both tests were to: (1) assess if it would be possible to perform all main tasks (download and upload data, create an experiment, search, check profile), (2) understand if the interface was user-friendly and the structure self-explanatory and (3) identify the existence of obstacles, errors or ambiguities when interacting with the database.

The first small-scale test was conducted at TUC. A total of four students (two male and two female; mean age of 24 years old) took part. The functionalities tested included: login area, main menu and view data. Prior to the test, researchers uploaded parts of empirical data collected in [22] and also dummy data, in the database. Participants were then asked to perform two tasks: (1) go to a specific experiment and search for information about when and how the experiment was conducted (metadata related to a specific experiment) and (2) perform two distinct data downloads: first, an entire dataset (all the variables from a questionnaire), and second, certain variables with the application of a filter (e.g. select data only from male participants). The interaction with the database was recorded with a camera and at the end of the test a final interview was conducted to acquire more information about problems and difficulties that were experienced.

Results confirmed that information about the experiments (metadata) was easy to find. Participants were able to download an entire dataset. Nevertheless, applying a filter to select certain variables was identified as the most difficult task. For this test the main problems identified were inconsistent sizes of certain GUI elements; locating the download button; making a table preview if another table was already open; applying a filter.

Before performing the second small-scale test, adjustments were made in the sizes of the GUI elements, the location of the download button and in the table preview behaviour. The second small-scale test was done after all functionalities were implemented in the database. Available ESR of the ADAPTATION network ($N = 9$) were selected to take part in the test. As it was not possible for all participants to meet in one place, ESR were instructed to interact with the database from their workplaces and then report their experiences. The outcome of their actions could be viewed in the database (e.g. successful uploading of a dataset). Interaction problems and further suggestions were described in a word document. An email was sent with a detailed explanation of the test's main goals. The main task was to upload three datasets and respective metadata of one experiment. Text files with the three different datasets to be manipulated were sent to participants via email. After performing this task subjects could view their experiment and download data (with and without the application of filters).

In the second test, participants identified specific improvements as follows: increase the size of images in the help file and experiment description, reduce the number of steps required for saving a complete experiment, provide additional information on how to apply a filter, need for more feedback (positive and negative) after performing certain actions and eliminate bugs. Though it was clear that improvements are still needed, a success rate of 100% indicated that all participants were able to create an experiment and upload the datasets.

The results obtained in both first and second small-scale tests can be used for the improvement of the tool and serve as support in the development of similar systems in the future.

15.3 Conclusion

Using a database to store empirical data collected by different research institutes is a novel concept in the Human Factors in transportation domain. There are few databases in which such data are shared amongst the research community.

Rare examples are the ones developed by VTTI. Databases bring several advantages to the work of researchers, aspects that can be multiplied if a large community shares the database. The present project allowed the development and implementation of a database for the storage and joint access of empirically collected behavioural data. The development of technology and the expertise of technicians make the creation of such tools a reality. The major challenges faced were testing the concept of sharing empirical data among researchers and investigating the possibility of developing a system that could be utilised by non-database experts (i.e. interaction without technical training). At the time of writing this chapter, the ADAPTATION database could be accessed exclusively by project partners. Access apart from this, that is, for the research community or general public, has been discussed within the consortium, but no further action has been taken as it is beyond the scope of the project.

ESR have favourable opinions about the concept of sharing data collected in empirical experiments through a database setting. However, exclusivity in data sharing was preferred by industry partners due to greater legal constraints than those of research institutes. The legal feasibility of the database was assured through fulfilment of specific requirements, which included authorisation of participants for recording and storing data, prohibiting storage of personal data and registering all transfers performed in the database.

Limited technical knowledge of handling databases was apparent in ESR of the ADAPTATION network. As knowledge of specific database languages is sparse, the uptake of such systems would greatly benefit from an interface that promotes problem-free interactions. However, an adequate interface is not only important for non-database experts – it also improves the structure of the entire database and makes the information contained therein understandable. This is a crucial aspect of secondary analysis and is independent of the background knowledge of the database users. As evidenced by ESR, a precise definition about the study' design and the type of information contained in the database is needed to optimise use. This was made possible in the ADAPTATION database through the option of uploading text descriptions (previously defined as metadata).

The main functionality and interface problems of database development were highlighted in two small-scale tests, which further helped to identify areas in need of improvement. Further analyses are required, mainly of data preparation processes, for subsequent upload in the database. Though it is technically possible to transform the data to a format supported by the database, researchers still need to be adequately guided by the tool to do so. Although it was not possible to verify this in the present work, it is considered important. Only through adequate design of all tasks (from data preparation to data upload/download) can a database support researchers in managing its content without the constant need for technical support.

In conclusion, DBMS offer researchers several advantages in the field of Human Factors in road transportation research. The knowledge gathered in the present work showed that sharing empirical data among researchers can be accepted by researchers if certain conditions are met. It is possible to develop a system dedicated to non-database experts, but future systems should have a clear structure in which users can easily understand the database content and its organisation. A user-friendly interface is also crucial to boosting its uptake. And last but not least, future systems might also have to overcome legal constraints, which are dependent on the type of data stored and on the respective countries involved. In line with experiences developing the ADAPTATION database, it is expected that certain parties will have greater difficulties sharing data, which provides all the more reason why flexible solutions must be found to make data sharing inclusive for all potential partners.

Acknowledgement

This research received funding from the European Commission Seventh Framework Programme (FP7/2007-2013) under grant agreement no. 238833 (Marie Curie Initial Training Network ADAPTATION: 'Drivers' behavioural adaptation over the time in response to ADAS use).

References

1. Engström J., Hollnagel E. 'A general conceptual framework for modeling behavioural effects of driver support functions'. In Cacciabue P. C. (ed.). *Modeling Driver Behavior in Automotive Environments: Critical Issues in Driver Interactions with Intelligent Transport Systems*. London: Springer; 2007. pp. 61–84
2. Haigh T. 'Charles W. Bachman: Database software pioneer'. *IEEE Annals of the History of Computing.* [online]. 2011. Available from http://www.tomandmaria.com/tom/Writing/BachmanBio.pdf [Accessed September 2012]
3. Date C. J. *An Introduction to Database Systems*. Boston, MA: Addison-Wesley; 2004
4. American National Standards Institute/Standards Planning and Requirements Committee (ANSI-SPARC). *Study on Data Base Management System.* Interim Report nr 7514TS01. Bulletin of ACM SIGMOD. [online]. 1975. Available from http://www.citeulike.org/user/voj/article/7766192 [Accessed 29 September 2012]
5. ScienceDirect. Available from http://sciencedirect.com
6. European Commission. *CARE: Community Database on Accidents on the Roads in Europe Database*. 1993. Available from http://ec.europa.eu/idabc/en/document/2281/5926.html [Accessed 1 October 2012]
7. European Commission. *Projects Financed by the Commission*. Available from http://ec.europa.eu/transport/road_safety/specialist/projects/index_en.htm [Accessed 1 October 2012]
8. FOT-Net. Available from http://www.fot-net.eu/en/welcome_to_fot-net.htm [Accessed 1 October 2012]
9. Kielcolt K., Nathan L. *Secondary Analysis of Survey Data. Quantitative Applications in the Social Sciences*. Beverly Hills, CA: Sage Publications; 1985

10. Lietz H., Petzoldt T., Henning M., Haupt J., Wanielik G., Krems J., Mosebach H., Schomerus J., Baumann M., Noyer U. *Methodische und Technische Aspekte einer Naturalistic Driving Study*. Report BASt FE 82.0351/2008. 2008
11. Virginia Tech Transportation Institute Scientific Data Warehouse. Available from http://forums.vtti.vt.edu/index.php?/files/ [Accessed 3 October 2012]
12. Strategic Highway Research Program 2 (SHRP 2). Available from http://forums.shrp2nds.us/ [Accessed 3 October 2012]
13. Virginia Tech Transportation Institute. *2011 Annual Report*. Available from http://www.vtti.vt.edu/publications/annualreports/FY11%20Annual%20Report_Final.pdf [Accessed 3 October 2012]
14. Dingus T. A., Klauer S. G., Neale V. L., Peterson A., Lee S. E., Sudweeks J., et al. *The 100-Car Naturalistic Driving Study, Phase II – Results of the 100-Car Field Experiment*. NHTSA Report DOT HS 810 593. 2006
15. Olson R., Hanowski R., Hickman J., Bocanegra J. *Driver Distraction in Commercial Vehicle Operations*. U.S. Department of Transportation Federal Motor Carrier Safety Administration Report FMCSA-RRR-09-042. 2009
16. SHRP 2. Available from http://www.trb.org/StrategicHighwayResearchProgram2SHRP2/Blank2.aspx [Accessed 3 October 2012]
17. Barnum C. *Usability Testing Essentials: Ready, Set... Test!* Burlington: Elsevier; 2011
18. Hox J. J., Boeije H. R. 'Data collection, primary vs. secondary'. *Encyclopaedia of Social Measurement*. 2005;1:593–599
19. Boslaugh S. *Secondary Data Sources for Public Health*. [online]. 2007. Available from http://www.cambridge.org/gb/knowledge/isbn/item1174141/?site_locale=en_GB [Accessed 19 September 2012]
20. International Organization for Standardization. *Human-Centred Design for Interactive Systems*. ISO 9241-210. 2010
21. EuroFOT Project. Available from http://www.eurofot-ip.eu/
22. Beggiato M., Krems J. F. 'The effects of preliminary information about adaptive cruise control on trust and the mental model of the system: a matched-sample longitudinal driving simulator study'. In de Waard D., Merat N., Jamson H., Barnard Y., Carsten O. M. J. (eds.). *Human Factors of Systems and Technology*. Maastricht, the Netherlands: Shaker Publishing; 2012. pp. 63–74

Chapter 16

Synthesis of new knowledge on adaptation processes in response to ADAS and associated advances in research methods

Véronique Huth[1], Mercedes Bueno[1], Alexandra Fort[1] and Corinne Brusque[1,2]

Abstract

Within the ADAPTATION project a series of individual research studies was carried out on the basis of a Joint Conceptual Theoretical Framework. An overview of the findings that were obtained is provided in this chapter, combining the results so as to conclude on three topics that are discussed in the context of implementation and deployment of Advanced Driver Assistance Systems (ADAS): the roles of system characteristics, driver populations and user support strategies. Factors that increase or diminish the expected benefits of ADAS are inferred both from changes detected in driving behaviour (induced by ADAS use) and from subjective system evaluations by drivers. Furthermore, a synthesis of the critical review of methods to study behavioural adaptation in the driving context is given and the two tools developed within the project are discussed. The synthesis of new knowledge allows for conclusions that go beyond the individual studies and confirms the benefit of carrying out research under a joint framework.

16.1 Introduction

In the framework of the ADAPTATION project, a series of individual research activities has been carried out. This book sheds light on the range of subjects that have been approached within the research performed by Early Stage Researchers and by Experienced Researchers involved in the project. The results of this work have been presented at numerous conferences and published in scientific journals, and they constitute or support the PhD theses of the Early Stage Researchers. The aim of this concluding chapter is to review all these results and synthesise them into an integrated picture of the project outcome.

[1] IFSTTAR-LESCOT, Bron, France
[2] University of Lyon, Lyon, France

The ADAPTATION studies have considered different types of Advanced Driver Assistance Systems (ADAS). Depending on their interaction with the driver, ADAS can be classified into informing, warning and intervening systems. Within ADAPTATION two informing systems have been studied: a Satellite Navigation (SatNav) system that provides the driver with route guidance [1] and an Intersection Assistant (IA) that supports the driver's decision making at an upcoming intersection (Chapter 7). The warning systems assessed within the project were a Forward Collision Warning (FCW) system that alerts the driver when a risk of collision due to critically small headways is detected (Chapter 12), a Visual Distraction Alert (VDA) system that emits a warning when the driver looks away from the road scene for too long or too often [2], and a Blind Spot Information System (BLIS) that warns the driver who is about to cut into the trajectory of another vehicle during a lane change [3]. Concerning intervening systems, two individual research projects have studied Adaptive Cruise Control (ACC), an ADAS that maintains a maximum speed and headway set by the driver (Chapters 4 and 5). A further project worked on ACC in a modified form that takes into account driver workload (WACC, Chapter 10). Other studies tested the effects of an FCW coupled to ACC (brake-capacity FCW) that requires the driver to take over control when the brake capacity of the ACC is exceeded [4].

The ADAS mentioned above have been tested in real driving conditions or adapted for simulator tests, and focus group discussions have been carried out with users of some specific systems. In addition, a questionnaire survey and focus group studies have been conducted with users of a range of ADAS (24 systems). Therefore most studies have focused on drivers equipped with ADAS. ADAS users share the road with unequipped drivers, who can also be influenced by the presence of ADAS in other vehicles. For this reason, some research activities have focused on how unequipped drivers adapt their behaviour in the vicinity of automated vehicles, considering the example of platoons of electronically coupled vehicles (Chapter 13). Furthermore, the effects of secondary tasks related to in-vehicle technologies have been examined with regard to use of mobile phones while driving (Chapter 9) and to performance of a visual search task considered as a surrogate informing ADAS (Chapter 6).

The studies have analysed behavioural adaptation of drivers to ADAS. In order to generate research questions and design the experiments within the ADAPTATION project, a Joint Conceptual Theoretical Framework (JCTF, Chapter 2) was established at the beginning of the project, based on a review of theoretical and empirical research related to the concept of driver behavioural adaptation over time. As Chapter 2 highlights, theories about learning and skill acquisition, driver risk models, as well as more recent integrative models of behavioural adaptation, have been considered for the JCTF development. In addition to providing insights concerning the research questions, within the JCTF each set of studies contributes to a more complete picture of drivers' behavioural adaptation. Thus, the project in its entirety can reach more generalizable conclusions based on the patchwork of studies that it comprises. The results largely consist of detected changes in driving behaviour or subjective system evaluation by the drivers following ADAS use.

This behavioural change (adaptation) is presented in relation to three topics. First, this chapter discusses the roles of ADAS characteristics, regarding the functionality and design of the system on one hand, and its reliability on the other hand (Section 16.2). Then, specific driver populations are addressed, including older drivers, Parkinson's drivers, ADAS experienced drivers and professional drivers (Section 16.3). Furthermore, user support strategies for deployment of ADAS are suggested, referring to system knowledge, driver awareness and training (Section 16.4). This synthesis of knowledge is illustrated by specific examples, taken from the individual studies.

The activities of the ADAPTATION project were not only confined to generating new findings about drivers' behavioural adaptation to ADAS, but also included methodological work. The variety of available research methods was critically reviewed regarding advantages and drawbacks, and two tools were created during the ADAPTATION project in order to deal with two specific difficulties frequently found by researchers. The first was developed in order to compensate for a lack of programming skills of many researchers. It consists of a new approach and user interface for the modelling of driving scenarios within driving simulators (Chapter 14). Another issue researchers frequently have to deal with is the storage and handling of large amounts of data collected during the life of a project. Responding to this need, the second tool is a database that allows all members of the project to store, access and share data (Chapter 15). Following a brief conclusion on the methods of research on behavioural adaptation to ADAS, these two research tools are presented in Section 16.5 of this chapter.

16.2 The roles of system characteristics

The assessment of the effects of ADAS comprises the users' response to the system characteristics in terms of behavioural changes and subjective appreciation. Generally speaking, the ADAS studied within ADAPTATION have proven to fulfil their function satisfactorily and are well accepted by drivers. Positive effects on driving behaviour that can be attributed to system use have been found for the informing system SatNav [1], as well as the warning systems FCW and VDA [2, 5–7]. For example, the use of SatNav led to visual search behaviour that was focussed on potential hazards (other road users) rather than on orientation points (e.g. road names) as observed during the use of a paper map [1]. Moreover, the intervening WACC showed a safety advantage in deceleration and braking parameters compared with ACC. That is, drivers benefit from larger headways that are automatically adjusted by the WACC under high cognitive workload conditions. Their reaction time to a critical braking event is the same for WACC as for ACC, but they use the advantage of the WACC to decelerate more smoothly. These two factors could lead to better management of a potentially critical situation if the drivers have to brake in case of a sudden object or braking manoeuvre in front of their car [8].

Concerning the subjective evaluations by drivers, positive satisfaction ratings have been registered for the warning systems BLIS [3] and IA [9], as well as the intervening systems ACC [10, 11] and its variety WACC [8]. For example, the BLIS was judged as useful, comfortable and safe by the participants despite some limitations presented by the system. It was especially appreciated by the users for its contribution to avoiding dangerous situations during a lane change [3].

These results, however, should not be interpreted as absolute findings. Instead, drivers' response to ADAS has been found to depend on a range of factors related to system characteristics. The first part of this section discusses the role of ADAS functionality and its design, while the second part presents results concerning influence of reliability on driving performance and driver opinion.

16.2.1 Functionality and design

In the context of the functionality of an ADAS, its adaptability (i.e. how flexibly the system parameters adjust to specific conditions) plays an important role. In this regard, ADAPTATION studies have shown that the distraction level of the driver influences the effectiveness of the FCW system; drivers benefit less from the support of the warning system in cases of high mental workload [7]. This finding suggests that considering the attentional state of the driver can improve system functionality. Within ADAPTATION, the evaluation of the WACC that adapts to the workload state of the driver in real-time [8] and a VDA system that considers the driver's glance history [2] have confirmed the benefits of such adaptability. In addition, controllability of the system perceived by the drivers may have an influence on its acceptance. For example, referring to ADAS in general, the possibility of switching off the system has been requested by users [1].

Beyond the effects intended by the system functionality, unexpected effects can appear because the design of the system can be a source of driver distraction. An unfavourable presentation of information that promotes driver distraction can reverse the intended benefit of the ADAS. On that score, ADAPTATION studies demonstrated that visual information of a brake-capacity FCW system should not be presented in the instrument cluster as this might cause misallocation of attention [4]. Similarly, route guidance information should be presented in a way that supports the driver keeping their eyes on the road scene as far as possible. A field driving study detected that, proportionally, drivers made more glances towards the SatNav than towards a paper map [1]. By contrast, the design of the BLIS has been confirmed as a positive example of information presentation. The position of the warning light allows the drivers to perceive it without taking their eyes off the road and the system thus supports drivers in avoiding dangerous overtaking while keeping focus on the road scene [3]. The review of the state of knowledge on mobile phone use while driving presented in Chapter 9 suggests that interface design is crucial in this context as well. Among all interactions with in-vehicle technologies, the use of a mobile phone while driving is one of the most distracting. Considering that this behaviour is frequently shown by a large number of drivers, efforts need to be made regarding interface ergonomics, so as to facilitate use of the phone that is compatible with driving (Chapter 9).

Furthermore, unexpected effects can occur when the drivers adapt their driving behaviour in a potentially inappropriate way. The ADAPTATION studies observed such forms of unintended driver behaviour related to use of functionality offered by the system, mostly concerning ACC. Findings from focus groups suggest that there may be an increase in secondary task engagement (e.g. mobile phone use) while driving with a version of ACC that is adapted to stop-and-go traffic and drivers were observed to be using the ACC speed buttons instead of the accelerator or brake for specific manoeuvres [12]. A further aspect of unintended usage behaviour of ACC could be found in the choice of the shortest headway option by most drivers [12]. This effect might be due to flexibility limitations of the system studied; since drivers are constrained to use the same headway in a variety of situations they might thus prefer a headway that is less disturbing in specific situations, e.g. in heavy traffic or overtaking situations. This might not necessarily correspond with the headway that is safest and most comfortable in the situations for which the ACC has mainly been designed. Alternatively, the predefined settings might not adjust closely enough to real needs, such that the drivers do not consider the longer headways as practical. By contrast, unintended driver behaviour that could have been expected in response to WACC has neither been observed during simulated nor on-road driving. Short-term studies have not detected any adaptation to WACC in terms of risk homeostasis, which would have predicted later braking in critical situations when the headway margins are increased by the system [8]. Finally, a behavioural adaptation effect to the IA was found for older drivers and drivers with Parkinson's disease, who crossed the intersection more quickly and carried out less visual checks when using the ADAS in a driving simulator. It is yet unclear whether this behaviour is unsafe and thus inappropriate (due to insecurity/age or health related constraints) or whether there was no necessity of more checks and the drivers were thus adapted to the context [13]. A similar effect was found for navigation assistance: drivers crossed intersections more quickly when they used the SatNav system than when they used a paper map [1].

16.2.2 Reliability

When analysing the effects of system characteristics, it has to be considered that ADAS have technical limitations which can impair their reliability. For instance, physiological measures that are used to detect driver workload as in the case of the WACC can be biased by the age or by possible diseases of the driver. Moreover, systems rely on sensors that can encounter limitations in certain conditions, thus producing false alarms and misses. For instance, ACC and the FCW can fail in cases of adverse weather conditions; the VDA system may be influenced by inaccurate head or gaze tracking; and the BLIS has some difficulties in detecting rapidly approaching vehicles. For this reason, system reliability has been considered as a variable when studying drivers' adaptation to ADAS. Not only is the actual reliability of the system important, but its perception by the users also has a consequent effect on driver acceptance. Focus group discussions within ADAPTATION revealed that drivers are more sceptical towards newer ADAS compared with those having a more established market position [1]. Further studies showed that ACC is considered reliable by its users [11, 12]. The example of the BLIS illustrates that a system can receive a positive evaluation even though it was not considered completely reliable due to known limitations [3] and the FCW that was reliable at 75% also received positive satisfaction and usefulness ratings [7]. Similarly, the acceptance of ACC was not affected by some unreliability, as long as the drivers had a mental model of the system that matched their practical experience [10]. By contrast, unexpected automation failures experienced in a simulator setting had a particularly negative effect on driver acceptance and trust [10]. In line with this result, the VDA system was less accepted during a driving simulator test when it presented a higher rate of false alarms that were not previously explained to the drivers [14]. Thus, transparency regarding system reliability is of great importance for user acceptance.

In addition, it has to be considered that system reliability can also affect driving performance. Interestingly, variations in the reliability of the VDA system did not lead to differences in glance behaviour nor driving parameters [2]. Thus, objective driving performance is not necessarily affected by some unreliability of the ADAS. This finding was confirmed by results on a 70% reliable FCW system, which still improved driver performance compared with baseline driving [5–7]. However, possible negative behavioural adaptation effects may appear in the longer term if drivers over-rely on such systems. In the case of the BLIS, for example, drivers could cease checking the mirrors before lane changes when they get used to relying on the system's warnings and do not take into account system limitations [3].

16.3 Considering the specificities of driver populations

The interaction of the driver with the ADAS is not only shaped by the system characteristics but also by characteristics of the user. Different driver populations have different needs with respect to assistance functions and interface designs. Consequently, their preferences regarding ADAS may vary as well as their behaviour in response to system use. While most studies within ADAPTATION did not focus on a particular driver population, some research has been undertaken which explicitly studied specific driver groups. These included older drivers, drivers with Parkinson's disease, ADAS experienced drivers and professional truck drivers. The four parts of this section present the results concerning these specific user groups.

16.3.1 Older drivers

Older drivers tend to have difficulties in managing complex driving situations, such as intersections. They can have physical limitations that may influence their driving; e.g., more restricted head and neck movements may impair their ability to make rapid checks on the driving environment. Additionally, retrieving and processing information can be more challenging for older drivers, which can lead to difficulties in selecting the most important traffic signs in order to judge a traffic situation correctly. In particular, older drivers can have trouble dealing with situations that require divided attention and decision making under pressure (Chapter 7). For these reasons, the IA tested within ADAPTATION appears to be particularly applicable for this user group. Simulator studies showed that healthy older drivers learned to drive with the IA and might benefit from this assistance system [15].

It could be argued that using an ADAS increases demands on the driver, at least during those moments when interaction with the system is required. Interacting with an ADAS interface could be challenging for older drivers, since it represents a secondary task and requires attention sharing. Comparisons of secondary task interference in the driving performance of older and younger drivers, using a visual search task as a surrogate informing ADAS, demonstrated that initial differences between 'younger' and 'older' age groups diminish with practice (Chapter 6). Therefore older drivers might need more time to adapt to system use, but can equally benefit from it after some practice. The ADAPTATION studies also found that feedback on driving performance helps older drivers to prioritise the driving task and reduce interference from the secondary task, that is the interaction with the ADAS [16].

16.3.2 Drivers with Parkinson's disease

The effects of ageing in terms of physical and cognitive limitations are more pronounced in older persons with neurodegenerative disorders such as Parkinson's disease; hence, in its more severe form this disorder is incompatible with driving. However, patients with milder forms of Parkinson's disease can keep driving safely, while adequately compensating for their impairments. Often, they manage by severely restricting the driving situations they choose. For example, they may drive only under good weather conditions, during daytime and in familiar driving environments. Under most circumstances, drivers with Parkinson's disease avoid complex traffic situations. Thus, this compensation behaviour can represent a limitation in the patients' mobility. Assisting these drivers with an ADAS may help them compensating for their shortcomings and, as a consequence, extend the range of situations in which they feel comfortable and safe. ADAPTATION studies found a positive bias in trust and acceptance ratings of the IA for drivers with Parkinson's disease [13]. This result could be due to drivers' awareness of their cognitive and motor limitations associated to the disease. Yet, drivers who suffer from Parkinson's disease might also encounter particular difficulties when interacting with an assistance system. Unlike healthy older drivers, they did not improve their performance when using the IA over time [13]. This result might be explained by the difficulties with procedural learning that patients with Parkinson's disease often experience. In order to confirm this result and to further analyse in which way drivers with Parkinson's disease could effectively benefit from IA, further safety indicators such as time-to-collision and safety margins would need to be considered in addition to intersection crossing time.

16.3.3 ADAS-experienced drivers

It can be expected that interaction with an ADAS evolves with use of the system. Having passed an initial learning phase and having witnessed a range of situations while using the system, ADAS-experienced drivers may respond to the system in different ways than novice users. Once drivers are familiar with an ADAS, support by the system can reduce the driver's workload. This, in turn, may encourage secondary activities, as has been reported in focus group studies with experienced ACC users [12].

A questionnaire survey conducted within ADAPTATION revealed that experienced ADAS users were more aware of the risks related to secondary task engagement, at least regarding the specific task of writing a text message [1]. However, these users indicated a greater intention for secondary task engagement in the case of hands-free phone use than other drivers [1]. On a related point, Chapter 9 provides insights into the discrepancies between risk awareness and usage behaviour related to mobile phones.

Long-term use of ADAS might also lead to changes in perception of the system and its utility. The questionnaire survey demonstrated that the perceived need to use ACC for safety reasons when a child is travelling in the car was positively correlated with ADAS experience [17]. This effect persists when taking into account a possible influence of age, driving experience and annual mileage, which demonstrates the relevance of ADAS experience regarding the appreciation of the safety benefit provided by the use of the ADAS.

16.3.4 Professional truck drivers

The task of professional truck drivers differs from that of other drivers. Professional truck drivers usually travel much longer distances and predominantly drive in less complex environments, such as the highway. The characteristics of this kind of driving task increase the chances of experiencing monotony, which in turn may encourage the truck driver to engage in secondary tasks. This may engender visual distraction from the roadway. Within ADAPTATION, studies on VDA system use while performing a secondary task were therefore carried out with professional truck drivers. The results indicate that VDA systems have a positive effect, comprising an improvement in driving performance and an increase of attention-on-road time [2]. Furthermore, professional truck drivers participated in a naturalistic driving study investigating eye movement and brake reactions to a real world brake-capacity FCW system. The results show that truck drivers initiate braking responses when the warning is issued and reallocate their attention to the road if necessary. However, once the threat is over, eye movements are directed towards the instrument cluster (Section 16.2.1). This particular glance behaviour occurred when truck drivers were distracted or during unpredictable warning events, suggesting that the truck drivers might need to identify the warning source [4]. Given that all collision warning systems in trucks on the market present visual warnings in the instrument cluster, this result should be taken as advice to reconsider the interface design so as to ensure professional truck driver safety.

16.4 User support strategies for the deployment of ADAS

Market introduction of ADAS may be accompanied by support strategies, so as to guide users in the introductory phase of system use. In doing so, drivers' adaptation to ADAS can be positively influenced in terms of subjective reactions and also regarding the way drivers use the system. ADAPTATION studies have dealt with three highly relevant factors in this context: knowledge the driver has about the system, awareness of system utility and learning processes related to system use. Correspondingly, the three parts of this section deal with the effects that have been found concerning driving behaviour and driver opinion.

16.4.1 System knowledge

From their first contact with an ADAS, drivers start building a mental model of the system and its functioning. An initial mental model is formed from preliminary information and is later updated with system experience, until it eventually converges with a realistic appreciation of the ADAS [10]. However, the influence of experience on development of the mental model implies that aspects of the system that are not experienced might not be retained and represented in the model. For example, it has been found, both in simulator and on-road studies, that limitations of ACC (such as system failure during adverse weather conditions) drop out of the mental model if they are not experienced, and that the model shifts towards an overly positive representation as a consequence ([10] and Chapter 4 of this book). This practical experience with a system has a strong influence on the mental model that drivers have about the ADAS. However it has been found in a focus group study using the example of ACC that even experienced users are not always fully aware of critical situations linked to ADAS use [12]. Considering this finding, information given to the driver initially, e.g. at the time of the system acquisition or during use, gains crucial importance. Misuse cases of ACC that have been reported during focus group interviews [12] might also be explained by an incorrect mental model of the system. This provides the opportunity to prevent unintended usage behaviour by correcting the mental model.

System knowledge may also be important for unequipped drivers who share the road with ADAS users. For instance, drivers travelling behind an ACC user can be confused by the longitudinal control determined by the ADAS of the preceding car, e.g. abrupt braking, which differs from the expected driving behaviour in some situations [12]. Moreover, a negative behavioural effect, which could possibly be attributed to incomplete system knowledge of unequipped drivers, has been found regarding platoons (automated closely coupled vehicles). The short time headways maintained in the platoon induced unequipped drivers in its vicinity, who did not have any information about it, to reduce their own headway [18]. Yet, this effect cannot be ascribed to missing system knowledge with all certainty, given that the occurrence of the effect may be influenced by many variables, such as the penetration rate of platoons, the exposure time, the frequency of exposure and the conspicuity of platoons.

16.4.2 Driver awareness

In order to successfully implement ADAS and have drivers use them, it is necessary to make drivers generally aware of the utility of ADAS. The ADAPTATION studies have revealed that drivers tend to perceive ADAS as useful for close relatives, e.g. their parents or children, rather than judging them as necessary for themselves [1]. In focus group discussions, drivers expressed high levels of confidence in their own driving, and the feeling of control even makes some of them believe that they can compensate for other drivers' errors [1]. Likewise, drivers tend to overestimate the quality of their driving performance and their ability to perform secondary tasks [1]. Thus, there is a need to improve drivers' awareness of their own capability limits and how they can benefit from the support of ADAS.

Driver awareness of a system has been found to be related to its time on the market, with acceptance and usage rates increasing over time [1]. Knowledge about the limitations of an ADAS has not proven detrimental to the recognition of its benefits in the case of ACC and the BLIS [3, 12], which corresponds with the findings concerning mental models and acceptance mentioned earlier in this chapter. As already discussed in the context of system reliability (Section 16.2.2), transparency regarding system functioning also seems to be beneficial for the user experience. For example, an increase in comfort ratings on WACC has been registered after having been informed about its functioning in a simulator setting and partial increases have been found in on-road settings depending on system awareness [8].

Research activities on the use of mobile phones while driving revealed that driver awareness needs to be improved concerning some aspects of the distracting effects of phone interactions (Chapter 9). Another aspect of driver awareness that should be further addressed relates to effective exposure regulation strategies the driver can apply to minimise the risk of phone use. The ADAPTATION studies showed that drivers' reported intentions to regulate their phone use sometimes differ from what can be observed in their actual behaviour [19].

16.4.3 Learning and training

The uptake and use of ADAS involves drivers' learning processes, which will be reflected, on one hand, in driving performance and, on the other hand, in subjective evaluation of the system by users. These may be specific to user groups. Within ADAPTATION, learning processes related to driving performance have been studied with older drivers who used the IA in a simulated setting. Not only did they learn to travel with the ADAS, they also transferred the knowledge they acquired though the IA support when driving later without the IA [15]. Further studies are needed to clarify whether this behavioural adaptation is favourable in terms of driving safety, but the results do show that learning processes need to be considered when introducing an ADAS. By contrast, a positive transfer effect was not observed for the FCW when the system was switched off after 45 minutes' driving [7]. Still, it was not clear whether this result was due to the absence of the system or to other factors such as practice or fatigue effects. The studies on the FCW confirm the need to extensively investigate the consequences of learning processes during system use on subsequent driving without assistance.

A comparison of older and younger drivers' performance when using a surrogate informing ADAS detected a positive effect of practice on system use for both groups (Chapter 6). Although this effect might be confounded with a simulator training effect, the results point towards a beneficial impact of training on older drivers' behavioural adaptation to ADAS. A further example of learning processes that are reflected in driving performance was found for driving assisted with a VDA system. In addition to the immediate adaptation effect of eyes-on-road time, a constant improvement of this parameter was detected during the 18 minutes of test driving when the VDA system was using glance history for the generation of warnings [2]. Immediate improvements of driving performance have also been found for the FCW: drivers adapt their behaviour positively as soon as a brake-capacity FCW is issued, showing faster reaction times to braking events [4]. Similarly, when an independent FCW system was introduced, drivers immediately reacted faster and adopted safer headways, and these effects seemed to be stable over time during the 45 minutes of testing [7].

Learning processes related to subjective evaluation were studied with ACC. Drivers who did not receive any information about the system functioning did not develop stable trust and acceptance regarding the ADAS [10]. Thus, a trial-and-error strategy of learning is insufficient for successful deployment of this assistance system. By contrast, giving initial information on the system leads to a steady (logarithmic) increase of trust and acceptance ratings with practice of system use [11]. After a few hours of ACC use and less than 200 km driven, the learning curve stabilises at a favourable level of system perception [11].

An important issue concerning training and behavioural adaptation is raised in the DO-IT BEST Driver Feedback Model [20]. For learning processes to evolve, it is essential that drivers receive feedback about their behaviour ([2] and Chapter 11 of this book). Here a comprehensive feedback strategy on safety behaviour has been proposed within ADAPTATION [2, 20]. This approach emphasises the importance of driver training and education in order to achieve positive safety effects of ADAS that persist for the long term. It includes behavioural check-ups in order to analyse the safety benefit of these interventions. In short, the DO-IT BEST Driver Feedback Model presents a comprehensive user-centred strategy on how to manage attention-related failures involving driver distraction countermeasures before, while and after driving and thus preventing those failures in future trips.

16.5 Tools for research on drivers' behavioural adaptation

Research on behavioural adaptation to ADAS can be carried out in laboratory settings and naturalistic environments, providing different types of data and allowing complementary insights. The research tools and environments have been described in Chapter 3, and their benefits and drawbacks have been discussed in detail. In short, simulator studies offer a controlled and safe setting for hypotheses testing with good data quality, but are constrained to an artificial setting, which results in potentially low external validity of the findings. Naturalistic studies deliver findings that are closest to reality but are inappropriate for a range of research questions, in particular those concerning causality, due to the lack of experimental control. The variety of research methods, including laboratory studies and experimental field studies, represents a set of complementary tools which need to be carefully chosen by the researcher depending on the specific research questions that are to be studied regarding the aspect of behavioural adaptation. Using combinations of study types can provide insights that go beyond the potential of each individual method. As discussed in Chapter 3, behavioural adaptation extends over time and different effects of ADAS use can be observed on the short, medium and long term. Series of simulator sessions at several points in time or a naturalistic driving study allow investigation of aspects of behavioural adaptation from learning processes to transfer effects. Although several challenges are linked to long-term studies, presented in Chapter 3 and further discussed in Chapter 17, they are essential for the examination of drivers' behavioural adaptation to systems.

Responding to difficulties often encountered in research on behavioural adaptation to ADAS, two research tools have been developed and tested within the ADAPTATION project: an adapted user interface for simulator studies and a common database for storage and usage of behavioural research data. The motivation for creating these tools, related challenges and the main characteristics of the instruments are described below.

Researchers often lack programming skills and do not have enough time to deal with simulator scenario development and, besides, there is no standardised way of describing scenarios. So, in most cases, researchers have to rely on help from technicians or scenario developers, which consumes a considerable amount of time and resources. A study was conducted with researchers (end-users) in order to get more insight into their requirements, problems and profiles ([21], Chapter 14 of this book). With the aim of overcoming these difficulties, a new user-centred methodology and interface has been created that allows relatively easy modelling of scenarios on different driving simulators. It has been proposed that this instrument will enable the end-user to develop an experimental protocol using the skills they already have regarding the typical programming process. A preliminary evaluation of the new approach and the interface performed with end-users has been positive.

Identical driving scenarios are often tested on different simulators within European projects. Given that the execution platforms of driving simulators are rather heterogeneous and frequently no standardised specification for the scenario descriptions exists in the experimental protocol, a Scenario-Meta Language (SML) and an interoperability framework have been developed within ADAPTATION. This proposed approach and interface are generalizable, and are expected to be usable on most existing driving simulators due to the interoperability of the scenarios generated from the new tool.

Behavioural studies often produce large amounts of data which need to be stored and accessed by various researchers in different parts of the world. Databases and data analysis protocols have to render individual and integrated analyses. To serve these purposes, a unique tool was created to enable the storage, joint access, individual analysis and integration of different types of data collected during driving simulator and field studies (Chapter 15). The idea of creating such a tool was motivated by researchers' past experience. The development process focused on two main challenges: testing the concept of freely sharing empirical data among researchers and developing a system for use by non-database-experts, in which users can not only benefit from stored information but also contribute to the database without requiring technical support. The researchers (end-users) were integrated in the development process by contributing their knowledge level regarding databases and by sharing their needs and expectations. After the first development stages, two small-scale tests were conducted to identify the main database problems. Although the researchers recruited to test the database were not IT specialists, they recognised the advantages of working with Database Management Systems for sharing empirical data. The possibility of developing a system dedicated to non-database-experts was demonstrated through the implementation of a user-friendly design and a clear organisation of the database contents. Future systems will benefit from an unambiguous structure, in which users can easily understand what type of data is stored in the database, and how is it arranged. A user-friendly interface is also crucial to boosting its uptake.

It is anticipated that future systems might have to overcome legal constraints, which are dependent on the type of data stored and on the respective countries involved. In order to avoid such legal constraints in sharing data, flexible solutions must be found to make data sharing inclusive for all potential partners.

16.6 Conclusion

The results of the series of individual research studies shed light on a range of aspects of drivers' behavioural adaptation in response to ADAS use and in this chapter they are integrated into one picture of the project outcome. The JCTF established at the beginning of the project has been found useful for the development of research questions related to behavioural adaptation. Findings have been generated that, taken together, have allowed for conclusions that go beyond the individual studies. However, the limitations of the studies need to be considered: in particular, some results are very specific and might not be generalizable over different systems, driver populations and usage situations. Nevertheless, the findings give interesting insights, which need to be confirmed by future research. Although it cannot be considered all-embracing, the JCTF will serve as a starting point to obtain further insights in new projects and thus to enlarge the range of validity of the findings. Above all, the ADAPTATION project confirmed the usefulness of an approach that began with the creation of research hypotheses from a model in a top-down process, that then produced individual results and integrated them bottom-up into a joint framework.

In general, informing, warning and intervening ADAS induce positive effects on driving behaviour and on driver opinion. The project results have demonstrated the benefits of adaptability of system functionality, for instance, by taking into account the driver's attentional state. The importance of a system design that does not induce driver distraction has been evidenced, for example, by an appropriate location of warnings or information provided by the ADAS. Moreover, the system functionality and design can lead to unintended usage behaviour, e.g. misuse of the system controls, and the system use can have unexpected side effects, e.g. engagement in secondary tasks or a decrease in checking behaviour when crossing intersections or changing lane. Interestingly, it seems that ADAS do not need to be completely reliable to be accepted by the drivers and to improve driving performance. However, the driver's mental model plays a crucial role in this context. Unexpected failures have been shown to have a detrimental effect on system acceptance, and a mere trial-and-error strategy might be insufficient to properly learn how to drive with an ADAS. Although system experience helps the mental model converge with a realistic system appreciation after some time, providing initial information has been shown to be necessary in order to make drivers fully aware of the system functioning and its limitations (which might not be frequently experienced). Transparency in system functionality has also proven to be favourable for drivers' appreciation of the system. Furthermore, the research findings hint towards relevance of system information for unequipped drivers who share the road with ADAS users, both in terms of their own driving performance and their perception of the ADAS users.

The findings also reveal the importance of considering learning processes and their effects when introducing ADAS. Overall, it seems that drivers do not need a long learning phase in order to benefit from assistance by a warning system. The results of the ADAPTATION studies show that positive effects on driving behaviour appear to persist over time, but do not necessarily improve further. It can be expected, however, that systems with more complex interfaces require a training phase so as to fully reach their assistive potential. Similarly, the ADAPTATION studies suggest that specific driver populations such as older drivers may need some time to practice with the system in order to benefit from its assistance. Performance feedback helps reducing the interference of ADAS use on their driving. In the case of neurodegenerative disorders, more research should be conducted before determining whether these systems can be effective. Although immediate benefits of system use have been found for healthy drivers, their learning process has been shown to improve when performance feedback was provided. An integrated approach including driver training and education has been suggested so as to maximise the positive safety effects of ADAS. Whilst a need to improve drivers' awareness of the benefits they can obtain from using ADAS has been detected, drivers with some degree of impairment, e.g. patients with Parkinson's disease, seem to be particularly suited to derive benefit from assistance systems.

The tools developed within the ADAPTATION project (simulator scenario generator and database assistant) are expected to considerably facilitate future research work and improve the generation and exchange of scientific knowledge. Applying the tools in a range of different studies will reveal aspects that need to be further improved.

In a further step, conclusions from the project results can be used to produce recommendations on further research questions that should be addressed, so as to fill current knowledge gaps. These are presented in the following chapter, along with methodological advice derived from lessons learnt during the research activities of the ADAPTATION project and with more operational suggestions towards improved design and deployment of ADAS.

Acknowledgement

The research leading to these results has received funding from the European Community's Seventh Framework Programme (FP7/2007-2013) under grant agreement no. 238833/ADAPTATION project (www.adaptation-itn.eu).

References

1. Haupt J., Van Nes N., Risser R. 'Changes in motivational factors due to ADAS experience'. Presented at ADAPTATION meeting, Lyon, France, 2012
2. Wege C. 'Adaptive eyes: Driver distraction and inattention prevention through advanced driver assistance systems and behaviour-based safety'. Doctoral Thesis, Chemnitz University of Technology, Germany, and Volvo AB, Sweden, 2013
3. Piccinini G. F., Simões A., Rodrigues C. M. 'Focusing on drivers' opinions and road safety impact of Blind Spot Information System (BLIS)'. In Stanton N. A. (ed.). *Advances in Human Aspects of Road and Rail Transportation*. CRC Press, Boca Raton, FL, United States; 2012. pp. 57–66
4. Wege C., Will S., Victor T. 'Eye movement and brake reactions to real world brake-capacity forward collision warnings: A naturalistic driving study'. *Accident Analysis and Prevention*. 2013;50:259–270
5. Bueno M., Fabrigoule C., Deleurence P., Ndiaye D., Fort A. 'An electrophysiological study of the impact of a Forward Collision Warning System in a simulator driving task'. *Brain Research*. 2012;1470:69–79
6. Bueno M., Fort A., Francois M., Ndiaye D., Deleurence P., Fabrigoule C. 'Effectiveness of a Forward Collision Warning System in simple and in dual task from an electrophysiological perspective'. *Neuroscience Letters*. 2013;541:219–223
7. Bueno M. 'Impact of a collision warning system on information processing and driver behaviour'. Doctoral Thesis, University of Bordeaux and IFSTTAR, France, 2013
8. Hajek W. 'Workload adaptive cruise control and the potential of physiological based driving assistance systems'. Personal communication from thesis in preparation, Technical University of Chemnitz, Germany
9. Dotzauer M., Brouwer W. H. 'How does repeated exposure to ADAS effect older drivers' workload, trust, and acceptance ratings'. Poster presented at European Conference on Human Centred Design for Intelligent Transport Systems, Valencia, Spain, 2012
10. Beggiato M., Krems J. F. 'The evolution of mental model, trust and acceptance of adaptive cruise control in relation to initial information'.

Transportation Research Part F: Traffic Psychology and Behaviour. 2013; 18(0):47–57

11. Beggiato M. 'Short-term and long-term changes in drivers' higher level cognitive processes when driving with ADAS'. Presented at ADAPTATION meeting, Lyon, France, 2012
12. Bianchi Piccinini G. F., Simões A., Rodrigues C. M. 'Effects on driving task and road safety impact induced by the usage of adaptive cruise control (ACC): a focus groups study'. *Human Factors and Ergonomics*. 2012;1(3): 234–253
13. Dotzauer M. 'Longer-term effect of ADAS use'. Presented at ADAPTATION meeting, Lyon, France, 2012
14. Wege C., Victor T. 'The DO-IT BEST feedback model: Distracted driver behaviour management and prevention before, while and after driving'. Proceedings of the Third International Conference on Driver Distraction and Inattention, Göteborg, Sweden, 2013
15. Dotzauer M., Caljouw S. R., De Waard D., Brouwer W. H. 'Intersection assistance: A safe solution for older drivers?'. *Accident Analysis and Prevention*. 2013;59:522–528 doi:10.1016/j.aap.201307.024
16. Berthon-Donk V., Grosjean M., Rinkenauer G. 'Effect of feedback on performance in the Lane-Change Test'. Presented at Sixth International Driving Symposium on Human Factors in Driver Assessment, Training and Vehicle Design, Olympic Valley, Lake Tahoe, CA, 2011
17. Haupt J. 'Drivers' changes of intentions to pay attention to the driving task related to the use of ADAS'. Presented at ADAPTATION meeting, Lyon, France, 2012
18. Gouy M. 'Behavioural adaptation of drivers of unequipped vehicles to short time headways observed in a vehicle platoon'. Personal communication from thesis in preparation, Open University, UK
19. Huth V., Lancelle V., Gabel C., Bonnard A., Brusque C. 'Exploring mobile phone usage and its context with naturalistic driving observations'. Presented at HFES Europe Chapter Conference, Toulouse, France, 2012
20. Wege C. 'The DO-IT BEST feedback model – distracted driver behaviour management and prevention before, while and after driving'. Presented at 3rd International Conference on Driver Distraction and Inattention, Gothenburg, Sweden, September 4–6, 2013
21. Bhatti G., Brémond R., Jessel J.-P., Dang N.-T., Millet G., Vienne F. 'Filling the user skill gap using HCI techniques to implement experiment protocol on driving simulator'. Presented at 6th International conference on Advances in Computer-Human Interactions (ACHI), Nice, France, 2013

Chapter 17

Recommendations for research on drivers' behavioural adaptation and for the design and deployment of ADAS

Alexandra Fort[1], Véronique Huth[1] and Corinne Brusque[1,2]

Abstract

Based on the knowledge gained within the ADAPTATION project and synthesised in Chapter 16, this concluding chapter gives three kinds of recommendations. First, it deduces future research prospects concerning drivers' behavioural adaptation in response to ADAS use and on its underlying processes. Next, methodological recommendations are given following the identification of challenges related to this kind of research. Finally, the outcome of the ADAPTATION project is translated into suggestions concerning how to increase the expected benefits of new or existing ADAS on driving safety.

17.1 Introduction

Within the framework of the ADAPTATION project, 10 early stage researchers (PhD students) and 2 experienced researchers (post-docs) performed research works on the topic of drivers' adaptation processes in response to ADAS use. An overview of the findings has been synthesised in Chapter 16. The current chapter complements the previous one by presenting three types of recommendations based on these results and the experience gained within the project.

First, recommendations for future research are framed, starting from open research questions that have arisen from the ADAPTATION studies. These deal with the aspects that should be considered in research on ADAS, as well as the knowledge gaps that should be filled by future research. Second, methodological issues are identified that represent a challenge in research in behavioural adaptation and suggestions are made as to how to address these in future studies, building on lessons learnt within the project. Third, more operational suggestions towards improved design and deployment of ADAS are derived from the acquired

[1] IFSTTAR-LESCOT, Bron, France
[2] University of Lyon, Lyon, France

knowledge of drivers' adaptation in response to ADAS use and of the underlying processes.

17.2 Future research prospects on drivers' behavioural adaptation

Research performed in the different studies within the ADAPTATION project raise new issues and needs for investigations that would shed more light on the different processes involved in the use of ADAS and on the way different groups of drivers adapt their behaviour in using ADAS according to the system characteristics. This knowledge is crucial to support the development and improvement of ADAS in the coming decades.

The identified research needs to focus on integration of the temporal dimension in the investigation of behavioural adaptation. Investigating behavioural adaptation through different underlying processes is then discussed, differentiating between motivational, cognitive and energetic processes. The question of countermeasures to mitigate the effect of the systems' technical limitations and specific issues related to behavioural changes induced by automation are also addressed.

17.2.1 Integration of the temporal dimension in the investigation of behavioural adaptation

In the context of ADAS use, behavioural adaptation suggests an *unexpected* or *unanticipated* modification of behaviour following the introduction of a new system designed to impact the driving task in a positive way by facilitating it or by increasing safety and comfort. When drivers face a new system, they have to learn to drive with the support given by the system. This learning process takes time, and different phases can be distinguished [1] – the first encounter phase, when the driver first learns how the system works and how to handle it; the learning phase, when the driver gains control over handling the system; the trust phase, when the driver gains trust in the system and then the adjustment/readjustment phases once the driver has experienced the system in different driving contexts and has better identified its limitations, and also the way to deal with them. During these two last phases, the driver's trust in the system fluctuates before it becomes steady.

The existence of an ADAS learning process provides an argument for the integration of the temporal dimension in the investigation of behavioural adaptation. Within the framework of the ADAPTATION project, several studies pointed out the lack of research concerning the long-term effects of ADAS use and the progressive behavioural changes during the different phases of the learning process. So, behavioural adaptation has not merely to be considered as a result of the system introduction, but as a process including different temporal phases that will be affected by the level of expertise and the level of trust the driver has in the system. This investigation, over the time, has to take into account the specificities of the studied system and the specificities of its user population.

Indeed, the duration of each phase of the learning process varies according to the characteristics of the system: the kind of support provided by the system (informing, warning or intervening), the more or less self-explanatory functioning and interfaces of the system and its frequency of activation or use. Another point to consider is the quality of the information related to the system given to drivers when they start to use it. As stated in Chapter 16, the learning process is shortened if the driver is immediately aware of the functioning and the limitations of the system instead of having to discover them through a trial-and-error process. These factors related to the system characteristics should be considered during further investigations.

Furthermore, the characteristics of the driver are worth taking into account, as each driver has different capacities for learning and some drivers can take more time to deal with the system and to build a correct mental model of it. Chapter 6 illustrates this issue by showing that older drivers can safely interact with technology, but they need more time than younger drivers to learn how to deal with it. Consequently, when investigating behavioural adaptation in response to use of an ADAS, it seems necessary to take into account the physical, perceptive and cognitive skills as well as the motivations of the user population and their learning capacity. In order to adapt to a broad range of users, the design of the system should be intuitive and, possibly, new users should be accompanied during their initial usage to educate them about system functioning and precautions to be taken. This could prevent some drivers abandoning the ADAS because of critical situations encountered during its initial use.

Integrating the temporal dimension of behavioural adaptation also means investigating what happens when the driver stops using a system after an extended period of use. This mostly occurs when a driver switches from an equipped vehicle to a non-equipped vehicle (e.g. car rental, system breakdown, change of fleet vehicle). Within the concept of behavioural transfer, that has been introduced to describe the behavioural changes after system removal, three kinds of transfer have been distinguished: neutral, positive and negative [2]. Neutral transfer means the absence of behavioural transfer after the end of the system use, whereas positive transfer occurs when the system has a long-term benefit on driver behaviour and on road safety after its removal. Here the system appears to support drivers when it is active and also trains them to modify their behaviour in order to enhance safety. Negative transfer represents the main situation of interest in terms of behavioural adaptation. It occurs when the performance of the driver is worse after removal of the system than before its use. It can be postulated that, by substituting for the driver for some of the driving tasks, the system can introduce a risk of loss of skills and a decrease of attention towards some critical aspects of the driving task. There is a research gap to fill regarding these issues. The kind of transfer (neutral, positive or negative) can depend on the support provided by the system and on the driver population. In order to guarantee the expected benefits of existing and emerging in-vehicle technologies, systematic evaluations of transfer effects should be undertaken as a matter of course, both during ADAS conception and implementation phases.

17.2.2 Investigation of behavioural adaptation through different underlying processes

The originality of the ADAPTATION project has been addressing behavioural adaptation by investigating the motivational, cognitive and energetic processes underlying behavioural changes in response to ADAS use. The knowledge gained through the different studies confirms the relevance of this approach.

Chapter 8 points out the role of motivational factors on driver behaviour. For the authors 'Motivation drives people to act, behave, evaluate, judge, feel and to believe the way they do'. In the context of driving, motivational factors can influence perceptive and cognitive processes related to vehicle control and to the management of driving situations, and thus modify driving performance. Beyond acceptance and trust, which are usually considered in research addressing behavioural adaptation in response to ADAS use, it is crucial for further research to deal with other motivations factors, such as risk perception and risk homeostasis, the feeling of security and protection, social needs and self-actualisation needs.

In addition, it would be valuable to gather further information about the cognitive mechanisms involved in system use or those underlying the behaviours in response to support provided by an ADAS. Results obtained in the research performed within ADAPTATION demonstrate the need for more information on this issue. Beyond the learning processes involved in the use of ADAS discussed above, several studies within ADAPTATION have also emphasised the role of cognitive processes during use of ADAS. In this regard, methodological progress has been made in the assessment of mental models and situation awareness (see Chapter 4). For instance, the importance of users obtaining a realistic mental representation of the system to ensure trust and acceptance has been shown (see Chapters 4 and 5). On this point, further research would be needed in order to better understand the parameters that can influence this mental representation. Concerning the auditory alerts used in warning systems, it has been found that they require attentional resources in order to be processed (Chapter 12). This raises the question of whether people with cognitive deficiencies (e.g. age-related ones) are able to benefit from such warning systems. It would be interesting to extend research with the aim of determining which resources are necessary, and for which ADAS, in order to optimise system designs and, possibly, to adapt them according to user characteristics.

Energetic processes such as mental workload, emotion, stress and vigilance are particularly important for consideration when examining ADAS use and associated behavioural adaptation. Indeed, some ADAS have been designed in order to mitigate the impact of distraction (implying an increase in mental workload) or vigilance decrement. A better understanding of how these factors impact on driving behaviour is crucial to optimise assistance.

In addition to traditional behavioural approaches, physiological techniques are especially promising for examination of different processes in driving. These techniques comprise the recordings of cardiac, respiratory or electro-dermal activities as well as neuroimagery techniques (such as electro- and magneto-encephalography and functional magnetic resonance imagery). First, these techniques can bring new insights on these processes during driving, as has been illustrated in Chapter 12. Second, they can be used to develop adaptive technology by providing on-line indices of the driver state. As illustrated in Chapter 10, an ECG (electrocardiogram) can provide evidence concerning the driver's mental workload in real time. Nevertheless, knowledge of these techniques has to be deepened since a wide variety of parameters can impact on the measures (e.g. emotion, stress, workload, temperature). For this reason, further research has to be pursued to identify the best indicators of the driver's state, or their combinations.

17.2.3 Countermeasures to mitigate the effects of ADAS limitations

Most ADAS on the market are not perfectly reliable due to technical limitations, choice of thresholds or the environmental conditions of use. But, as has been concluded in Chapter 16, the effect of system unreliability on efficiency and drivers' acceptance can be compensated for by users having a correct mental representation, including the predictability of limitations of the system. Nevertheless, further research is necessary to get a clearer picture of the impact of the different systems' limitations on their efficiency and acceptance by users. It is likely that the impact of these limitations will differ according to the system and the level of automation (informing, warning or intervening). While such system limitations can be viewed as negative, we should also consider that the limitations can avoid over-reliance by the users (see Chapter 12). Another important issue in the context of system efficiency and acceptance concerns identification of different user population needs, their abilities, preferences and motivations. Driver populations do not only encompass differences in age and expertise of driving, but also differences in terms of driving styles (confident, careful, anxious, aggressive, etc.), which can be culture-bound or related to personality traits. If an ADAS is particularly useful for one kind of users, this is not necessarily the case for all users.

Therefore, it is important to assess the efficiency of each system throughout the wide range of potential users and the potential misuses of each system that might occur depending on these user types. For example, drivers with neurodegenerative disorders may have a particular need for assistance in order to compensate for their impaired driving-related capacities (Chapter 16). However, the extent to which an ADAS can be effective for this driver population has to be examined, since the impaired capacities might also lead to difficulties in dealing with the system. Systems that require divided attention can be particularly critical and the expected benefits should be contrasted with possible costs regarding attention allocation (Chapter 7). The controllability of system function parameters and the individualisation of human machine interfaces are promising approaches for improving driver acceptance, which should be developed further in future research. According to personal preferences and motivation, users could adjust their systems individually; for instance, some commercial ACC already offer several headway settings.

17.2.4 Behavioural change induced by automation

As has been suggested in previous chapters, one of the main benefits of automation is to facilitate driving by taking one or several specific sub-task(s) away from the driver. However, the driver often needs to continuously monitor the status of the system and to maintain sufficient situational awareness in order to take control of the vehicle in situations that exceed the system's functional limits. Even though automation aims to facilitate the driving task, it can also result in three potentially negative effects: (1) driver under-load, monotony and fatigue; (2) driver distraction due to engagement in other tasks not relevant to driving and (3) driver overload due to some focus of attention on what the system is doing. This third effect is especially likely during the first encounter phase and the learning phase. However, whatever the effect produced, it can be highly critical in some situations when the system fails or reaches its limits and the driver is required to take over control of the vehicle in order to avoid an accident. More research is needed to better understand the impact of automation on driver attention mechanisms following first use of the system as well as after a longer exposure.

Another issue to take into account during assessment of road safety aspects of ADAS concerns the interaction between equipped and non-equipped vehicles. As this book illustrates, the use of ADAS systems can induce behavioural adaptation and the behaviour of an automated vehicle can surprise drivers of non-equipped vehicle or can insidiously influence their behaviour (Chapter 13). The consequences of the introduction of equipped vehicles on the behaviour of drivers of non-equipped vehicles are dependent on many factors such as the penetration rate, the conspicuity of equipped vehicles, the exposure frequency and time, the length of platoons, etc. Before the ADAPTATION project, these questions had not been considered, although the issue can be quite critical as some ADAS can induce unexpected behaviour for non-users confronted with them. As illustrated in Chapter 13, the introduction of automated vehicle platoons could impact on the driving behaviour of non-users who unconsciously begin to mimic the behaviour of drivers in the platoon by reducing their safety margins. The factors mentioned above that might moderate this behaviour need to be further analysed and it needs to be investigated whether the effects found in response to platoons are generalizable to other ADAS. If this kind of behaviour for non-equipped drivers is observed on the road, some countermeasures may have to be conceived before wide deployment of certain systems. Some of the possible countermeasures are detailed in Section 17.4.1. Prospective research is also needed in this domain, considering the penetration rate of the different systems, the demographic evolution, the infrastructure and vehicles evolution. This research should also take into account the potential development of communication with other vehicles and communication with the infrastructure, which seems to be a likely future step in the evolution of ADAS.

17.3 Methodological challenges in research on drivers' behavioural adaptation

The work performed in the ADAPTATION project has allowed identification of methodological challenges that are specific to the research topics addressed.

Based on the project experience, this section presents the lessons learnt that are considered relevant for similar research. Although the list of methodological and organisational issues does not claim to be exhaustive, the recommendations based on identified challenges in research can be of great value for the design, planning and realisation of further studies.

Detecting and analysing drivers' behavioural adaptation to the use of in-vehicle technologies implies considering multiple processes over time. As described in Chapter 3, driver adaptation effects can appear over the short, medium or long term. Learning processes that take place during the introduction of an ADAS need to be observed, focussing especially on effects at the operational level of the driving task. In the medium term, the integration of the ADAS into the mental model of driving needs to be analysed, mainly at the manoeuvring level of driving, while long-term effects can also be reflected at the strategic level. Post-use transfer effects can appear when driving without the ADAS or with other systems. Studies thus need to be fitted to these adoption phases of ADAS. For instance, an A-B-A study design (i.e. a test drive with the system that takes place in between two baseline drives) will allow detection of initial learning effects as well as possible transfer effects once the system is turned off. Repeated training trials, in turn, offer the possibility to fit learning curves as an indicator of the speed and ease of adoption of the ADAS by the driver.

Given that behavioural adaptation can evolve over time, longitudinal studies examining long-term effects of ADAS are essential. However, researchers face a range of difficulties when carrying out studies over an extended period. First and foremost, they imply a huge organisational load and special care regarding the management of the participant sample. Avoiding participant drop-out in long-term studies is a particular challenge that needs to be resolved by well-organised agendas, participant reminders and motivational efforts or incentives. In addition, long-term studies require the storage and handling of large data sets. The database for behavioural studies developed within the ADAPTATION project is a useful tool that significantly improves this task by enabling storage, joint access, individual analysis and integration of different types of data collected during driving simulator and field studies (Chapter 15).

The variety of processes that are involved in behavioural adaptation need to be considered by implementing appropriate study designs and data collection methods. It is hence necessary to understand benefits and drawbacks of the variety of existing methods and to choose appropriate method for each research question (Chapter 3). Certain research questions, such as the analysis of response times to critical events when assisted by a warning system, require experimental control of variables so as to avoid confounding effects by third variables. Fatigue and carry-over effects need to be controlled for by appropriate counterbalancing of the experimental conditions in such studies. During experiments that last several hours, participants are prone to vigilance loss and boredom. It is therefore recommended to consider how to keep the participants in the loop during the experiment. Planning of breaks, task variety and participant motivation by the experimenter can be helpful in this context.

Simulator studies present the opportunity to study behavioural adaptation in a strongly controlled setting that is free of risk, but on-road studies are imperative so as to study the effects of an ADAS on driving in more realistic settings. In cases of new systems, it is recommended that simulator studies first be carried out and then replicated in real traffic. In the design of simulator studies, the creation of scenarios has represented a challenge in the past. The tool created within ADAPTATION (Chapter 14) provides a promising solution to this issue. As compared to studies with an experimental protocol, naturalistic studies are much less controlled but offer the possibility of getting insights into the effects of systems use in their natural context, including behavioural adaptation at the strategic level. Chapter 3 provides detailed information concerning the advantages and disadvantages of these study types.

Given that the behavioural reactions provoked by an ADAS can depend on the driver state, corresponding conditions or variables have to be included in the study design or measurements. For instance, different levels of driver distraction should be defined and included within studies that test the effectiveness of a warning system. In this regard, physiological measures such as electroencephalography (Chapter 12) and electrocardiography (Chapter 10) have proven useful for the quantification of distraction. Although experimental precautions have to be taken in order to record a clean signal, Chapter 12 also illustrated that electroencephalography is a promising approach in order to better understand how a warning system impacts on the drivers' information processing during driving. These studies have demonstrated the feasibility of using such physiological techniques in the driving context. Physiological measures have also been used to study the adaptability of ADAS depending on driver workload at the example of a Workload Adaptive Cruise Control (Chapter 10). Although positive results have been obtained (Chapter 16), further research is needed on the detection and differentiation of driver emotions that could impact on physiological measurements and thus interfere with the workload measurement. Moreover, advances in technologies could be targeted at developing physiological measurement sensors that allow collecting data without interfering with the normal procedure of driving (e.g. with measurement devices that work without physical contact).

A qualitative approach can be more appropriate to address some research questions. Above all, research on driver opinion and strategies related to the use of technology while driving can benefit from qualitative data that complements results on behavioural adaptation obtained in quantitative studies. For instance, focus group discussions have been carried out within ADAPTATION, with the aim of exploring the importance of motivational factors in driving and ADAS use and of proposing a model on their interplay (Chapter 16). This model was later tested in a questionnaire study, designed on the basis of the qualitative information provided by the focus group study.

Driver needs regarding ADAS functionality and the effects of system use can differ considerably according to the driver population. The motivation and ability of specific driver populations to adopt ADAS should therefore be studied further and, in doing so, challenges in participant recruitment and motivation need to be met. Efforts are necessary to recruit a sufficiently large sample of participant that meets previously defined relevant driver characteristics such as having reached a certain age, suffering from a disease or being a professional driver. The level of ADAS experience is also worth considering, since drivers' attitudes and behavioural responses evolve with system experience. Again, finding ADAS experienced drivers might be challenging when studying relatively new systems. As mentioned above, participant motivation plays an important role because dropouts need to be minimised in long-term studies. Changes in participant motivation during a (shorter or longer) study that could bias results should be accounted for by counterbalancing the conditions in the study design.

Finally, studies on behavioural adaptation are particularly sensitive to data-related difficulties. Handling large data amounts or especially delicate data types, such as physiological data, is challenging and requires procedures that include back-ups and sufficient time margins to check and fix the recording and storage. The database developed within the ADAPTATION project and described in Chapter 15 is expected to considerably facilitate the work with different types of behavioural data. However, more analyses are required so as to further improve the data preparation process that is part of this tool. Security and privacy of data represent another critical issue in the context of behavioural adaptation research. They need to be warrantied to the participants of the studies; legal and ethical guidelines that should be respected by the researchers need to be established.

17.4 Towards improved design and deployment of ADAS

Just as the findings from the ADAPTATION project raise further research prospects in order to refine knowledge on how ADAS impacts driving behaviour, they also highlight some operational aspects relating to the design and deployment of ADAS so as to promote their expected benefits. In the following section, recommendations are given regarding the role of system-provided information and the resulting driver awareness of the ADAS functionality and limitations. With regard to functionality and design, recommendations focus on issues related to presentation of warnings to drivers on the one hand, and the adaptability of system based on real-time driver monitoring on the other hand. Additionally, driver training is suggested to accompany the introduction of ADAS and evidence of the need for reminders at later stages of system use.

17.4.1 System information and driver awareness

Findings from the ADAPTATION project highlight the relevance of the information on the form of assistance system provided to users (Chapters 4 and 5). It has been shown that, regarding trust and acceptance, the mental model drivers have about the system and its functioning is particularly important. Also, an inappropriate mental model could explain some misuses of ADAS and it is thus recommended that enough information is given to users to allow them to build an adequate mental model of the system.

It is recommended that information about the system should include details of the assistance function and any specific functional limitations. Provided that the drivers are exposed to this information (by reading the manual or being otherwise instructed) and comprehend it, this will help prevent drivers from learning about the system through trial-and-error which has proven insufficient for the development of stable trust and acceptance. At the same time, drivers will be more efficient in any monitoring tasks since, based on realistic expectations, they can more easily detect system limitations and failures. It can be anticipated that development of an adequate mental model would speed up the process of resuming control by the driver.

It is important to avoid drivers 'idealising' the system (i.e. anticipating perfect and intelligent performance). Idealisation has been shown to lead to strong decreases in acceptance of the system if unexpected failures then occur (Chapter 16). Instead, the information provided should be objective and should include all limitations of the system. Far from diminishing confidence in the system, the ADAPTATION studies have shown that drivers can accept errors and system limitations as long as they know about them before encountering them (Chapters 4 and 5).

Although user manuals can give a substantial amount of important information concerning an ADAS, its benefits, its limitations and its implications, they are not always sufficient. Indeed, not all drivers read manuals systematically and, perhaps, dealers should compensate for this behaviour by giving the necessary information to their customers. In particular, they should be aware of the effects of information mentioned earlier in this section so as to achieve optimal system acceptance by the drivers.

ADAS that take over aspects of vehicle control may induce driving behaviour that differs from driving without the ADAS. This can have an effect on unequipped drivers who share the road with system users. On the one hand, they can be confused or irritated by this unusual behaviour (e.g. regarding acceleration or braking behaviour and headway choice); on the other hand, drivers might show mimetic behaviour, for example by reducing their safety margins (headways), and consequently drive in a way that is less appropriate for non-ADAS-equipped driving. It is therefore recommended that information on system functioning is provided to non-users as well, which might be combined with information on the penetration rate of the system. Ideally, information on ADAS in use by vehicles could be presented individually so that non-equipped drivers become aware of their presence. This could be achieved by a specific signal, which would need to be designed to be easily understandable and either visible on the exterior of the vehicle or presented inside the unequipped vehicle by car-to-car communication.

Whenever misuse or negative behavioural adaptation to a system is detected, efforts should be made to raise driver's awareness of the implied risk. For instance, focus-group data have revealed that the use of certain assistance systems (like ACC) could increase the propensity to engage in secondary tasks where assistance by the system frees resources, which can be reallocated to activities that are not directly related to the driving task. However, this can imply a misjudgement of the effort necessary to monitor the driving situation and the system status. Knowing that drivers tend to overestimate their ability to perform secondary tasks and to misjudge the compatibility of some secondary tasks with driving, decreases in situation awareness are to be expected. For this reason, it seems important to raise awareness of the risk related to engagement in secondary tasks in general, and specifically during ADAS use.

Mobile phone use has been discussed as a popular secondary activity while driving, which is expected to further increase as smartphone capabilities develop and the everyday use of such technologies keeps augmenting (Chapter 9). Although drivers are generally aware of the risk that mobile phone use while driving can imply, Chapter 9 has identified a range of obstacles that hinder the implementation of corresponding regulation strategies. This knowledge should guide the design of driver education actions, which seem indispensable to induce them to use the phone in a risk-minimising way by regulating their exposure to phone use or appropriately adapting their driving behaviour.

17.4.2 System design: Warning presentation and adaptability based on driver monitoring

Behavioural adaptation to ADAS use can include unintentional driver distraction. In this regard, a detrimental effect of ADAS that reverses its expected benefit has been illustrated with the example of a brake capacity FCW system within ADAPTATION (Chapter 16). The visual warning that was presented in the instrument cluster diverted the visual attention of professional truck drivers away from the road scene in a critical driving situation. The identification of the warning source competed with the safe accomplishment of the driving task. This finding underlines the importance of an appropriate interface design for warning ADAS and attracting the drivers' visual attention by a warning presented outside the visual field for safe driving should be avoided.

Positive results for the interface design have been obtained for a Blind Spot Information System (Chapter 16), where the warning was located in the peripheral field of view of the drivers and did not require them to take their eyes off the road. This location also corresponded with the meaning of the warning and was thus especially appropriate for the functionality of the ADAS. Considering these findings, a location for warnings that is compatible with driving, such that it does not distract the driver's attention, should be the rule. Additionally, using warning locations consistent with the ADAS function attract attention towards relevant elements of the driving situation; this approach seems to be especially promising.

The ADAPTATION studies have shown that the efficiency of warnings provided by ADAS can depend on the driver state. For instance, the efficiency of a Frontal Collision Warning system depended on the driver's mental workload (Chapter 12). This result suggests benefit in adapting the alert according to the mental workload in order to keep an equal level of efficiency for the warning ADAS. The alerts could be designed to be more or less intrusive depending on the drivers' attentional state, just as the timing of the warning presentation could be varied. Similarly, safety margins adopted by intervening ADAS could be adjusted according to the driver state so as to improve the efficiency of the system. By way of example, the Workload ACC developed and tested within the ADAPTATION project is an efficient and well-accepted solution for how driver workload, based on physiological measures, can be taken into account to adjust the parameters of the assistance (Chapter 10). As another example, ADAPTATION studies have pointed out the benefits of Visual Distraction Alerts that take into account the glance history of the driver for generation of warnings (Chapter 4).

Thus, in the ADAS development process, it is important to test the system efficiency according to the driver state and, in particular, according to different levels and types of distraction (cognitive, visual, auditory or motor). These examples suggest that real-time monitoring of the driver state (measured through physiological parameters or driver behaviour) should be considered when developing new ADAS or improving existing ones. System adaptability based on such driver monitoring is expected to increase the benefits of ADAS by continuously optimising their functioning in varying conditions.

As has been discussed in Chapter 9, efforts need to be made in the design of the possibilities to interact with mobile phones while driving, taking account of the limited use of effective regulation strategies by drivers themselves. Likewise, solutions that support the driver in appropriately managing the dual task situation of driving and using the mobile phone should be developed and tested for effectiveness. For instance, interaction opportunities with the phone and its interface design could be adaptive according to the situational demands and driver workload, which are monitored in real-time. The driver could thus perform the secondary task whenever it is compatible with driving, while excessively complex dual task situations and consequent driver overload would be prevented.

17.4.3 Driver training and regular reminders

Given that information that allows building an adequate mental model is of such importance for acceptance of an ADAS and prevention of system misuse, educating drivers about ADAS functionality and practically introducing the system should be the rule. Driving schools should be encouraged to integrate classes on use of the most common ADAS into their course offer, if they have not yet done so. As the results of the ADAPTATION studies suggest, drivers tend to recognise the usefulness of systems for other drivers but are often not sufficiently aware of how they can benefit personally from the system use (Chapter 16). In order to modify this perception, driving classes should focus on presentation of the most common systems available on the market and the way their use can influence driving behaviour, and they could possibly include a practical session where the systems can be experienced. In order to create specific training programmes around these issues, it would be valuable to foster collaborations between car dealers, research centres, universities and driving schools. Beyond the education of novice drivers, these training programmes should also be made available to experienced drivers who have to learn how to manage driving with new ADAS installed in their cars. In particular, training programmes can be adapted to professional drivers, given that their intensive road exposure creates a specific need for assistance systems.

As illustrated in Chapter 16, mental models are influenced by experience. Aspects of the system that are not experienced drop out of the mental model. As a solution, to sustain a constant mental model, tutoring systems on the ADAS could be incorporated in the on-board computer of the car. Such tutoring could remind the driver about the availability of the ADAS, their benefits and their limitations. Research is needed on the design and timing of such reminders. However, it seems reasonable that they pop-up when the driver enters the car and do not appear during driving, so as to avoid driver distraction. Messages such as 'do you remember the system does . . . ' could be presented at regular intervals that still need to be defined. Alternatively, analysis in combination with other systems and sensors could allow the provision of real time information about probabilities of failure of the ADAS in the current driving situation. For instance, the drivers could be kept informed about limited sensor functioning within an ACC on a winding road, detected by a series of sharp turns in the navigation map.

Finally, it could be valuable to give feedback to users on their driving performance in order to improve behaviour. By means of warnings, ADAS could provide real time feedback on critical situations during driving activity, inciting the driver to initiate corrective action. Providing post-trip feedback on the driving performance could further increase the benefits of such warning systems. The warning history could be used to make the driver aware of the flaws in her/his performance and to advise of the need for driver training. Providing this kind of feedback represents one of the actions of the Behaviour-Based Safety (BBS) programme (Chapter 4). The DO-IT BEST strategy that has emerged from ADAPTATION studies and consists of a combination of BBS approach (post-trip feedback) with ADAS systems (immediate feedback) appears promising. It raises driver alertness to safety critical situations, speeds up driver response in these situations, raises awareness of safety-critical behaviour and coaches the drivers towards their target behaviour. The strategy of giving post-trip feedback has already been applied in some companies involving fleets of commercial vehicles. Its extension to individual drivers, coupled with immediate feedback from ADAS, could be beneficial for road safety in the long term.

17.5 Summary and conclusions

The advances made in research on drivers' behavioural adaptation in response to ADAS use within the lifetime of the 4-years ADAPTATION project comprise insights on drivers' behavioural adaptation and the underlying cognitive, motivation and energetic processes that have allowed three kinds of recommendations to be presented in this concluding chapter.

From the synthesis of the knowledge gained within the ADAPTATION project, future prospects for research into drivers' behavioural adaptation have been identified in order to shed more light on the different processes involved in the use of ADAS. This research concerns four main issues.

1. The consideration of the temporal dimension related to the learning phases when facing a new system, from short-term use to long-term use. This research should take into account the characteristics of the system, the information users have about it and the characteristics of the users. It includes also the transfer effects when driving with a system off after an extended period of use.
2. Research within the ADAPTATION project has shown the value of addressing behavioural adaptation through different underlying processes: motivational, cognitive and energetic processes. Efforts on this topic should be pursued taking advantages of the different methodological approaches available (subjective, behavioural and physiological data).
3. Because ADAS typically have technical limitations, countermeasures to mitigate the effects of these limitations have to be examined. The performance of each system has to be further studied according to its limitations and taking into account the wide range of potentials users.
4. The impact of automation on driver behaviour has to be further examined considering the potential change in user attention mechanisms, as well as the interaction between equipped and non-equipped drivers.

Following the experience gained within ADAPTATION project, methodological challenges specific to the research topics have been identified. They concern long-term studies (e.g. management of the participant sample over the time in longitudinal studies, storage and handling of large data set, vigilance loss, boredom) that are essential so as to examine long-term effects of ADAS used. Context studies (e.g. simulators, on-road studies), experimental design (e.g. long or short studies, counterbalancing), studied variables (e.g. subjective, behavioural, electrophysiological data) and population recruitment (e.g. number of participants, characteristics, motivation to participate) have to be carefully designed, taking account of advantages and limitations of each methodological approach depending on the research question.

Finally, some more operational advice for the design and deployment of ADAS can be proposed so as to increase their expected benefits and promote road safety for ADAS users and non-users. Such advice first concerns system information given to users so as to allow them to build an appropriate mental model of the system. Some information should also be diffused to non-users to allow them to better understand the behaviour of equipped vehicles. In addition information concerning potential misuses of the system and their consequences has to be given when necessary. Advice concerning system design should be formulated so as to avoid distractive effect and to increase efficiency by designing adaptive system using, for instance, real-time monitoring of the driver attentional state. A third kind of advice concerns driver training and reminders. Indeed, it seems valuable to educate drivers about ADAS and to introduce the systems practically to users. In addition tutoring and reminders on the benefits and limitations of the systems could be valuable so as to sustain an appropriate and constant mental model of